How and Why Do Policies Change?

A Comparison of Renewable Electricity Policies in Belgium, Denmark, Germany, the Netherlands and the UK

P.I.E. Peter Lang

Bruxelles · Bern · Berlin · Frankfurt am Main · New York · Oxford · Wien

Isabelle DE LOVINFOSSE

How and Why Do Policies Change?

A Comparison of Renewable Electricity Policies in Belgium, Denmark, Germany, the Netherlands and the UK

"Public Action"
No. 3

Published with the financial support of the Fondation universitaire

No part of this book may be reproduced in any form, by print, photocopy, microfilm or any other means, without prior written permission from the publisher. All rights reserved.

© P.I.E. PETER LANG S.A.
Éditions scientifiques internationales
Brussels, 2008
1 avenue Maurice, B-1050 Brussels, Belgium
www.peterlang.com; info@peterlang.com

ISSN 1783-6077
ISBN 978-90-5201-398-5
D/2008/5678/30

Bibliographic information published by "Die Deutsche Bibliothek"
"Die Deutsche Bibliothek" lists this publication in the "Deutsche Nationalbibliografie"; detailed bibliographic data is available on the Internet at <http://dnb.ddb.de>.

Table of Contents

Introduction 9
    I.   Research Question 9
    II. Methodology 12

CHAPTER 1. Theoretical Framework 17
    I.   Definition of Policy Change: A Conceptual Framework 17
    II. Explaining Policy Change: A Theoretical Framework 35
    III. Conclusion 60

CHAPTER 2. European Policy 63
    I.   Introduction 63
    II. European Liberalisation Policy 64
    III. State Aid for Environmental Protection 69
    IV. The Preussenelektra Case against Schleswag 70
    V. European RES-E Policy 71

CHAPTER 3. RES-E Policy in Belgium 77
    I.   Introduction 77
    II. Overview of the Electricity Sector 79
    III. Overview of the RES-E Policy History
        in Belgium (1970-1999) 83
    IV. Liberalisation of the Electricity Sector (1999) 89
    V. RES-E Policy change in Belgium (1999-2004) 92
    VI. Interpretation of RES-E Policy Change in Belgium 99
    VII. Explaining RES-E Policy Change in Belgium 101
    VIII. Conclusion 112

CHAPTER 4. RES-E Policy in Denmark 115
    I.   Introduction 115
    II. Overview of the Electricity Sector 116
    III. Overview of the RES-E Policy History
        in Denmark (1976-1999) 122
    IV. RES-E Policy Change in Denmark (1999-2004) 131
    V. Interpretation of RES-E Policy Change in Denmark 141
    VI. Explaining RES-E Policy Change in Denmark 143
    VII. Conclusion 151

CHAPTER 5. RES-E Policy in Germany 155
   I.   Introduction 155
   II. Overview of the Electricity Sector 156
   III. Overview of the RES-E Policy History
        in Germany (1974-2000) 159
   IV. RES-E Policy Change in Germany (2000-2004) 165
   V.   Interpretation of RES-E Policy Change in Germany 171
   VI. Explaining RES-E Policy Change in Germany 173
   VII. Conclusion 182

CHAPTER 6. RES-E Policy in the Netherlands 185
   I.   Introduction 185
   II.   Overview of the Electricity Sector 187
   III. Overview of the RES-E Policy History
         in the Netherlands (1974-1998) 191
   IV.   Liberalisation of the Electricity Market
         in the Netherlands (1998-2002) 198
   V.   RES-E Policy Change in the Netherlands (2002-2003) 208
   VI.   Interpretation of RES-E Policy Change in the Netherlands ... 213
   VII. Explaining RES-E Policy Change in the Netherlands 214
   VIII. Conclusion 224

CHAPTER 7. RES-E Policy in the United Kingdom 227
   I.   Introduction 227
   II.   Overview of the Electricity Sector 228
   III. Overview of the RES-E Policy History
         in the UK (1979-1997) 231
   IV.   RES-E Policy Change in UK (1998-2004) 240
   V.   Interpretation of RES-E Policy Change in the UK 248
   VI.   Explaining RES-E Policy Change in the UK 250
   VII. Conclusion 258

CHAPTER 8. Comparison 261
   I.   Comparison of the Policy Changes 261
   II.   Test of the Hypotheses 265
   III. Explanations of Policy Changes 283
   IV.   Conclusions of the Comparison 286

Conclusion 289

References 297

Introduction

I. Research Question

Today, energy policy issues dominate the political agenda in Europe and worldwide. Different but interrelated issues are at stake. Firstly, our energy supplies are threatened by the volatility on the oil and gas markets due to limited offer, geopolitical conflicts in the main producing countries (e.g. the Middle East, Russia), soaring energy demands, limited flexibility (e.g. the transport sector) and the lack of substantial energy alternatives. Therefore, the priority of most governments is expected to focus on increasing support to research and development in alternative energy sources, and the diversification of energy supplies in terms of energy sources and geographic origin. In this strategy, the renewable energy sources (RES) are playing an important role, as they offer a sustainable and decentralised alternative to the traditional fossil fuel energy sources. Secondly, the threat of the climate change issue has a significant impact on the energy sector, the main source of CO_2 emissions. The current pattern of energy production and consumption in the industrialised countries is expected to be substantially modified if climate changes are to be mitigated in the future. This is one of the purposes of the main inter-governmental solution addressing this issue so far, the Kyoto Protocol, adopted in 1997 and entered into force in February 2005. In this context, the renewable energy sources appear to be a solution full of promises in the future, since they are CO_2-free (or, at least, they emit significantly less CO_2 than the other energy sources). Thirdly, nuclear power was put on hold in most European countries after the Chernobyl disaster in 1986, but today the future of nuclear power is gaining renewed attention with the climate change issue and the concerns about the security of supplies. In some countries, the nuclear power plants are expected to be phased-out progressively (e.g. Germany and Belgium), but in others, the development of new nuclear facilities are being discussed (e.g. the UK, Finland) mostly as part of the solutions to reduce CO_2 emissions. So, the future of nuclear energy is still uncertain in Europe, with very different political priorities among the member states, but the revival of nuclear power is not a taboo anymore. Fourthly, the electricity and gas sectors have been going through significant reforms in Europe during the last decade due to the progressive liberalisation of the electricity and gas markets and international concentration among the energy companies. As a consequence of the liberalisation process, a stronger emphasis on the competitiveness of the energy supplies has emerged, which threatens the pursuit of social and environmental objectives in the energy sector. This is why the formulation of appropriate social and environmental policies at the European and national levels appear even more critical today than in the past. In conclusion, the role of the public authorities in the energy sector is critical today in order to address the current energy challenges and ensure sustainable and secured energy supplies in the future.

In this context, the RES happen to be at the core of the debate as one of the main energy solutions for both the energy security and the climate change issue. The political priority in favour of RES differs from one European country to another, but a similar revival of this solution can be observed, especially in the electricity sector. In the context of the liberalisation of the electricity sector, the issue of renewable electricity (RES-E) is on the political agenda of all member states. The question of how and how much we should support the RES-E on the electricity market has been addressed by all member states in previous years, with different results. However, several obstacles handicap the development of RES-E. Most RES-E technologies are hardly competitive on the electricity market compared to the traditional power plants (nuclear, gas, coal plants) because the external costs of these technologies are not reflected in the electricity prices or because they benefited from intensive public support in the past (nuclear plants). So, as long as the market does not consider these costs, it is the responsibility of the state to ensure that the electricity prices internalise the external costs and benefits of the various technologies as much as possible in order to make the competition more transparent. In addition, improvements are still necessary regarding most RES-E technologies (e.g. solar PV, wave and tidal plants, small hydropower, biomass plants) in order to improve their effectiveness and competitiveness. Therefore new RD&D programs in favour of RES-E are needed to improve the RES-E technologies. Another important obstacle to RES-E is the NIMBY (not in my backyard) problem, which hampers the development of RES-E (especially wind turbines and waste plants) in the rural areas. Indeed, the development of RES-E tends to clash with other environmental concerns: landscape protection, nature protection, noise or odour pollution, etc. So, it is crucial to the future of RES-E that the RES-E projects are developed in consultation with local interests and take into consideration surrounding environmental impacts, especially in those countries where local protests proved to be very powerful in the past (e.g. the UK, Germany). Finally, we observe that the RES-E lobby is still very weak and divided in most European countries as well as at the European level compared to the lobby of the traditional energy companies, which reduces its influence in the political debates. In conclusion, RES, and RES-E in particular, are more and more considered a promising solution to the energy issues of today and tomorrow, but significant obstacles have to be addressed to make this solution come true.

This research looks more specifically into the RES-E policies in Europe, namely at how the member states intervene in the electricity sector to remove the obstacles to RES-E and support its development in the electricity market. We observe that RES-E policies have already been implemented in some member states for decades (e.g. Denmark, the Netherlands, Germany, the UK), while others have only considered it recently (e.g. Belgium); hence the diversity of the RES-E policies. Besides, the development of an RES-E policy at the European level started very recently, at the end of the 1990s, and it is still in its infancy, which explains the absence of harmonisation between the RES-E policies in Europe. Due to the emergence of the new energy issues that we described above, we observe that the RES-E policies have changed more or less significantly in all member states in previous years. These changes usually reflect an increased political priority in favour of the RES-E policy as well as new approaches to implement it, but the patterns of policy change differ between the European countries. This is precisely the empirical question addressed in this research: "How and why have the RES-E policies changed in Europe recently?". The empirical basis of the research includes five EU member-states with diverging RES-E policy change patterns: Belgium, Denmark, Germany, the Netherlands, and the United Kingdom. In each country, a specific turning point has been identified lately in the RES-E policy: 1999 in Belgium, 1999 in Denmark, 2000 in Germany, 2003 in the Netherlands and 2000 in the UK. The "how" question considers the patterns of policy change in the five countries and investigates whether the policy objective and/or instrument of the policy has changed compared to the past. As a result of the comparison of the five countries, we will then be able to observe whether the RES-E policies tend to converge or diverge among these countries. The "why" question analyses the reasons why the RES-E policies have changed and tries to identify recurrent explanatory factors between the different countries.

The related theoretical question of this research is: "How and why do public policies change?". Based on the political science literature about the public policy analysis, especially the policy change literature, we will develop, on the one hand, a conceptual framework that specifies what we mean by policy change (the "how" question); then, on the other hand, a theoretical framework that identifies the main explanatory variables to policy changes and formulates specific hypotheses for each variable (the "why" question). The conceptual and theoretical frameworks of this research are based on the work of Peter Hall on the tree orders of policy change (Hall, 1993), but we complement and improve it with additional approaches to policy change, such as the Europeanisation literature (for an overview: Featherstone and Radaelli, 2004), the partisan theory (for an overview: Schmidt, 1996 and 2002), and the veto players theory (for an overview: Tsebelis, 2002). In addition, we specify operational details (indicators for policy changes) and theoretical details (specific hypotheses) about how Peter Hall's approach can be applied. Therefore, we rely on an existing approach to address the theoretical question of this research, but we adapt and improve it both theoretically and operationally.

In conclusion, the originality and the added value of this research are twofold. Firstly, we describe, explain and compare the RES-E policy changes in five European countries, based on a common analytical framework. Unlike most of the existing literature on the RES-E policies in Europe, our research goes beyond the description of individual cases, it relies on a strong analytical framework and it results in a systematic and rigorous comparison of five countries that includes not only the various RES-E policies, but also the larger political and institutional context in which those policies are formulated. Therefore, this research seeks to make sense of the RES-E policy changes in Belgium, Denmark, Germany, the Netherlands and the UK, from an analytical perspective that goes beyond the description of the policies. Secondly, although it does not bring any theoretical breakthrough, this research contributes to the political science literature, or more specifically to the public policy literature, by discussing existing hypotheses and applying them to a new empirical background. On the one hand, our theoretical framework aims at questioning and extending Peter Hall's approach of the policy changes. On the other hand, we confront it with an original empirical background, the RES-E policies, which have hardly been explored by political scientists so far (unlike economists or engineers), especially from a comparative perspective. Finally, we will ultimately derive from this research some prescriptive policy advice that could potentially guide the formulation of future RES-E policies at the domestic and European levels.

II. Methodology

This research is based on a combination of induction and deduction. Like in the grounded theory method (Glaser and Strauss, 1967; Strauss and Corbin, 1990; Corbin and Strauss, 1990), the hypotheses of our theoretical framework are partly based on empirical observations and partly on the theoretical literature. The interaction between the theoretical literature and the empirical framework helped us to formulate scientific and general hypotheses that are at the same time relevant and empirically workable (chapter 2). Then, following the hypothetico-deductive research process, we confront our theoretical framework with the empirical background in order to validate or invalidate the hypotheses. In addition, we also assess the empirical relevance and significance of the hypotheses in order to classify them according to a hierarchical system (chapters 4 to 8). Finally, still according to the hypothetico-deductive process, we revise and reformulate the hypotheses in order to improve and generalise the theoretical framework (chapter 9).

Figure 1: Research process

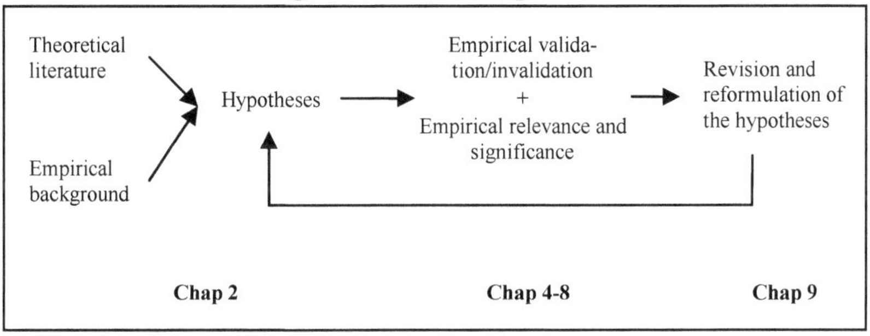

```
Theoretical
literature  \                        Empirical valida-
             \                       tion/invalidation       Revision and
              →  Hypotheses    →          +               →  reformulation of
             /                       Empirical relevance and  the hypotheses
Empirical   /                            significance
background
                   ↑_____|

        Chap 2                        Chap 4-8                Chap 9
```

This research relies on a comparative policy analysis approach. The main benefit of a comparative approach is twofold (Hassenteufel, 2000; Smith, 2000): to stand back from the individual cases and give them another look, which improves the eventual in-depth understanding of the cases; and to test the theoretical framework in more cases, which enables to increase its eventual generalisation. In order to avoid falling into the common traps of the comparison, we have used the same predefined conceptual and theoretical framework in all cases. This makes sure that we work with comparable concepts (the comparability issue) and that we do not concentrate on case-centred descriptions (e.g. canada-dry comparisons, Hassenteufel, 2000).

Comparative policy analysis relies on in-depth case studies using qualitative data. The choice of a case study method is explained by the small number of cases that we decided to investigate (Five cases) as well as the ambition to capture the details of each case with an historical perspective. Indeed, the benefit of the case study method is that it allows to understand complex issues and to carry out detailed contextual analysis, and if carefully and systematically implemented, the results of these case studies can offer interesting grounds for generalisations (Hamel, 1997; Dente, 1999; Yin, 2003), especially comparative case studies. We selected five cases according to a dissimilar system design, since our purpose is to analyse common explanatory factors in different contexts.

Based on their differing configurations on a number of key aspects (the RES-E policy change (dependent variable), their past RES-E policies, their policy actors configuration, their political and institutional context, and their relations with the European policies (independent variables), we decided to analyse the RES-E policy changes in Belgium, Denmark, Germany, the Netherlands, and the UK. Other countries could have been selected, but this sample of European countries provides a good idea of the diversity of the RES-E policies in Europe and at the same time these cases are similar enough to allow the comparison (e.g. the same geographical area, the same period of time for policy change, similar cultural background).

In order to make sure that we were collecting comparable data in each case, we developed a common case study framework that we applied in every case and used the same kind of sources (primary and secondary sources) to collect the data. The case study framework we used to describe the RES-E policy changes in the five countries includes: an overview of the electricity sector (statistics on the electricity supplies over the last decades), a chronology of the RES-E policies in the past, an overview of the electricity market liberalisation process and its impact on the RES-E sector, an analysis of the actors involved in RES-E policy making, and finally a description of the RES-E policy changes and the current state of the policy. Each case study then ends with the interpretation of the policy change (dependent variable) and the explanation of the policy change (independent variables). The empirical research on each case is based on qualitative data collected from primary and secondary sources. Primary sources cover, on the one hand, all sorts of available documents (legislation, administrative decisions, working papers, press releases) and, on the other hand, information collected directly at the policy actors by means of interviews or questionnaires[1]. Secondary sources cover publications and reports elaborated by private, public or academic actors specialised in the RES-E sector (publications, scientific reports) or interviews with policy experts. Finally, the implementation of the case studies in this research has benefited heavily from a research project conducted in the period 2002-2004 in coordination with the Thermodynamic Department of the University of Louvain, which resulted in two publications (de Lovinfosse and Varone, 2004; Tchouate Heteu, 2004). This research project involved the participation of external academics who conducted in-depth case studies of the RES-E policies in Denmark (Asano, 2004), Germany (Lauber and Pesendorfer, 2004) and the UK (Connor, 2004).

---

[1] In this research, the use of questionnaires concerns only the German and UK cases, where interviews with actors turned out to be more difficult to organise.

Introduction

When doing comparative research, two strategies are available to structure the paper (Hassenteufel, 2000): a structure by cases with a comparison at the end, or by analytical variables, with a continuous comparison. In this paper we decided to adopt the structure by cases, which makes it easier to apprehend for an outside reader, but to combine it with the adjunction of analytical sections at the end of each case. In this way, the reader who is not familiar with the cases will find it easier to understand the paper, and the reader who is familiar with the cases can focus on the analytical and comparative sections of each case study and focus directly on the analysis. Therefore, the paper is composed of ten chapters beyond this introduction. Chapter 2 presents the theoretical framework of the research. Chapter 3 briefly describes the European RES-E policy in order to give an overview of the context in which the domestic policies are being formulated. Chapters 4 to 8 present the five case studies: Belgium, Denmark, Germany, the Netherlands and the UK. Then, chapter 9 presents the comparison of the five cases and the revision of the theoretical framework. And finally, chapter 10 concludes the paper.

CHAPTER 1

Theoretical Framework

The theoretical question addressed in this research is "How and why do policies change?". In order to investigate this question empirically, we choose to look at the RES-E policy changes that have recently been implemented in five European countries. Looking at the RES-E policy changes will help us to better understand the policy change process: validating/invalidating hypotheses on the causes of policy changes and, eventually, discussing the grounds for a more general explanatory framework for policy changes.

The first section of this chapter clarifies what we mean by "policy change" and how our definition relates to the literature (the "how" question). In addition, we will show how to operationalise our definition in the empirical investigation about RES-E policies. In the second section of this chapter, we propose a theoretical framework and formulate hypotheses about the causes of policy changes, and we apply them to the particular case of RES-E policy changes in Europe (the "why" question).

I. Definition of Policy Change: A Conceptual Framework

Since the first call for the "policy sciences" from Lerner and Lasswell in the 1950s (Lerner and Lasswell, 1951), the literature and research on public policy have developed in different directions. One must distinguish two different dimensions of the public policy analysis: the knowledge "of" and knowledge "in" the public decision process (Lerner and Lasswell, 1951). The second dimension (knowledge "in", or knowledge "for") refers to a more operational approach of the policy analysis which seeks to use the knowledge on policy for political or professional purposes (Bobrow and Dryzek, 1987). Policy analysis aims at providing policy actors (public or private) with advice on and recipes for the "right" policies to adopt in order to achieve the "right" goals. The first dimension (knowledge "of") is concerned with explanation rather than prescription. It refers to the scientific understanding of the causes and consequences of public policy. This research participates in the first dimension since our purpose is scientific and not political. Even if political prescriptions might certainly be drawn from the conclusion of the research, this is another job.

In addition, policy analysis (as knowledge "of") can be used with different ambitions: to describe policies (information), to explain policies (causes or determinants) or to evaluate policies (consequences or impacts) (Dye, 1998). Our purpose in this research is twofold. Firstly, we describe the RES-E policies and therefore give information and data about a specific policy area in specific locations. Secondly, we seek to go beyond description by uncovering the determinants and causes of the RES-E policies. But we do not specifically look at the consequences of the policy in this research. Our focus is on the policy formulation and not on the policy outcomes.

Besides these two dimensions of the "policy sciences", no such thing as a commonly agreed definition of "public policy" is to be found in the literature. Most researchers understand public policy as "whatever governments choose to do or not to do" (Dye, 1998), but many questions remain about the scope of the concept and how to apprehend it empirically. As Heclo noticed, "the term policy is usually considered to apply to something 'bigger' than particular decisions, but 'smaller' than general social movements" (1972, p. 84).

Indeed, if a policy refers to all a government (or whatever public authority) does or does not do, it covers a very large scope of action, inaction, decisions, absence of decision, that may be very difficult to identify empirically. Where does a public policy begin and finish? It is relatively easy to identify the actions or decisions of the government, but it is much less easy to identify its inactions/non-decisions. Furthermore, it is often very hard at once to apprehend empirically all the actions/decisions that compose a public policy, especially because a public policy is hardly ever either conceptually or empirically given.

Therefore, many authors insist on the constructed aspect of a public policy. A policy is understood as a social and research construct (Meny and Thoenig, 1989; Muller and Surel, 1998; Thoenig, 2004). One of the most difficult tasks of the policy analyst is thus to define the boundaries of the policy, namely which actions/inactions, decisions/non-decisions of the public authorities compose the public policy to be analysed. To do so, several authors give a more detailed and pragmatic definition of public policy which can help us to delimitate the boundaries of a public policy.

Thoenig (Meny and Thoenig, 1989; Thoenig, 2004) defines public policies as the interventions of a public and legitimate authority on a specific domain of the society or of the territory. These interventions can take three different forms: the policy conveys contents, materialises in outputs and produces outcomes. Accordingly, the policy analyst must look at what the public authority actually does and not at what it ought to do.

Muller and Surel (1998) have a slightly different approach since they focus on three other fundamental aspects of a policy: the normative action framework that the policy produces, the public authority and expertise resources on which it relies, and finally the local order that it seeks to create. The visible content of a policy (the normative action framework) is composed of resources (financial, intellectual, and regulatory) and outputs (regulatory, financial, and physical). The problem for the policy analyst is on the one hand to identify the relevant outputs of a given policy (actions and decisions) and on the other hand to reconstruct the meaning, the objectives of the policy (coherence between outputs).

Knoepfel, Larrue and Varone (2001) give a more "operational" definition of public policy that brings together the main consensual elements of the literature. In short, public policy is defined as a series of intentionally coherent decisions or activities taken or carried out by different public – and sometimes – private actors, whose resources, institutional links and interest vary, with a view to resolving in a targeted manner a problem that is socially and politically defined as collective in nature. Thus there is no public policy unless several elements are found: a series of decisions or actions (no single decision or action), with an intended coherence (model of causality), taken by public actors (private actors in specific circumstances, such as delegation), at different levels (constitutional, legislative, administrative etc.), that result in concrete actions (implementation of the decisions), aiming at modifying the behavior of specific target-groups (causes of the policy problem). Building on the policy cycle approach of the policy process (May and Wildavksy, 1978), Knoepfel *et al.* divide the policy process into four stages (agenda-setting, policy formulation, implementation, evaluation), and each stage produces specific products: the agenda-setting stage defines the policy problem, the policy formulation stage elaborates the politico-administrative program and the politico-administrative arrangement, the implementation stage produces the action plan and the outputs, and finally the evaluation stage focuses on the policy impact (effects of the policy on the target groups) and policy outcomes (effects on the policy beneficiaries). This definition is very useful and handy for empirical research on public policies since the authors go into detail when describing each stage and each product. However, the scope of their definition is still large since empirical research hardly ever considers all stages and all products.

In conclusion, building on these definitions, we look at public policies as what public authorities do or choose not to do, and we do so by emphasising the formulation stage of the policy process (politico-administrative programs). Indeed, we choose to address the question of policy change neither from the agenda-setting perspective, nor from the policy implementation or evaluation perspectives, but with a focus on the formulation of the policy programs. Therefore, we assume that public policies materialise in public decisions and that by analyzing their content, the main dimensions of the policy can be revealed (policy goals, policy instruments, and settings of the instruments).

A. Overview of Policy Change Literature

The literature on policy change is rich and diverse. In order to put some order in it, we decided to look at the position of the different authors or approaches on three fundamental questions: *"What in the policy changes?"*, *"How does the policy change?"*, and *"Why does the policy change?"*. The first two questions ("what?" and "how?") are addressed in this section and the third one ("why?") will be discussed in section 2.2 of this chapter. However, this distinction between the three questions is very useful so as to make sense of the literature as well as elegant to present, but it involves some simplifications that must be acknowledged beforehand. The theoretical literature is still under construction and is based on perpetual questioning of one approach by another, so the different approaches are not independent and interact. I try to highlight some of these interactions in this overview, but the reality is certainly more complex than it appears. In addition, the questions "how" and "why" are usually addressed together in the literature, which is why there are many overlaps in my review. But I believe it does make sense to attempt to distinguish the author's answers to those questions individually (even if the answers are closely linked), just to help understand the literature on policy change from a more systematic perspective.

The *"what?"* question aims to identify which aspect of the policy actually changes. The literature is full of different focuses on what changes. Some authors investigate the policy agenda and look at how new policy problems arise on the policy agenda (Kingdon, 1984; Baumgartner and Jones, 1993). Others put the policy instruments at the centre of their investigation about policy changes (Hood, 1983; Salamon, 2002; Hall, 1993; Linder and Peters, 1989; Jobert, 1994; Palier, 2002; Eliadis *et al.*, 2005). Therefore, policy instruments are more and more being considered not only to be technical or functional tools of government but as political choices. Finally, the policy change literature also considers the larger cognitive dimension of the policy, namely the policy paradigm (Hall, 1993; Surel, 1995; Howlett and Ramesh, 1995) or referential (Jobert and Muller, 1987) or core belief system (Sabatier and Jenkins-Smith, 1993). All three concepts refer to the stable characteristics of a policy model that determine individual policy choices over a period of time. Nevertheless, some authors do not pay particular attention to "What changes?", but concentrate on "How do policy changes happen?" (e.g. policy cycle approach, incrementalism, path-dependency, policy learning).

The *"how?"* question looks at the ways policies change in terms of extent (degree of change) and type (characteristics of change). Most of the literature on policy change makes the distinction between major and minor policy changes (incrementalism, path-dependency, policy learning, punctuated equilibrium, etc.) with different assumptions and hypotheses about the process and characteristics of the change itself. Other authors consider the life cycle of a policy and identity different types of policy change according to the stage of the cycle where the change appears (Bardach, 1976; de Leon, 1978; Hogwood and Peters, 1983) without discussing the degree of change per se.

The proponents of the *policy cycle approach* consider that change is constantly taking place in the policy process. For instance, many authors have focused on the question of policy termination (end of a policy) and the conditions under which such a termination is expected or improbable (Bardach, 1976; de Leon, 1978; Hogwood and Gunn, 1984). In this vein, Hogwood and Peters have developed a comprehensive framework of policy change that distinguishes different kinds of policy change closely linked with the stages of policy cycle: policy innovation (new area of policy), policy succession (replacement of existing policies), policy maintenance (adaptive changes to existing programs) and policy termination (termination of an existing policy) (Hogwood and Peters, 1983). These types of policy change take the form of ideal-types in Hogwood and Peters' model. Indeed, any policy change in practice contains elements of one or more of these types of policy change. Therefore, any given policy change will be located on a dimension on which the ideal types form the polar extremes (innovation-succession, succession-maintenance, maintenance-termination, etc.).

In this approach, "how policies change" is closely related with the idea of the policy process as a cycle composed of different stages. Even if this conception proved to be useful as a heuristic tool to understand the policy process, it has important shortfalls when it comes to operationalise it systematically in empirical studies and understand complex policy changes.

Most other publications on policy change do not develop *ad hoc* typologies of change, but rather discuss the conditions under which policies continue (with minor change) or change substantially (major change). In this vein, most authors look at policy change according to these two modalities (minor versus major change), but with different meanings behind them.

According to the *incrementalist approach*, public policies change gradually by means of incremental adjustments which only marginally modify the status quo (Lindblom, 1959). Because of the bounded rationality of the decision-makers (Simon, 1957), the decision-making process is constrained by on the one hand the existing policies that limit the range of possible alternatives, and the need for consensus decisions that require bargaining and negotiations between actors. Therefore, this approach is more useful to account for policy continuity than for significant policy changes. In addition, compared to other approaches of policy change, incrementalism focuses more on the policy-making process than on the policy itself.

The *path-dependent* approach, which originates from economics (increasing returns), also focuses on the policy continuity within policy changes (Pierson, 1993, 2000). It is the central idea of the historic neo-institutionalist approach, which seeks to account for the continuity in public policy (Palier, 2004). Policy change is therefore determined on the one hand by the existing policies and previous policy choices and, on the other hand, by the existing institutions. What Pierson calls "policy lock-in" effects therefore dominate the policy change process. This conception of policy change also relates to the "policy inheritance" and "policy legacies" concepts which were developed concurrently to account for past determinism on future policy choices (Rose, 1990; Rose and Davies, 1994). To sum up, this approach again focuses more on minor than on major policy changes.

The *punctuated-equilibrium* model of Baumgartner and Jones give a more comprehensive account to policy change, especially policy agendas change (Baumgartner and Jones, 1993; True, Jones and Baumgartner, 1999). Their model explains why a political system can be both incrementally conservative and subject to more radical phases of policy-making (Parsons, 1995, p. 203). Indeed, policy change is understood in terms of "punctuated equilibrium", which means that the policy process is characterised as having long periods of stability (equilibrium with minor policy changes), then followed by periods of instability and major policy change (critics of the policy monopoly). During the periods of instability, the policy image (policy problem) and the institutional context (institutional policy venues) are challenged, reconsidered or destroyed, resulting in major policy changes. Therefore, this model is more comprehensive than the incrementalist and path-dependent approaches since it accounts for both minor (during stability periods) and major policy changes (during instability periods).

Another school considers policy change in terms of *policy learning*. Consistent with the idea "to bring the state back in" (Evans, Rueschemeyer, Skocpol, 1985), they focus on how the State learns from past experiences and adapts to the environment, regardless of social pressure. While it has been developed primarily as a partial corrective to theories of policy change based on the notions of power and conflict (Heclo, 1974; Bennett and Howlett, 1992), it proved to be not an alternative but a complementary hypothesis (Hall, 1993). Like the incremental and path-dependant approaches, policy learning is mainly relevant to explaining minor policy changes since policy changes occur through progressive learning processes. However, several proponents of this approach have developed a more general framework of policy change which also accounts for major policy changes (Hall, 1993; Sabatier, 1993). Therefore, while policy learning originally applied to minor policy changes, it has also proved to be relevant to understanding major policy changes under certain conditions and circumstances.

The policy learning literature is very rich and diverse (see Bennett and Howlett, 1992; May, 1992; Stone, 1999 for a review). Two authors have developed well-grounded frameworks of policy change (based on the policy learning approach) that deserve to be presented into more detail (Hall, 1993; Sabatier, 1993).

Sabatier and Jenkins-Smith, in their "advocacy coalition framework", suggest that the key to understanding policy change is the dynamics of policy-oriented learning which takes place within the advocacy coalitions (Sabatier and Jenkins-Smith, 1993). Furthermore, they distinguish the dynamic of policy-oriented learning which results in modifications of the secondary aspects of the policy (minor policy change, which is frequent) and changes of the core aspects of the policy (major policy change, which is very rare) that result from either a shift of the dominant advocacy coalition, or learning between coalitions due to pressures external to the subsystem. According to this framework, the policy process takes place within a specific policy subsystem composed of advocacy coalitions. The policy subsystem is defined as "the set of actors who are involved in dealing with a policy problem" (Sabatier and Jenkins-Smith, 1993, p. 24). It contains a large and diverse set of actors who are grouped in advocacy coalitions. Each advocacy coalition shares a particular belief system which is characterised by "core" (deep normative core and near policy core beliefs) and "secondary" elements (secondary aspects of the policy). The core elements of the belief system are hypothesized to be relatively stable over time (one decade or more), while secondary aspects can be easily reconsidered by learning. The public policies (empirical governmental programs) are conceptualised in the same way as the actor's belief systems, that is, as sets of value priorities and causal assumptions on how to achieve their objectives (see table 1).

Table 1: Structure of belief systems and governmental programs[1]

	Deep (normative) core	Near (policy) core	Secondary aspects
Defining characteristics	Fundamental normative and ontological axioms	Fundamental policy positions concerning the basic strategies for achieving normative axioms of deep core	Instrumental decisions and information searches necessary to implement policy core
Scope	Part of basic personal philosophy. Applies to all policy areas	Applies to policy area of interest (and perhaps a few more)	Specific to policy area/subsystem of interest
Susceptibility to change	Very difficult; akin to a religious conversion	Difficult, but can occur if experience reveals serious anomalies	Moderately easy; this is the topic of most administrative and even legislative policymaking
Illustrative components	Nature of man	Scope of governmental vs. market activity	Decisions concerning administrative rules, budgetary allocations, disposition of cases, statutory interpretation, statutory revision
	Ultimate values	Identification of social groups whose welfare is most critical	Information concerning program performance, seriousness of the problems, etc.
	Basic criteria of distributive justice	Orientation on substantive policy conflicts	
		Basic choices concerning policy instruments	

Source: Adapted from Sabatier and Jenkins-Smith, 1993, p. 31.

Therefore, there is a direct link between modifications in the belief system of the policy actors and policy change. That is, the process of policy-oriented learning usually results in minor policy changes (secondary aspects of the policy) and major policy changes (core aspects of the policy) occur only rarely, viz. when the core aspects of the belief system are altered. Finally, this framework suggests that policy learning is relevant to explain minor policy changes, but that major policy change rather responds to larger factors and struggle for power logics.

Peter Hall also reaches a similar conclusion to Sabatier's in his work on social learning when he states that

---

[1] Governmental programs translate public policies in practical and perceptible units. Note that the American meaning of "program" is larger than the traditional European understanding.

the processes of first and second order change in policy correspond quite well to the image of social learning. (...) However, the process of third order change (...) was quite different (Hall, 1993, pp. 287-288).

Both authors acknowledge the usefulness of learning to account for minor policy changes, but recognise that other processes are working when major shifts in policy occur. However, they do not share the same approach to learning.

Building on Heclo's work (1974), Peter Hall defines social learning as

a deliberate attempt to adjust the goals or techniques of policy in response to past experience or new information (Hall, 1993, p. 278).

Learning is assumed to take place inside the state (autonomy of the state from social pressure), which means that the key actors pushing forward the learning process are

the experts in a given field of policy, either working for the state [civil servants] or advising it from privileged positions at the interface between the bureaucracy and the intellectual enclaves of society (Hall, 1993, p. 277).

Hall further develops the social learning approach by stating that the learning process can take different forms depending on the kinds of policy change that are involved. Accordingly, he identifies three orders of policy change (see table 2): first order change (the most frequent one) relates to modifications of the settings of the policy instruments; second order change refers to circumstances when the policy goals remain largely the same but the instruments used to attain them are modified; and third order change (the least frequent one) entails simultaneous changes in all three components of the policy (policy goals, policy instruments and settings of instruments). Even if they represent distinct kinds of change, each order of change is linked to the others hierarchically. As Hall states,

this classification reflects a lexical ordering: first order learning represents a change in the simplest of policy variables; second order learning changes both policy instruments and their settings; and third order learning entails a change in all three sets of variables (Hall, 1993, p. 293).

Table 2: The dimensions of policy and corresponding policy change orders

Dimensions of policy	Policy goals	Policy instruments	Settings of instruments
Policy changes			1st order change
		2nd order change	
	3rd order change		

Hall associates change in the policy instruments or the settings of instruments (1st and 2nd order changes) with social learning and perceive it as normal policy-making. However, 3rd order change only happens in specific and rare circumstances associated with shifts in policy paradigms, namely the interpretative framework of ideas and standards shared by the policymakers which specifies not only the goals of the policy and the kind of policy instruments that can be used to attain them, but also the very nature of the policy problem. Unlike other orders of change, the process of 3rd order change spills beyond the boundaries of the state into the broader political arena and involves not the adjustment of the goals and techniques of the policy (social learning), but the revision of the existing policy paradigm (paradigm shift). To sum up, Hall distinguishes minor policy changes associated with social learning inside the state (1st and 2nd order changes) and major policy change associated with a shift in the overall policy paradigm (3rd order change).

The approaches of policy change developed respectively by Sabatier and Hall have several characteristics in common. Firstly, they focus on the learning process that takes place inside a dominant coalition of actors of a specific policy area (change in secondary aspects of a belief system, first and second order change). Consequently, in both perspectives, major policy change (change in core beliefs and 3rd order change) must be analysed in the long term (over a decade, according to Sabatier), while minor policy change occurs more frequently in the normal policy-making process.

Secondly, both describe the policy change process in three different degrees based on the dimensions of the policy to be altered (the link to "what?" and "how?). However, their dimensions do not perfectly match since they do not use the same definition of and approach to policy change (see table 3). Changes in the secondary aspects of the policy in Sabatier corresponds to the 1st order change in Hall (settings of the policy instrument), but it also covers modifications in the policy instruments that go beyond the settings in Hall (2nd order change) (see table 3). Indeed, Sabatier's definition of the settings of the policy (see table 1) involves "instrumental decisions necessary to implement policy core", which goes beyond the definition of the settings of the policy instruments in Hall because it concerns the choice of the instrument itself, at least partially. However, some aspects of the policy instruments are also included in Sabatier's "near policy core" (e.g. basic choices concerning policy instruments, see table 1), so the dimension of 2nd order change in Hall seems to overlap with Sabatier's secondary aspects and near policy core dimensions (see table 3). Finally, the idea of 3rd order change in Hall (change in policy goals) is addressed to a large extent in Sabatier's near policy core dimension (fundamental policy positions, applicable to the policy area, difficult to change; see table 1). The last dimension of policy change in Sabatier, deep normative core beliefs (fundamental normative and ontological axioms, personal philosophy, applicable to all policy areas, akin to religious conversion; see table 1), is not considered as such by Hall (see grey area in table 3). Some aspects of the policy paradigm idea in Hall relate to certain dimensions of the deep normative core of Sabatier (e.g. the Keynesian versus the monetarist vision of the basic criteria of distributive justice or nature of man), but in Hall they tend to apply to a given policy area (e.g. macroeconomic policy in Hall, 1993). In Sabatier they relate to the more general philosophical axioms of the policy actors in all policy areas (e.g. nature of man), so those two dimensions overlap only to a small extent.

Table 3: Sabatier and Jenkins-Smith and Hall's dimensions of policy change in comparison

Sabatier and Jenkins Smith (1993)	Deep (normative) core	Near (policy) core	Secondary aspects
Hall (1993)	3rd order change	2nd order change	1st order change

In conclusion, many different approaches of policy change coexist in the literature. Most emphasise the "How do policies change?" question, with many different answers being developed, without systematically looking at the "What changes in policy?". Furthermore, only few authors formally and systematically link the "what?" and the "how?" questions in a comprehensive framework. In the next section, we develop an original conceptual framework of policy change based on the existing literature and we apply it to a specific policy area (RES-E policies).

B. A Conceptual Framework for RES-E Policy Changes in Europe

We notice that most of the literature on policy change makes the distinction between "major" policy changes (less frequent, more difficult, involving fundamental aspects of the policy) and "minor" policy changes (more frequent, less difficult, involving more operational aspects of the policy). But even if this distinction dominates, the actual definition of what major versus minor change covers differs from one author to another.

The purpose of this research is to make sense of why policies change over time. Before looking at why they change, we must define what we mean by "policy change". Therefore, we present a conceptual framework of policy change that is based on Peter Hall's approach of distinguishing policy changes at the level of the policy goals, the policy instruments or the settings of the instruments. However, unlike Peter Hall, we consider only two modalities of policy change, in the line with most of the literature on policy change: change in the policy objectives/goals and change in the policy instruments.

We do not consider changes in the settings of the instruments as a particular category of policy change in our framework, and this for two reasons. Firstly, changes in the settings of the policy instruments are relatively easy to achieve and very frequent (Hall, 2003). They usually do not reflect significant modifications of the policy, but only marginal adaptations of the instruments. Having said that, sometimes modifications of the settings have a greater impact on the policy outcomes than the modification of the instrument itself. For instance, if the budget of an instrument is tripled (as a result of increased expected policy goals), the impact on the policy outcomes is expected to be significant, even in the absence of change in the policy instrument itself. The settings of the instruments can therefore result from changes in the policy objectives, in the absence of change in the policy instrument. So, we do consider changes in the settings of the policy instruments in this research, but we do not develop a specific conceptual framework about them. Secondly, Peter Hall considers changes in the settings of the instruments as a particular kind of policy change (1st order change), but in his explanatory framework, he tends to consider change in the policy instruments (2nd order change) and the settings together. Indeed, in his approach, 1st order or 2nd order policy changes represent normal policy making (corresponding to the image of social learning) as opposed to major policy change in case of 3rd order change; and the factors that he presents to explain the 1st and 2nd order changes are the same (past policy failures and no change in the locus of authority over the policy) as opposed to 3rd order change explanatory factors that differ (past policy failure, change in the locus of authority, and debate spilling beyond the borders of the state) (Hall, 1993, pp. 287-288).

In addition, unlike Peter Hall, we do not postulate a hierarchy between the two dimensions of policy change, namely that changes in the policy goals necessarily lead to changes in the policy instruments. We consider them as two related but not linear instances of policy change. This means that change in the policy objectives does not necessarily lead to change in the policy instruments. For instance, there are cases where the objectives of the policy will change in the light of new policy problems (e.g. climate change, liberalisation of the electricity market), but the existing instrument is still adapted to deal with the new objective (e.g. ecotax, feed-in tariffs for RES-E). Or there are cases of policy instruments that are looking for a new legitimisation by a new policy objective because the political agenda will change (e.g. subsidies to nuclear power with the climate change issue) (e.g. Cobb and Elder, 1983; Rochefort and Cobb, 1994; Kingdon, 1994). On the other hand, in some cases the policy objectives do not change, but the instrument used to attain them change because it failed to reach the goals (e.g. ecotax exemption will increase the demand for RES-E but not domestic RES-E production), or because it is not adapted to the context anymore (e.g. the liberalisation of the market). So, the relation between change in the policy objectives and instruments is not as simple as it appears at first sight, and certainly not hierarchical.

In order to operationalise this approach of policy change in empirical research about RES-E policy changes, specific variables for changes in policy objectives or policy instruments must be identified. Peter Hall does not clearly specify the variables of policy change in his work, but only gives some empirical examples from his analysis of the macroeconomic policies in Britain over the period 1970-1989. Table 4 summarises our conceptualisation of policy change as either a change of the policy objectives or the policy instruments. On the one hand, changes in the policy objectives refer to a modification of the target of the main policy objective that the policy seeks to address (e.g. increase of the RES-E share in the electricity sector). On the other hand, a policy change at the policy instrument level can involve a change either in the target group[2] (e.g. demand actors or supply actors), or in the incentive used to reach the policy objective (e.g. favourable prices for RES-E or

---

[2] Note that there is a fundamental distinction to make between the "target" of the policy objective (sometimes referred as "policy target"), namely the quantitative target pursued by the policy; and the "target group", namely the actors that are targeted by the policy (actors whose behaviour is expected to be modified by the policy instrument).

guaranteed quantity of RES-E), or in the resources used to finance the instrument (e.g. public budget or private funds).

Table 4: Definition and operationalisation of policy change

Policy change	Policy objectives	Policy instruments
Variables	Target of the main policy objective	– Target group – Incentive – Resources

Firstly, a public policy can pursue different *objectives* at the same time, which are more or less complementary or conflicting. Usually, there is a hierarchy between these objectives, and some of them are given more political priority over the others. This hierarchy is not always formally acknowledged, which makes it difficult sometimes for the policy analyst to observe. The hierarchy between the objectives may change over time (Hall, 1993). Besides, not only the hierarchy but also the content of the policy objectives, namely their specific target, what they aim to achieve, change over time. So changes in the policy objectives can take two forms: a modification in the hierarchy of the objectives, or changes in the objectives themselves.

In his analysis of the macro-economic policy in Britain over the period 1970-1989, Peter Hall looks primarily at the change in the hierarchy of the policy objectives (reduction of unemployment versus command of inflation). In our research, we look not so much at the hierarchy among the overarching objectives of the RES-E policies in Belgium, Germany, the Netherlands, Denmark and the UK; but at the main objectives their RES-E policies seek to achieve and how these objectives may change over time. We observe that the main objective of the RES-E policy in all five countries is to increase the share of RES-E in the domestic electricity market. This objective is usually formulated through quantitative targets (e.g. percentage of total electricity supplies/ generation, quantity of kW/MW installed capacity) to be reached within a given period of time, usually a medium to long-term deadline (e.g. by 2010/2020). So, according to our framework, when the quantitative targets that the policy aims to reach are modified, it represents a change in the policy objectives (see table 4).

Of course, the RES-E policy in those five countries is also driven by other, non-mutually exclusive, policy objectives, such as the development of new RES-E technologies (R&D policy) or the promotion of domestic RES-E industries (industrial policy); but we observe that these objectives are often complementary to the main objective of the RES-E market penetration. In the past, the R&D or industrial goals have been the main objectives of the RES-E policies in most countries, and they still remain significant (e.g. Germany, the UK, Belgium), but today the priority of those countries appears to be the increase of the RES-E market shares in the liberalised electricity market. The main reasons behind this priority to market development include: the national and international commitments to the reduction of CO_2 emissions (replacement fossil fuel electricity generation by RES-E generation), the liberalisation of the electricity market (new market regulation, new market players), the European indicative targets adopted in the 2001 RES-E directive (in percentage of total electricity supplies), and the availability of new and competitive RES-E technologies (e.g. wind turbines).

Table 5 gives a summary of the RES-E policy objective changes in the five European countries analysed in this research. The details of those changes can be found in the next chapters. Since some countries have experienced more than one RES-E policy change during the last decades, we had to make a decision about which change we will look at. We decided to focus on the most recent significant change in each country, which leads us to identify a specific turning point in each case. This turning point is the moment when the policy change is formally initiated, but the subsequent policy change process continued afterwards in several cases (Belgium, Germany, Denmark, and the UK).

Table 5: Changes in the policy objectives

Country	Turning point	Target before	Target after	Change of policy objective
Belgium	1999	None	3% by 2004 and 6% by 2010	Yes
Denmark	1999	20% by 2005, 29% by 2010 and 68% by 2030	20% by 2003, 29% by 2010 and 68% by 2030	No
Germany	2000	None	12.5% by 2010 and 20% by 2020	Yes
Netherlands	2003	9% by 2010	9% by 2010	No
UK	2000	1,500 MW installed capacity (+- 3%) by 2000	10% by 2010 and 15% by 2015	Yes

The other modality of the policy change that this research addresses is change in the *policy instruments*. The literature on policy instrument is abundant and participates in both approaches of public policy, namely policy science (knowledge for policy) and policy analysis (knowledge of policy) (Eliadis *et al.*, 2005). Most authors tend to argue in favour of an increased importance of the instruments in the public policy process (Hood, 1983; Varone, 1998; Salamon, 2002; Lascoumes and Le Galès,

2004; Eliadis *et al.*, 2005). Policy instruments appear not only as instrumental devices determined by policy objectives (Hood, 1983), but also as the expression of political choices (Varone, 1998, Salamon, 2002; Peters, 2002; Ringeling, 2002; Lascoumes and Le Galès, 2004).

In Hall's approach of policy change, the policy instruments may change even in absence of changes in the policy objectives (second or first order changes). It appears in cases when the policy objectives are not modified, but the policy actors are dissatisfied with the instruments implemented to reach them. In those cases, either the policy instruments themselves are replaced or the settings of the instruments are adapted. Since changes in the settings of the instruments are very frequent and usually less controversial than instrument changes, our research focuses only on changes in the policy instruments themselves. Changes in the settings are investigated in the case studies, but they are not the subject of this theoretical framework.

Most of the time, the governments use a mix of instruments to achieve their policy objectives. This raises a lot of questions about the policy mixes and their effectiveness in reaching the goals of the policy (Howlett, 2005). However, the policy mix issue is not the subject of this study. We recognise that a mix of instruments is often used in the RES-E policies to increase the share of RES-E in the electricity market, but we also notice that there is usually a central instrument specifically adopted for this purpose and the other instruments of the mix either aim at filling the gaps the central instrument does not cover (e.g. supporting less competitive technologies), or were primarily designed for achieving other policy objectives (e.g. R&D in new technologies, support to a specific industry, CO_2 emissions reductions). So, even if in practice a mix of policy instruments contribute to the policy, our research does not look at changes in this policy mix, but at changes in the central policy instrument that has been adopted to reach the policy objective. Identifying this instrument is usually quite straightforward as we analyse the policy documents of the government as well as the legislation.

When taking a closer look at what a policy instrument is and how to characterise it, we face a large number of different typologies with a wide inventory of criteria in the literature (Hood, 1983; Varone, 1998; Salamon, 2002; Lascoumes and Le Galès, 2004; Howlett, 2005). Indeed, a policy instrument can be characterised by many different attributes: the nature of the coercion (e.g. Howlett and Ramesh, 1995), target groups (e.g. Ingram and Schneider, 1991; Varone, 2000; Knoepfel *et al.*, 2001; Howlett, 2005), impact on target groups (e.g. Padioleau, 1982), state control mechanisms (e.g. Hood, 1983; Howlett and Ramesh, 1995), the resources mobilised (e.g. Hood, 1983; Varone, 1998; Howlett, 2005), etc. Out of all these criteria, we believe that three attributes of the policy instruments will help us best to characterise and understand the RES-E policy changes in Europe: the actors targeted by the policy instrument (target group), the incentives used to reach the policy objective, and the financial resources mobilised by the instrument (see table 4).

Firstly, policy instruments aim to influence the behavior of a specific group of actors (target group). Target groups are the actors whose behavior needs to be modified in order to reach given policy goals. The choice of a specific target group refers to a particular causal assumption about the policy problem (who caused it and how?) which may be very controversial among the policy actors and subject to intensive lobbying from the potential target groups (Knoepfel *et al.*, 2001). The RES-E policy instruments that seek to increase the shares of RES-E in the electricity market usually target one of those two main categories of actors: the demand actors (demand side of the electricity market, e.g. electricity suppliers, consumers) or the supply actors (supply side of the electricity market, e.g. electricity generators). However, since the electricity market is vertically fragmented, several actors act as intermediaries between the electricity generators (supply) and the final electricity consumers (demand). The question of where to put the boundary between the supply actors and demand actors is thus potentially controversial. We decided to look at the electricity market from the RES-E generators' point of view and thus we adopt a strict definition of supply (electricity generators and electricity traders) and an extended definition of demand (from network operators to suppliers and final electricity consumers). So, if the target group of a policy instrument switches overtime, we consider that the policy instrument changes. For instance, the switch of target group from the electricity generators (supply) to the electricity suppliers (demand) means that the policy instrument has changed (see table 6).

Secondly, policy instruments aim to reach the objective of the policy by using specific incentives designed to induce a modification in the actor's behavior. In the RES-E policy, the policy instruments traditionally aim to increase the share of RES-E in the electricity market or the number of RES-E installations installed in the country. To do so, two types of incentives tend to be used: prices or quantities. In the first case, favourable prices are offered to RES-E under specific conditions (e.g. connection to the grid, type of RES-E technology, location) and under different modalities (e.g. €/kWh generated, €/kW installed, percentage of investment costs). In the second case, a minimum quantity of RES-E (capacity or generation) is enforced (e.g. quota, call for tenders). So, if the type of incentive of a policy instrument changes from prices to quantities or vice versa, we consider that the instrument has changed (see table 6).

Thirdly, another important attribute of a policy instrument is the resources the instrument mobilises. Different criteria can apply about how to characterise the resources: the nature of the resource (e.g. information, organisation, authority, legitimacy, money), the effect of the resources (redistribution among the actors), and the origin of the resource (public versus private). In the RES-E policy, the last criterion makes a lot of sense to characterise the policy instruments. Indeed, in the context of the privatisation and liberalisation of the electricity sector in Europe, the question of where the resources of the RES-E policy come from is highly relevant. In the past, most of the RES-E instruments were financed by the public budget, but for the last decade, most RES-E policies have been funded by the electricity sector (in the end, the electricity consumers). Changes in the origin of the financial resources behind a policy instrument from public (public budget) to private (electricity consumers) funds represent significant changes for the policy instrument (see table 6).

Table 6 presents a summary of the changes in the RES-E policy instruments in the five European countries analysed in this research. The details of those changes can be found in the next chapters.

Table 6: Changes in the policy instruments[3]

	Before policy change			After policy change		
	Target group	Incentive	Resources	Target group	Incentive	Resources
Belgium	demand	price	private	demand	quantity	private
Denmark	supply	price	public	demand	quantity	private
Germany	demand	price	private	demand	price	private
Netherlands	demand	price	public	supply	price	private
UK	supply (demand)	quantity (price)	private	demand	quantity	private

Finally, table 7 integrates the information of table 5 and table 6 and gives an overview of the policy changes in our five countries.

---

[3] The grey cells emphasise the characteristics of the instrument that change.

Table 7: RES-E policy change in Belgium, Denmark, Germany, the Netherlands, and UK

Country	Turning point	Change policy objective	Change policy instrument
Belgium	1999	Y	Y
Denmark	1999	N	Y
Germany	2000	Y	N
Netherlands	2003	N	Y
UK	2000	Y	Y

In conclusion, following Meny and Thoenig (Meny and Thoenig, 1989), we believe that a public policy does not exist in abstracto, but is constructed by the analysis. In this section, we have presented our definition of what policy change is and how we suggest addressing it. Of course, the literature abounds in other approaches of policy change, but we believe this approach is consistent with the literature and will best help us to understand the latest RES-E policy changes in Europe.

II. Explaining Policy Change: A Theoretical Framework

Now that the "what?" and "how?" questions about policy change have been discussed and the conceptual framework of this research has been presented, this section addresses the "why do policies change?" question. As we mentioned above, the "how" and the "why" policy change questions are closely linked in the literature, so several references will be made here to the previous section of this chapter. Indeed, the way a theoretical approach characterises "how" policies change determines to a large extent the way it shall explain "why" changes occurs.

After having briefly reviewed the literature on policy change with a focus on the different approaches to "why" question, we present our own theoretical framework and formulate four sets of hypotheses to explain RES-E policy changes in Europe.

A. Overview of Policy Change Literature

Four main factors of policy change can be identified in the literature: the experience from the past, the policy actors, the international context and external shocks. Each factor has been developed by different approaches, which results in a pretty heterogeneous understanding of why policies change.

Firstly, many different theoretical approaches emphasised the importance of the *experience from the past* on policy change. The policy cycle approach, for instance, looks at the feedback process following the evaluation of past policies on policy changes (May and Wildavsky, 1978). As May and Wildavsky noted, "policies intended as solutions to certain problems are themselves the causes of new problems for which policy solutions must be designed" (1978, p. 13).

The path-dependent approach also focuses on the past to explain the direction of policy change. Pierson notes that "preceding steps in a particular direction induce further movement in the same direction", which is well captured by the idea of increasing returns imported from economics (Pierson, 2000, p. 252, 1993; Palier, 2004). Because of increasing returns processes, policy changes tend to follow pre-existing paths and therefore remain largely limited (minor changes). A very similar idea is also at the core of the historical neo-institutionalist approach (Steinmo, Thelen and Longstreth, 1992; Hall and Taylor, 1996; Steinmo, 2004), which asserts that policy changes are dependent upon pre-existing institutions and policies. To sum up, these approaches (policy cycle, path-dependency, historical neo-institutionalism) seem more useful to explain policy continuity than major policy change. However, some authors emphasise the influence of policy failures on policy change (Hall, 1993; Jobert, 1994; Kingdon, 1984; Dolowitz and Marsh, 2000). For Peter Hall and Bruno Jobert, policy change is expected only if past policies have failed, or are perceived by the policy actors as if they have failed. Therefore, policy failure is seen as a necessary (but not sufficient) condition for policy change in their work. Moreover, John Kingdon considers policy failure to be one possible mechanism through which a "window of opportunity" for policy change would open (a problem window), but even without that actual policy change happens, since additional conditions need to be present (political receptivity and available policy alternatives).

In conclusion, the experience from past policy (success or failure) is without a doubt an important explanatory factor for policy change. However, it proved to be useful especially to understand changes in policy instruments, and less to account for shifts in policy objectives, except in specific circumstances and in association with other explanatory factors (Hall, 1993).

Secondly, many different approaches and authors emphasise the role of the *policy actors* in policy change. Both collective actors (advocacy coalitions, policy networks) and individual actors (policy entrepreneurs, policy brokers) are seen as crucial factors of change in the literature. In the last decades, research on public policy departed from the traditional focus on the iron triangle of actors (interest groups, civil servants and elected politicians) to steadily increase the number of actors involved in policy-making (e.g. academics, think-tanks, researchers, journalists, etc.). This phenomenon has been linked with the development of new forms of governance characterised by a more open decision-making process, and the increased use of private expertise to make decisions or delegate decision competences (Salamon, 2002; Grossman, 2004). Therefore, the question "who is a policy actor?" has become more complex in the last decades. The concept of policy networks emerged in the literature to account for the increasing openness of the state towards private actors and the resulting complexity and fluidity of the policy actors' structure. So more and more authors have developed new concepts to characterise the policy actors (e.g. policy community, issue network, advocacy coalitions, and epistemic communities). But, in order to study actor networks, it is necessary to first prove the existence of a strong and sustainable interdependence between the actors and/or shared beliefs and ideas, which is far from being trivial and may turn out to be tricky empirical work. In addition, the utility of these concepts to explain policy changes has been questioned by many authors who consider the policy network approach to be very useful as a heuristic device, but not as an explanatory framework (Dowding, 1995; Le Galès and Thatcher, 1995; Kassim, 1994; Dudley, 2003).

Among the rich literature on policy actors and policy change, two main dimensions of how actors induce change can be identified: policy learning and power struggle approaches.

The latter approach refers to how policy actors struggle to dominate the policy-making process and influence policy continuity or change according to their resource and interest. This literature is very rich, with many different studies focusing on, e.g. organised interests (corporatism) (Lehmbruch and Schmitter, 1982), the balance of ideas and interests in liberal democracies (pluralism) (Dahl, 1961; Dahl and Lindblom, 1976), party competition (Castles, 1982; Schmidt, 1996), veto players (Tsebelis, 1990), the rational choice of collective actors and veto players (Tsebelis, 1990; Scharpf, 1997), policy "networks" (Marsh and Rhodes, 1992; Haas, 1990; Sabatier and Jenkins-Smith, 1993; Dowding, 1993; Le Galès and Thatcher, 1995).

The former approach of policy actors focuses on ideas rather than power. Studies on policy learning look at how actors learn from past experiences and therefore change policies accordingly (Bennett and Howlett, 1992; May, 1992; Sabatier and Jenkins-Smith, 1993; Hall, 1993). Two theoretical frameworks on policy learning dominate the literature (Sabatier and Jenkins-Smith, 1993; Hall, 1993). They have already been presented in section 2.1. However, they both rely on different assumptions concerning which policy actors learn and how that needs to be developed here.

In Sabatier's work, the actors are aggregated into advocacy coalitions, which consist of both private and public actors from a variety of positions who share particular values and ideas (belief system, see Table 2) and show co-coordinated activity over time. Policy learning occurs when the belief system of the coalition is modified due to the influence of another coalition's beliefs, change in the context, or learning from past experiences. In this approach, major policy changes are unlikely as long as the prevailing coalition remains in power, except if its core beliefs are altered, which is highly improbable. Thus major policy change requires a redistribution of resources between the advocacy coalitions and the subsequent emergence of a new dominant coalition. On the other hand, minor policy changes are expected when the secondary aspects of the belief system of the dominant coalition are modified, but without a significant redistribution of resources among the existing coalitions. In brief, policy changes depend on the stability of the beliefs and resources of the advocacy coalitions who compete in a policy subsystem. Besides the "ideational" dimension of the coalitions (belief system), Sabatier also considers their resources and assumes that major policy changes result primarily from pressures external to the coalition that challenge its resources and power.

Unlike Sabatier, Peter Hall does not look at coalitions of actors, but at individual actors. He makes the distinction between the experts in a given policy field (civil servants or knowledgeable experts from society) and the politicians (without specific expertise, political actors and not bureaucrats), which is central to his framework. Indeed, policy learning is expected to take place among the experts of a policy field and result in first or second order changes (change in policy instruments or settings of instruments), which downgrades the role of politicians in policy change. But third order change (change in policy objectives), results from a shift in the locus of authority over policy from the civil servants (and other policy experts) to the politicians. So in third order changes, the key policy actors are the politicians, while in second or first order changes, the civil servants dominate the policy making process. In addition, Peter Hall also looks at the "openness" of the policy process to explain policy changes. He distinguishes between the cases in which the state (civil servants, state officials) acts rather autonomously from the pressures of external actors (first and second order changes), and the cases in which the policy making process spills over into the political arena and becomes the object of public debate (third order changes). So, Peter Hall makes the distinction between, on the one hand, policy changes in which the policy actors are experts in a given policy field and act autonomously from external pressures in a closed policy making process, and policy changes that involve the politicians as central policy actors and a larger public debate on the policy.

In conclusion, while two different approaches about how actors determine policy change exist in the literature (power struggle and policy learning), they benefit from being reconciled in order to account for major policy changes (Sabatier and Jenkins-Smith, 1993; Hall, 1993).

Thirdly, a large spate of the literature looks at how policy change is determined or influenced by the *international context* (e.g. policies from neighboring countries, international organisations, and supra-national institutions). Different approaches address this question, with a variety of perspectives: policy transfer (Dolowitz and Marsh, 1996; Dolowitz and Marsh, 2000; Evans and Davies, 1999; Stone, 2000; Page, 2000, James and Lodge, 2003), policy diffusion (Berry and Berry, 1999; Dolowitz and Marsh, 2000), lesson-drawing (Bennett and Howlett, 1992; Rose, 1993; Bennett, 1997; James and Lodge, 2003) and Europeanisation (Radaelli, 2000, 2004; Knill and Lehmkuhl, 2002; Featherstone and Radaelli, 2003; Cowles, Caporaso and Risse, 2001; Héritier *et al.*, 2001).

Policy transfer, policy diffusion and lesson-drawing tend to overlap in the literature. Therefore, Dolowitz and Marsh use policy transfer as the generic concept for the "process by which knowledge about policies, administrative arrangements, institutions and ideas in one political system (past or present) is used in the development of policies, administrative arrangements, institutions and ideas in another political system" (2000, p. 5). So lesson-drawing is seen as a sub-type of transfer, focusing only on voluntary processes, that is "action-oriented intentional activity" (Rose, 1993; Evans and Davies, 1999; Dolowitz and Marsh, 2000; James and Lodge, 2003). Like the policy transfer literature, the policy diffusion considers both voluntary and coercive processes, but with a focus on the process itself and not so much on the agents involved in the process. Thus the policy diffusion literature helps deepen the understanding of the diffusion process, but it reveals itself to be narrower than the policy transfer literature (Dolowitz and Marsh, 2000). Finally, the concept of policy convergence is often associated with policy transfer or policy diffusion, but they should not be confused since they address different questions. Indeed, research on policy convergence looks at the outcomes of policy-making processes (converging or diverging outcomes; Bennett, 1988, 1991; Seeliger, 1996), while policy transfer and diffusion literature focus on the process itself (Evans and Davies, 1999; Stone, 2000). For instance, Tews *et al.* use the policy diffusion approach to explain why more and more countries are converging in the adoption of new environmental policy instruments (Tews *et al.*, 2003). However, existing frameworks on policy transfer, diffusion or lesson-drawing are limited to conceptual frameworks used as a heuristic device to describe transfer processes (who?, what?, how?, why?, from where?), but not as a theoretical model. In short, policy transfer analysis does not have full explanation status since it would require the development of a causal model based upon a series of testable hypotheses, which has not been carried out successfully so far (Evand and Davies, 1999).

Europeanisation could be seen as a specific case of policy transfer, which includes transfer processes (more or less voluntary or coercive) from the European policy to domestic policy. The Europeanisation literature borrowed key ideas from the policy transfer literature, for instance: the degrees of transfer-lesson drawing (from copy to inspiration; Rose, 1993; Dolowitz and Marsh, 1996, 2000) inspired some typologies of Europeanisation outcomes (inertia, absorption, transformation, retrenchment; Schmidt, 1997 and 2002; Héritier, 1998; Héritier and Knill, 2000; Börzel, 1999; Green Cowles *et al.*, 2000; Radaelli, 2000); and the distinction between voluntary and coercive transfers, which is central to the policy transfer approach, has its counterpart in the Europeanisation literature that distinguishes vertical (coercive) from horizontal (voluntary) Europeanisation processes (Radaelli, 2003 and 2004). However, the Europeanisation literature differs from the policy transfer literature in so far as it looks specifically at the European institutions and policies, and considers not only the Europeanisation of public policies (e.g. Héritier *et al.*, 2001), but also the Europeanisation of political institutions and structures (e.g. Cowles *et al.*, 2001).

The European Union as a supranational institution benefits from exclusive or shared competences with the member states, which means that in some cases the transposition of the European policy is compulsory and in other cases it is not. In addition, the European policy can be more or less specified from highly specified rules, to less specified rules, suggested rules or even no rules at all (Schmidt, 2002). Starting from these characteristics of the European policies, several mechanisms of Europeanisation have been developed in the literature. Most make the distinction between vertical Europeanisation processes (with specified constraining European policies) and horizontal ones (with less specified and non constraining European policies). The vertical Europeanisation mechanisms usually look at whether the European and domestic policies fit (versus do not fit) and subsequently how strong the pressure on the domestic actors is to adapt their policies (Börzel, 1999; Cowles and Risse, 2001; Knill and Lehmkuhl, 2002; Börzel and Risse, 2003; Knill and Lehmkuhl, 2002; Radaelli, 2003; Héritier *et al.*, 2001). While some authors consider that a policy misfit is necessary to have Europeanisation (e.g. Cowles and Risse, 2001; Borzel and Risse, 2003), others consider other Europeanisation mechanisms without a misfit (e.g. Knill and Lehmkuhl, 2002; Radaelli, 2003). The horizontal mechanisms of

Europeanisation look at cases where there is no adaptation pressure from the EU (or weak adaptation pressure), but the domestic policy actors' power and resources or policy beliefs and preferences are modified by the European policies (Knill and Lehmkuhl, 2002).

The literature on Europeanisation identifies several variables that explain the divergence between the outcomes of the Europeanisation processes in different countries (called intervening variables), for instance: the characteristics of the domestic policy actors or advocacy coalitions (distribution of resources versus beliefs and preferences) (Héritier *et al.*, 2001; Cowles and Risse, 2001; Knill and Lehmkuhl, 2002; Schmidt, 2002; Börzel and Risse, 2003), the existence of veto players (Héritier *et al.*, 2001; Cowles and Risse, 2001; Börzel and Risse, 2003; Giuliani, 2003; Haverland, 2002), the policy structure and stage (Héritier *et al.*, 2001), and the policy legacies (Héritier *et al.*, 2001; Schmidt, 2002).

In conclusion, the literature on how the international context influences policy changes is very rich and diverse, with both general and specific frameworks. However, except in particular cases of European coercive policies, the influence of the international context on domestic policy changes remains largely voluntary and *ad hoc*, which explains why any attempt to formulate a general theoretical framework is very tricky indeed.

Fourthly, policy changes can also result from *external events* that modify the context of the policy and require it to be adapted to the new environment. Several authors raise the idea that policy changes are not only determined by endogenous factors (past policy, policy actors), but also by exogenous factors to be found in the context of the policy (Kingdon, 1984; Sabatier and Jenkins-Smith, 1993; Birkland, 1997). Sabatier distinguish two types of exogenous factors: relatively stable parameters (e.g. distribution of natural resources, constitutional structure) that are especially useful to explain policy continuity over time; and external events (e.g. changes in socio-economic conditions, changes in systemic governing coalition, impact from other subsystems) that cause perturbations in the subsystem (coalitions resources and beliefs) and result in major policy changes in certain circumstances. Indeed, major policy changes, Sabatier assumes, are unlikely in the absence of significant perturbations external to the subsystem. Therefore, external shocks are necessary (but not sufficient) conditions for major policy change in Sabatier's advocacy coalition framework. In Kingdon's multiple streams approach, dramatic events or crises in the context of a policy serve to bring problems to the attention of policy-makers and so open a "window of opportunity" for policy change. However, actual policy change only occurs if other conditions are observed, such as political receptivity (policy entrepreneurs) and available policy alternatives. Finally, following Kingdon and Sabatier, Birkland (Birkland, 1997) looks at how focusing events (man-made and natural disasters) cause advocacy coalitions to draw the attention to new policy problems and lead to new agenda setting and, ultimately, policy changes.

In conclusion, focusing events or shocks external to the policy subsystem tend to disrupt the status quo and challenge the existing policies. But, major policy changes only occur if the policy actors seize the opportunity to introduce major policy reforms.

B. A Theoretical Framework for RES-E Policy Changes in Europe

What stands out from this succinct literature review is that the four main factors of policy change that emerge from the literature can be addressed from a variety of approaches and with a variety of hypotheses. In order to remain coherent with our conceptual framework of policy change, we base our theoretical framework on Peter Hall's approach of policy change. Peter Hall considers three main variables to explain policy changes in the macro-economic policy in Britain between 1970 and 1989: the past policy (satisfaction versus dissatisfaction), the policy actors (civil servants versus politicians), and the autonomy of the state from the pressure of outside actors (autonomy versus public debate). What Peter Hall assumes in his work, is that, on the one hand, changes in the policy instruments or settings of the instruments are explained by the dissatisfaction of the civil servants in charge of the policy with the past policy in a context of relative autonomy of the state from outside pressures; and on the other hand, changes in the policy objectives result from a dissatisfaction with the past policy coming not only from the civil servants but especially from the politicians who take the lead over the policy and open up the policy-making process to a larger public debate. This approach is rich and will form the starting point of our theoretical framework.

However, there are two important variables that Peter Hall does not explicitly address in his work and that we will consider in our framework: the influence of the European Union on domestic policy changes (so-called Europeanisation), and the influence of the political and institutional context on policy changes.

Eventually, we will formulate hypotheses on the relation between the four explanatory factors and policy changes (see figure 1).

Figure 3: Theoretical framework

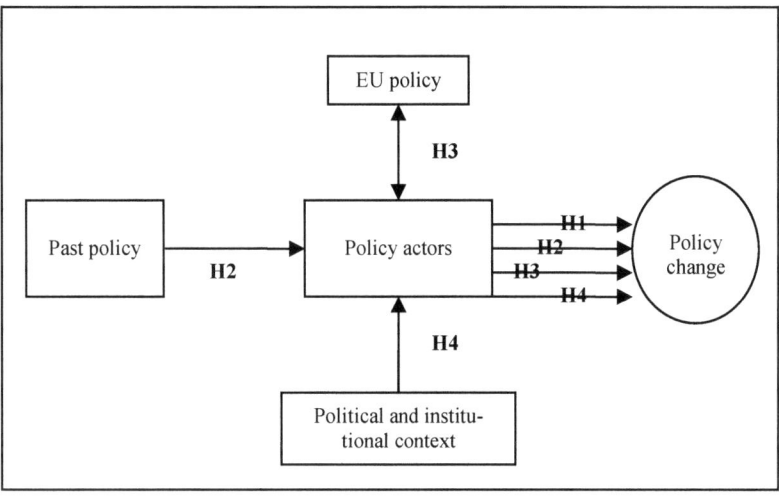

Firstly, the policy actors hold a central position in the framework (like in Peter Hall). The type of policy actors that dominate the policy-making process (civil servants versus politicians) and the influence of the actors of the sector in the policy making process (early versus late in the process, and in favour versus not in favour of policy change) determine to a large extent the type of policy change that we can expect (policy instrument versus policy objective) (H1). In addition, the other explanatory factors are intermediated by the policy actors in the process of explaining policy changes (past policy is evaluated by the policy actors, the policy transfer of EU policy is mediated by the policy actors, and the political and institutional structure determines the policy actors' opportunities and constraints).

Secondly, following Peter Hall, we assume that the performance of the past policy (policy success or failure) determines to a large extent whether this policy is going to change or not (H2). Peter Hall even considers the past policy failure as a necessary condition for any kind of policy change.

Thirdly, in the European Union, domestic policy changes are often influenced by the policies adopted at the European level (Europeanisation). Different processes of Europeanisation have been identified in the literature. Most depend upon whether the European policy and the domestic policy fit or not (misfit). Therefore, we will look at the fit or misfit between the policy objectives/instruments formulated at the EU level as well as the domestic policy objectives/instruments to explain policy changes (H3).

Finally, policy changes are determined by a series of political and institutional factors. We do not consider all the potential political and institutional factors that could explain policy changes, but we focus on two significant ones based on two well-grounded theories: the partisan theory (H4/1) and the veto players theory (H4/2).

1. Policy Actors

The first set of hypotheses (H1) of our theoretical framework is based on the common assumption that it is not possible to explain policy changes without looking at the policy actors, namely the actors involved in the policy-making process of a specific policy sector or subsystem. Many different approaches to who the policy actors are and how they explain policy changes have been developed in the literature (see section 2.2.1). In this research, we do not aim to add a new model to this literature, nor will we try to summarise the different approaches, but we focus on the work of Peter Hall and on his hypotheses on how the policy actors determine policy changes.

Peter Hall looks at which policy actors dominate the policy-making process: the civil servants (state officials with specific expertise in the policy) or politicians (elected officials without specific expertise in the policy). He postulates that the civil servants play the central role in processes of first and second order changes, but in case of third order change, the locus of authority over policy-making shifts towards the politicians. In addition, Peter Hall assumes that the policy-making process remains confined within the limits of the state in case of first and second order changes (no public debate over the policy), while it expands to a larger public debate with a confrontation between different ideas in case of third order changes. So, Peter Hall (1993) makes the distinction between, on the one hand, policy changes in which the policy actors are experts in a given policy field and act autonomously from external pressures in a closed policy-making process, and policy changes that involve the politicians (non-experts in the policy field) as central policy actors and a larger public debate over the policy.

Following his approach, we hypothesise that when the policy-making process is dominated by civil servants (administrative actors), changes in the policy instruments are expected. On the other hand, when the policy-making process is dominated by political actors (politicians), changes in the policy objectives are expected (H1/1).

In addition, we look at the participation of the actors of the sector (e.g. energy companies, RES-E interest groups, environmental associations, consumer groups) in the decision-making process and how they influence policy change. But, unlike Peter Hall, we formulate a distinct hypothesis about the participation of the external actors (H1/2). To do so, we first consider whether the actors of the sector participate in the policy process in its early stage or only in a later stage (see figure 2). Indeed, if the actors of the sector are consulted by the dominant public policy actor (civil servants or politicians) at the early stage of the policy process, when the basic choices concerning the formulation of the new policy are made, they are expected to have more influence on the policy change than if they are consulted only in a later stage, when the fundamental characteristics of the new policy have already been decided and only the settings are to be modified, or when the policy change has already been adopted and only the details of the implementation of the policy change have to be designed. So, the actors who are consulted early on are expected to have a stronger influence on the policy change (in favour of the change or not), while the actors who are consulted later on are expected to have a smaller influence (or no influence at all) on the policy change (the first proposition of H1/2), which means that they do not help us to explain the policy change. Then, if the actors of the sector who have a strong influence on the policy are in favour of the policy change, they support the policy change and make it more likely to be ultimately adopted (the second proposition of H1/2). However, if they are not in favour of the policy change, they are expected to oppose the policy change, which is likely to be abandoned (no policy change) or to be significantly revised so as to become more coherent with their policy preference (another policy change) (the second proposition of H1/2). Finally, the preferences of the actors who have a weaker or no influence on the policy-making process are assumed to have less importance in our framework since they are not relevant to explain the policy change. If the preference of those actors are congruent with the policy change, then those actors will help legitimise the policy change (but do not explain it); and if they are not congruent, they do not have enough influence to veto the policy change.

H1/1   If the civil servants dominate the policy process, then only changes in the policy instruments are expected.

If the politicians dominate the policy process, then changes in the policy objectives are expected.

H1/2   If the actors of the sector participate in the policy-making process at an early stage, then they are expected to have more influence on the policy change.

If the actors of the sector participate in the policy process at a later stage, then they are expected to have little or no influence on the policy change.

Besides, if the actors who have a strong influence on the policy change are in favour of the policy change, then it is likely to occur; but if they are not in favour, the policy change is not likely.

Figure 2: Policy actors and policy change

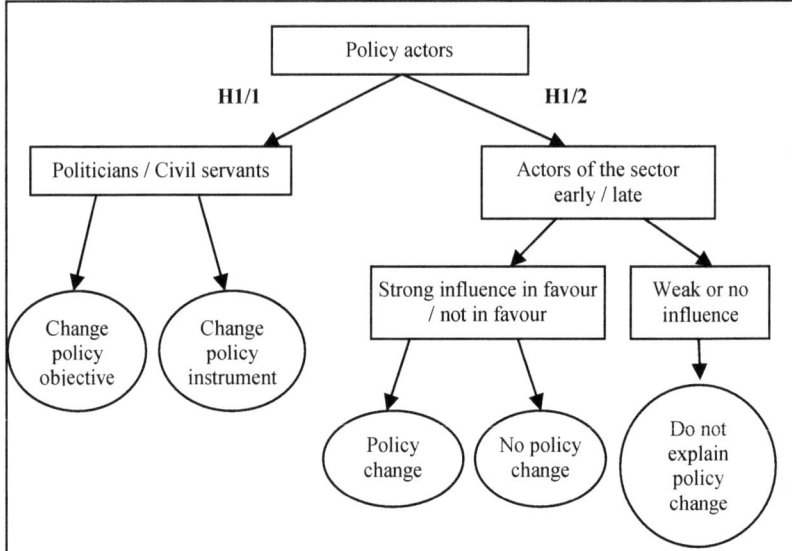

In the empirical investigations, one must consider that the dominant policy actor is not necessarily the actor who formally adopts the policy in the end (usually Parliament), but rather the central actor in the formulation of the policy (the administration or the political actors). The politico-administrative culture is different in every country regarding the relation between the political actors and the administrative actors, and this will influence our analysis of the policy actors. In the RES-E policies, the political actors who dominate the policy change process are expected to be: the Minister for Energy, Green Parties in Government or in Parliament, or a "green" majority in Parliament. On the other hand, the dominant civil servants are usually the ministry of energy or the energy division in a larger ministry (see table 8).

In order to observe the influence of the actors of the sector in the policy change, we must look at three different indicators for each actor (see table 8). Firstly, we observe if the actor participated in the policy making process at the early stage of the policy formulation or later, when the basic choices behind the policy change are already decided and no significant modifications are expected (except in the settings of the instrument). Then we investigate the actual influence of the policy actor in the policy making process (strong, weak or none). And finally, we consider the preferences of the actors about the policy change, namely if they are in favour of the policy change (as proposed by the dominant policy actor) or not, or indifferent to it. That information is available from three different sources: the documentation about the policy making process (administrative and legislative documents), interviews with the actors of the sector themselves, or interviews with experts of the policy.

Table 8: Indicators of policy actors

	H1/1	H1/2
Variables	Dominant policy actor	Influence of the actors of the sector
Indicators	- Civil servants: ministry of energy, department of energy in larger ministry - Politicians: Minister for Energy, Green Parties in government or parliament, "green" majority in parliament	- Early / Late participation in policy making process - Strong influence (+), weak influence (-), no influence (-) - In favour (+), not in favour (-) of the policy change, or indifferent to it (0)

2. Past Policy

In the literature, two main approaches describe how past policies determine policy changes. Firstly, policy changes are seen by certain authors as mostly incremental and path-dependent because they closely follow the past policy choices (e.g. Lindblom, 1959; Pierson, 1993 and 2000; Palier, 2004). So, these approaches are more useful to explain policy continuity than to explain policy changes. Secondly, several authors consider that the success or failure of the past policy explains why and how policies change (e.g. Hall, 1993; Jobert, 1994; Kingdon, 1984; Dolowitz and Marsh, 2000). Past policy is especially central in the work of Peter Hall, who assumes that past policy failure is a necessary condition for policy changes. This assumption questions the hierarchy between the explanatory factors of policy change and it shall be discussed it the comparative chapter of this paper.

Following these authors, we hypothesise that policy changes are explained by the failures (versus success) of the past policies. So if the past policy is evaluated as a success (regarding its objectives or instruments), no policy change is expected; otherwise, if past policy failed (regarding its objectives or instruments), it is expected to change (change of policy objectives or instruments, or both) (H2, see figure 3).

In addition, when one considers policy success or failure, it is important to keep in mind that the success or failure is not given nor objective, but rather constructed by the policy actors. Confronted with identical facts or data, different actors may interpret them differently according to their own interests or ideas. Looking at the policy actor's evaluation of past policy is the only way for policy analysts to account how past policy determines policy changes. Therefore, the policy actors have a central position in our hypotheses about past policy success or failure (see figure 3). But different actors may evaluate past policies differently, so the question the policy analyst is confronted with is: whose evaluation shall we take into account to explain policy changes? In this theoretical framework, we observed that some policy actors dominate the policy making process, they occupy a central position in the formulation of the policy. Accordingly, the evaluation of the past policy success or failure that determines subsequent policy changes, is the evaluation of the dominant policy actor in the policy making process (civil servants or politicians) (H2).

H2 If past policy is evaluated by the dominant policy actor as a success, then no policy change is expected.

If past policy is evaluated by the dominant policy actor as a failure, then policy change is expected.

Figure 3: Past policy and policy change

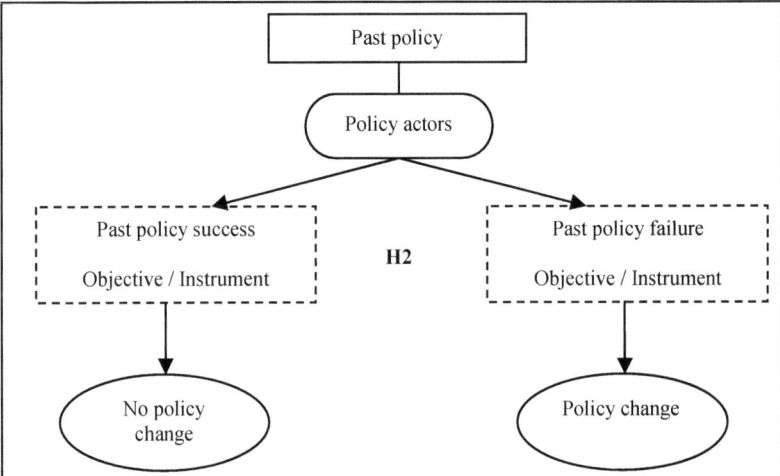

The empirical investigation of those hypotheses is very sensitive because it looks at subjective evaluations of past policies by policy actors. The policy analyst must give an objective account of the subjective evaluations, and therefore he/she needs to work with indicators (see table 9). The starting point of the empirical research is to analyse the papers, reports, discourses of the dominant policy actor (or ordered by the dominant policy actor) in the RES-E sector (e.g. the ministry of energy or the energy minister) in order to identify signs of evaluation of the past policy. We make the distinction between evaluations of the past policy objectives and of the past policy instruments because they can differ (success of the objective but failure of the instrument or vice versa), and they are at the core of our research. An in-depth historical analysis of the RES-E policies in each country is therefore very useful to test this hypothesis.

On the one hand, the evaluation of the past policy objectives discuss whether the past RES-E target is within reach, namely the objective is on the way to be achieved (success); or if the past RES-E target is too low (regarding the RES-E potential or the political priority towards RES-E) or too high to be reached (regarding the RES-E potential or the political priority towards RES-E). So past RES-E policy objective success or failure depends on the evaluation of the policy actor about the level of the RES-E target: within reach, too low or too high. If the target is within reach (success), it means that in the light of the past development of RES-E in the electricity market and the expected developments in the future, the target should be reached by the deadline, so no policy change is expected. If the target is too low (failure), it means that either it should be much higher considering the potential for RES-E in the country or it should be much higher considering the increased political priority towards RES-E (increased political ambitions for RES-E), so policy change is expected (increase of target). If the target it too high (failure), it means that it appears out of reach to the policy actors, either because they do not consider that the potential for RES-E in the country is sufficient, or because the political priority towards RES-E is not sufficient (e.g. political priority and subsequent financial support to nuclear versus RES-E in Belgium over the 1970s and 1980s), so policy change is expected (the target is less ambitious).

On the other hand, the evaluation of the past policy instrument looks at two indicators: the effectiveness of the instrument to help reaching the RES-E target, and the cost of the instrument for the public budget or the consumer bills (see table 9). Firstly, the instrument is successful if it has contributed in the past to increasing the share of RES-E in the electricity market (effectiveness), this means that it is effective in reaching the policy goal, and no policy instrument change is expected. If it does not contribute enough to reaching the policy goal, then change in the policy instrument is expected. Secondly, if the instrument represents a politically acceptable cost for the public budget or the electricity bills (depending on the type of policy instrument), no change is expected. But, if the instrument is questioned by the policy actors as being too expensive either for the public budget or the consumer bill, then change in the policy instrument is expected.

Table 9: Indicators of past policy success/failure

	H2
Variable	– Past policy success (objective or instrument) – Past policy failure (objective or instrument)
Indicators	Success: – Objective: RES-E target within reach – Instrument: effective (to reach the target) and affordable (regarding the public budget or consumer bills) Failure: – Objective: RES-E target too low or too high compared to RES-E potential or political priority towards RES-E – Instrument: ineffective (to reach target) or too expensive (regarding the public budget or consumer bills)

3. Europeanisation of Public Policies

The third hypothesis of our explanatory framework (H3) relates to the Europeanisation of public policies, namely the response of the domestic policies to the policies of the EU (minimalist definition in Featherstone, 2003). The definition of Europeanisation is still under discussion in the literature (Cowles *et al.*, 2001; Héritier *et al.*, 2001; Featherstone and Radaelli, 2003; Börzel and Risse, 2003; Radaelli, 2004) and it still covers a very broad range of theoretical and empirical issues. Europeanisation is often defined in the literature in opposition to other related but different concepts, such as: European integration (a prerequisite for Europeanisation), European convergence/divergence (the potential outcome of Europeanisation), European harmonisation (aims to reduce European diversity) (Radaelli, 2000, 2003, 2004; Vink, 2003). In his attempt to bring conceptual clarification, C. Radaelli defines Europeanisation as:

> Europeanisation consists of processes of construction, diffusion and institutionalisation of formal and informal rules, procedures, policy paradigms, styles, "ways of doing things" and shared beliefs and norms which are first defined and consolidated in the EU policy process and then incorporated in the logic of domestic (national and sub-national) discourse, political structures and public policies (Radaelli, 2000, 2003, 2004).

However, this definition encompasses a large range of processes covering a large number of elements, which makes it still very complex and vague and far from the conceptual clarification that it aims at. In the end, we prefer the definition used by Vink (Vink, 2003): "a process of change in national institutional and policy practices that can be attributed to European integration" It is certainly less comprehensive, but at least it helps to clarify the concept. Finally, most of the literature on Europeanisation focuses on the top-down aspect of Europeanisation

Theoretical Framework

(EU-member states), but we will see that the bottom-up processes (member states-EU) are also very significant.

The literature shows that the Europeanisation processes are mediated by the domestic policy actors according to different paths, depending on whether the European policy prescribes a policy model or not and if the European policy fits with the domestic policies. A significant stream of the Europeanisation literature, the so-called "goodness of fit" or "adaptation pressure" approach, starts from the assumption that there must be some degree of misfit between the European and the domestic policy, if policy change is to be explained by Europeanisation mechanisms (Börzel, 1999; Cowles *et al.*, 2001; Börzel and Risse, 2003). However, Europeanisation is also observed in policy areas where Europe did not adopt a specific policy model to be implemented in the member states, but just suggested some indicative policy objectives or instruments (e.g. RES-E policy) (Héritier and Knill, 2001; Knill and Lehmkuhl, 2002). So, in our framework we do not start from the distinction between compulsory versus voluntary European policies, because we believe that most of the European policies include both aspects, and because the literature shows that in the end it is not such a significant factor to explain domestic policy changes. However, we believe that the fit/misfit between the European and domestic policies is very relevant to explain policy changes, which is why we formulate the H3 hypothesis.

In addition, we believe that the influence of the EU policies on domestic policies can only be understood if one considers how the domestic policy actors are determined by (down-loading process; Radaelli, 2004) or make use of (up-loading process; Radaelli, 2004) the EU policy to pursue their goals (see bidirectional arrows in figure 4). Indeed, Europeanisation is often considered to be an interactive process between the European and domestic levels (top-down and bottom-up approaches; Radaelli, 2004). This means that the domestic policy actors are not only influenced by the EU policy to change or not to change the domestic policies (H3, down-loading the Europeanisation process); but that they also influence the formulation of the European policies in order to make them more congruent with the domestic policies or use the European policies to legitimise their policy choices (policy change or status quo) (up-loading the Europeanisation process). So, the policy actors are at the core of the relation between the EU policy and the policy changes and this relation is bi-directional (see bi-directional arrows in figure 4). For instance, when the domestic policy actors perceive that the European and domestic policies do not fit, instead of changing the domestic policies, the policy actors can attempt to modify the European policies in order to reduce the degree of misfit. Moreover, when the European policy fits with the domestic policy, the domestic policy actors can use the European policy to legitimise the policy change or the status quo they advocate. However, these so-called uploading Europeanisation processes are useful to explain the strategy of the domestic actors and legitimise the policy change or the status quo, but they do not explain the cause of the policy change. Therefore, we do not formulate a specific hypothesis on this idea in our theoretical framework, but still acknowledge its relevance by drawing bidirectional arrows between the policy actors and the EU policy in figure 4.

H3 If the European policy objectives or instruments fit with the domestic policy objectives or instruments, then no policy change is expected.

If the European policy objectives or instruments do not fit with the domestic policy objectives or instruments, then policy change is expected.

Figure 4: European policy and domestic policy change

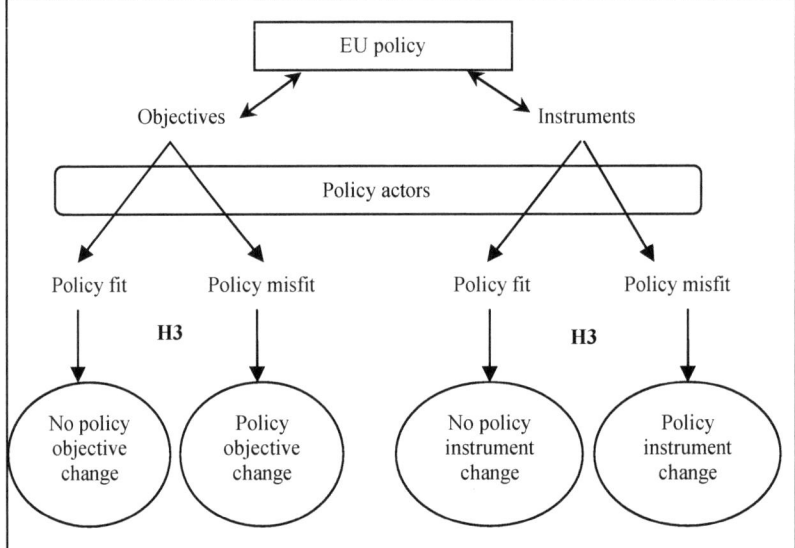

Hypothesis H3 assumes that the influence of the European level (policy objectives or policy instruments) on the domestic policy changes (policy objectives or policy instruments) depends on the fit (versus misfit) between the EU policies and the existing domestic policies. This means that if the objectives of the European RES-E policy are congruent with the objectives of the domestic RES-E policy, the domestic RES-E policy objectives are not expected to change. But if the objectives formulated at the European level differ from the domestic objectives, the domestic policy objectives are expected to change. We believe that this is true even if the EU policy is not compliant (e.g. indicative RES-E targets in 2001 directive on RES-E). Similarly, if the European RES-E policy favours one specific type of policy instrument over another and the RES-E policy instrument of a country does not fit with it, we expect that the policy instrument is going to change in the given country. The idea behind these hypotheses is that the European policies are looked at by the domestic policy actors as references to the fact that, even if they are not compliant today, are expected to become the rule in Europe in the short to medium term, so the policy actors had better adapt their policies as early as possible to take the lead on the policy issue (advantage of front runner) or reduce the future costs of policy change (cost of late runner). So, the evaluation of the fit/misfit between the European policy and the domestic policy is often more subjective than objective, since it is based on the perception of the domestic policy actors.

Table 10: Indicators of Europeanisation

	H3
Variables	– Fit between EU and domestic policy – Misfit between EU and domestic policy
Indicators	– Policy objective: domestic target congruent/ not congruent with European target – Policy instrument: domestic policy instrument congruent/ not congruent with European preferences

Héritier (2001) and Radaelli (2003) raise two important issues in the empirical investigation of Europeanisation. Firstly, Europeanisation, in order to produce domestic change, must precede domestic change. Therefore, it is necessary to place the main steps of the European RES-E policy-making process and the domestic RES-E policy change process on a time axis to make sure that the European policy precedes the domestic policy changes. Secondly, one must wonder if the domestic policy change would have taken place anyway without Europeanisation. Therefore, it is important to look at counterfactual cases and to consider the relative importance of the Europeanisation hypotheses compared to alternative hypotheses on policy change (see H1, H2 and H4 in our framework).

Finally, the European policies can take the form of formalised and constraining decisions (e.g. directives, regulations, judgements of the ECJ) or just policy proposals (e.g. green papers, white papers, working papers). In our empirical research about the European RES-E policy, we consider not only the European law (2001 directive on RES-E), but also the preliminary documents issued during the policy making process (RES green paper of 1996, RES white paper of 1997, working paper of

1999, directive proposal of 2000). Moreover, the judgments of the European Court of Justice (ECJ) in a specific policy area also participate in the related European public policy and sometimes even take the place of formal legislations (judgement in the PreussenElektra versus Schleswag case of 2001). This maximalist definition of the European policies is justified by the fact that even before the adoption of a formal decision, the policies suggested at the European level can be downloaded or uploaded by the domestic policy actors.

4. Political and Institutional Context

The last set of hypotheses of our theoretical framework considers the influence of the political and institutional context on the policy changes. To do so, it focuses on two predominant theories in political science: the partisan theory and the veto players theory. Both theories try to explain the relation between the characteristics of the partisan or institutional structure of a political system and its policy outputs. The partisan theory looks at how the public policies that a government adopts can be explained by the incumbent political parties in government (party manifesto) (Schmidt, 1996, 2002). The veto players theory looks at how the political institutions are characterised by a particular configuration of veto players and explains the policy stability or change in the policy status quo (Tsebelis, 1995, 1999, 2002). In the end, we observe that both theories are complementary and gain from being associated to explain policy changes.

The partisan theory, or parties-do-matter hypothesis, looks at the influence of political parties on policies, starting from the assumption that a major determinant in policy choices and policy outputs in democracies is the party composition of the governments. It puts forward the idea that politics, and especially governing political parties, matter to public policies. For instance, it assumes that the policies of a green government differ from those of a liberal or conservative government. In the literature, it has been used to explain differences between policy choices or outputs across countries and policy changes over time in a single country (e.g. Budge and Keman, 1990; Klingemann *et al.*, 1994; Schmidt, 1996). Some hypotheses have been formulated by the proponents of the partisan theory about how it performs in different political systems (e.g. majority versus non-majority democracies, sovereign versus semi-sovereign democracies). The partisan theory appears of limited significance in cases of non-majority democracies (e.g. all-inclusive coalitions or co-government of the opposition parties), since policies result from compromises between the government and the opposition (Schmidt, 1996, 2002). For instance, the Danish RES-E policy results from a large consensus between the governing and the opposition parties, which is not a surprise in a country characterised by minority governments. In addition, Schmidt acknowledges that the 'parties-do-matter' hypothesis performs better in sovereign democracies than in semi-sovereign democracies (Schmidt, 1996, 2002). Sovereign democracies are characterised by a largely unconstrained government majority (e.g. Westminster systems), while in semi-sovereign democracies the government is constrained by powerful checks and balances, such as federalism, local governments, and co-governing institutions. Therefore, the degree of semi-sovereignty, or the pattern of institutional pluralism, significantly determines the scope of action of the incumbent governing parties. Different indicators are used by Schmidt (Schmidt, 1996, 2002) to measure the degree of semi-sovereignty of a country: federalism versus unitarism (Lijpart, 1984), constitutional structures (Huber *et al.*, 1993), institutional pluralism, counter-majority constraints of central government (Schmidt, 1996), and veto players (Tsebelis, 1995, 1999; Schmidt, 2000). However, we observe that these indicators overlap to a large extent, since they tend to include the same characteristics, which are then over-represented in the end (e.g. number of veto players). In conclusion, the constitutional and institutional structure of a democracy determines the extent to which governing parties matter in shaping public policies (Schmidt, 2002). For instance, even if the governing coalition parties reach a consensus about the policy changes, the existence of counter-majority checks and balances (e.g. veto players) could prevent the policy changes from being ultimately adopted. So, in order to explain policy changes, one must look not only at the governing political parties, but also at the institutional context in which they govern.

Among the literature on political institutions, the veto players theory looks specifically at how institutions constrain or facilitate policy changes. The idea of veto players is not new in political science literature and has been used in many comparative politics research as an indicator of the institutional settings of democracies (e.g. Huber *et al.*, 1993; Schmidt 1996, 2002; Tsebelis, 2002). However, the veto players theory permits comparisons across different political and party systems, thus transcending the traditional typologies of the political institutions literature. In fact, regime types (presidentialism/parliamentarism), federal versus unitary countries, single party versus multiparty governments, majority versus consensus democracies, differ in terms of their veto player configuration, but the translation between those typologies and the feature of veto players is not direct, nor straightforward (Tsebelis, 2002). The same institutions may have different results in terms of veto player configuration in one country or another depending on the political and institutional context of this country. For instance, the Bundesrat and the Bundestag (the two chambers of the German parliament) will constitute two different veto players in Germany when they rely on different majorities; while the Chamber of Representatives and the Senate in Belgium are based on the same majority and thus in reality represent only one veto player.

The basic argument of the veto players theory is that in order to change policies (meaning change the legislative status quo), a certain number of individual or collective actors have to agree to the proposed change (Tsebelis, 1995, 1999, 2002). The veto players are such actors whose agreement is necessary for legislative change to take place. Tsebelis identifies two categories of veto players: the institutional veto players, those who are specified in the constitution (e.g. the King, the Senate, the Chamber of Representatives (with a specific 2/3 majority for certain issues) and the Regional parliaments in Belgium), and the partisan veto players, those who derive from the political system (e.g. the different political parties in government, the party groups in Parliament in Belgium). It is important to make the distinction here between the veto players whose agreement is necessary to modify the legislative status quo, and the policy actors who formulate the policies. Some policy actors can be veto players (e.g. political parties in government, members of parliament), but most of them are not (e.g. civil servants, private policy experts, interest groups). That is why Tsebelis talks about changes in the legislative status quo instead of policy changes. Policy changes do not amount to legislative changes, but they are usually formalised in changes in the legislative status quo, so the veto players theory does not provide an extensive explanation for policy changes, but contributes significantly to its understanding. The veto players theory hypothesises that significant departure from the legislative status quo is impossible when the veto players are numerous, there is a significant ideological distance between them and they are internally cohesive (Tsebelis, 1995). So the veto players configurations that matter for legislative change are the number of veto players (few or many), the ideological distance between them (consensus possible to reach or not), and their internal coherence (e.g. party discipline). In addition, Tsebelis (Tsebelis, 1999) assumes that legislative changes are more likely when the government stays in power for a longer period of time (governmental stability), or when the ideological difference between the current and the previous government increases (governmental change).

Building on the partisan theory and the veto players theory, we formulate two sets of hypotheses about how the parties in government (H4/1) and the veto players (H4/2) are expected to explain policy changes.

Theoretical Framework

Figure 5: Political and institutional context and policy change

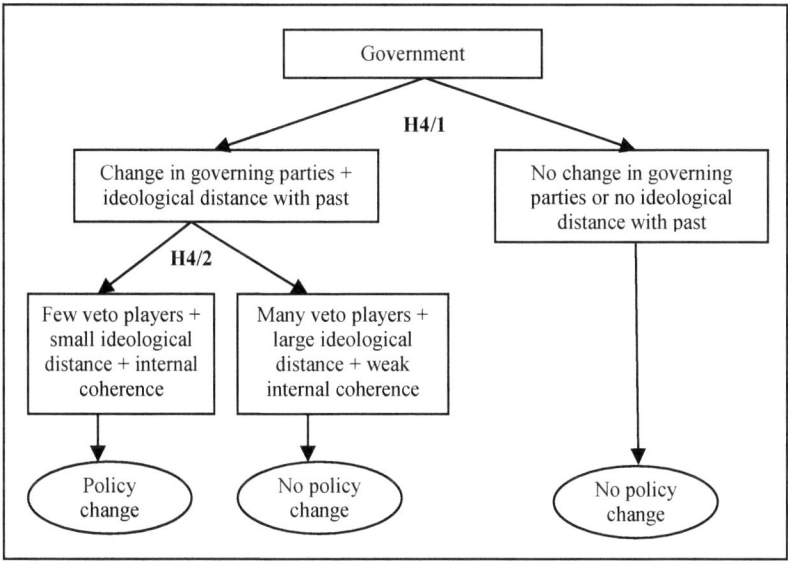

H4/1   If the political parties in government change taking significant ideological distance from the past governing parties, then policy changes are expected.

If the political parties in government do not change or they change taking no significant distance from the past governing parties, then no policy changes are expected.

H4/2   If the political and institutional context is characterised by many veto players, with a large ideological distance between them and weak internal coherence, then policy changes are less likely.

If the political and institutional context is characterised by few veto players with a small ideological distance between them and strong internal coherence, then policy changes are more likely.

On the one hand, we look at changes in the government that result from general elections and we assume (according to the partisan theory) that if the political parties in government change, then the policies are expected to change (H4/1). For instance, if a green party enters in a new government coalition, RES-E policy changes are expected. On the other hand, if the incumbent parties in government remain in place after a general election, RES-E policy changes are not expected. However, it is necessary to qualify those hypotheses with additional observations.

Firstly, the extent to which a new governing coalition is expected to produce policy change depends on the ideological difference between the past and new governing parties. If the new coalition includes one new party whose RES-E policy preferences are very similar to the other parties, then RES-E policy changes are less expected than if the new party holds very different RES-E policy preferences (e.g. a green party). Secondly, there is a difference between being in government (the governing party) and being in power. Indeed, within a governing coalition, smaller parties often have less power than big parties and this can be observed in the portfolio distribution (smaller parties get smaller portfolios) and in their overall influence on the decisions (except if small parties have a capacity to blackmail their coalition partners). For instance, it is not enough that a green party participates in the governing coalition to observe subsequent changes in the RES-E policies. It depends on which portfolio the green party holds (the sector in which the party can primarily introduce/change policies) and what power it has to impose its policies on the other governing parties (blackmail power).

So, we need to consider three indicators to account for the influence of the governing parties on RES-E policy changes: the entry of a new party in the governing coalition (e.g. a green party), the ideological difference between this party and the incumbent ones (e.g. over the RES-E policy preferences), and the power of the new party in the coalition (e.g. the new party occupies the energy portfolio or is able to blackmail the other governing parties) (see table 11).

On the other hand, building on the veto players theory, we hypothesise that when the institutional settings of a country are characterised by a large number of veto players (institutional veto players and partisan veto players), then policy changes are more difficult to achieve (H4/2). If we link this with the first hypotheses, it means that even if a new green party with different RES-E policy preferences and significant power enters in government, RES-E policy change is not guaranteed. Policy change will also depend on the number of institutional and political veto players that are able to prevent the government from changing the policy (see figure 5). In addition, besides the number of veto players, two other characteristics of the veto player configuration matter to explain policy changes: the ideological distance between the veto players (consensus possible or not), and their internal coherence (e.g. party discipline). If the veto players are ideologically close, they are expected to adopt similar positions about policy change, which ultimately reduces the number of veto players in practice and facilitate policy change. Moreover, when the collective veto players are internally coherent, they are expected to act as one veto player (political party with strong party discipline, e.g. Belgium), but when dissension appears within a collective veto player, it can account for different veto players in reality (parliamentary groups that support other positions than their political party, e.g. Germany).

Table 11: Indicators of political and institutional context

	H4/1	H4/2
Variables	– Change in governing parties – No change in governing parties	– Few veto players – Many veto players
Indicators	*Change:* new political parti(es) in the governing coalition (e.g. green party), with significant ideological difference from the existing political parties (e.g. RES-E policy preferences) and significant power in the coalition (e.g. energy portfolio or blackmail)	*Few veto players:* – Institutional veto players: monocameral parliament (or same political majority in both Chambers in a bicameral parliament), unitary and centralised state (versus federal and decentralised state) – Partisan veto players: one party government (versus coalition of parties), no counter majoritarian veto players (government governs without opposition) – Reduced ideological distance between the veto players and internal coherence (strong party discipline)
	No change: same political parties in the governing coalition, or new party with no significant ideological difference from the existing parties or no power in the coalition	*Many veto players:* – Institutional veto players: bicameral parliament with different political majorities, federal and decentralized state – Partisan veto players: multi-party government with coalition of political parties, counter majoritarian veto players (all party consensus or co-governing opposition parties) – Large ideological distance between the veto players and weak internal coherence

Several indicators can be considered to identify the actual number of veto players in a country (see table 11). The main indicators of the institutional veto players are: bicameral versus monocameral parliament, constitutional monarchies, and federal and decentralised versus unitary and centralised political systems. The main indicators of the partisan veto players are: one party versus multi-party government, and majority versus non-majority democracies. Then, the actual importance of the veto players depends on the ideological distance between them and their internal coherence (e.g. party discipline). Those indicators are closely linked. For instance, if the parliament is bicameral (two distinct chambers), it accounts for two institutional veto players, which increases the difficulty of adopting policy changes (e.g. Belgium, Germany, the UK). But this is only true if the political majorities in both Chambers of Parliament are different (e.g. the Bundesrat in Germany). If the majorities are the same in both Chambers, they account for only one veto player in reality (Belgium). Moreover, when the governing coalition includes a large number of political parties (multi-party government), policy change requires that a large consensus is reached, which makes it more difficult to achieve (many partisan veto players, e.g. Belgium), especially in case of a significant distance between the political parties. Finally, in federal and decentralised states, the regional (e.g. Belgium, Germany) and local authorities (e.g. Denmark) may have veto powers on the conduct of policies, which increases the number of veto players.

III. Conclusion

In the first part of this chapter, we presented our definition of what policy change is (conceptual framework) and how we suggest addressing it (empirical indicators). Our conceptual framework of policy change is based on Peter Hall's approach of distinguishing policy changes at the level of the policy goals, the policy instruments or the settings of the instruments. However, unlike Peter Hall, we consider only two modalities of policy change, in the line of most of the literature on policy change: change in the policy objectives/goals and change in the policy instruments. Then, in order to operationalise this approach of policy change in empirical research about RES-E policy changes, we identify specific variables for changes in policy objectives or policy instruments. On the one hand, changes in the policy objectives refer to a modification of the target of the main policy objective that the policy seeks to address (e.g. an increase of RES-E share in the electricity sector). On the other hand, a policy change at the policy instrument level involves a change in the target group (e.g. demand actors or supply actors), in the incentive used to reach the policy objective (e.g. price or quantity), or in the resources of the instrument (e.g. public or private). Of course, the literature abounds in other approaches of policy change, but we believe this approach is consistent with the literature and will help us best in understanding the latest RES-E policy changes in Europe.

In the second part of the chapter, we presented an original theoretical framework with four sets of hypotheses about policy changes. In order to remain coherent with our conceptual framework of policy change, we base our theoretical framework on Peter Hall's approach of policy change, which relies on two main variables: the policy actors (H1) and the past policy (H2). However, we also extend our framework beyond Peter Hall by introducing two additional variables: the Europeanisation of domestic policies (H3) and the political and institutional context (H4).

In the next chapters of the paper (chapters 4 to 8), we apply our definition of policy change on and test our hypotheses in five countries: Belgium, Germany, the Netherlands, Denmark and the UK. For each country, a systematic case study has been carried out based on the same framework divided into 3 parts: an overview of the electricity sector with significant statistical data (electricity generation by fuel and RES-E generation by fuel), an overview of the history of the RES-E policy before the policy change, and finally the description of the RES-E policy change itself. At the end of each case study, we analyse the policy change in the light of our conceptual framework, and then test the four sets of hypotheses. So the hypotheses are tested case per case, to evaluate if they make sense in the particular cases. But before the case studies, we present an overview of the European policies that are directly or indirectly related to the RES-E sector: the liberalisation policy, the community guidelines on state aid, the European Court of Justice ruling of 2001 in the case PreussenElektra, and the European RES-E policy (chapter 3).

CHAPTER 2

European Policy

I. Introduction

In order to explain the renewable electricity (RES-E) policy changes in EU member states, one of the most important variables (as identified in the introductory chapter) is the Europeanisation of domestic public policies. By the Europeanisation of domestic public policies, we mean the influence of European policy on the RES-E policy of member states. The aim of this chapter is to present the European policies that have potentially influenced the RES-E policy changes in Belgium, Denmark, Germany, the Netherlands and the UK: the liberalisation policy (Directives of 1996 and 2003), the community guidelines on state aid (1994 and 2001), the European Court of Justice (ECJ) ruling in the Preussen-Elektra case against Schleswag (2001), and the RES-E policy (the green paper of 1996, the white paper of 1997, the working paper of 1999 and the directive of 2001) (see figure 1).

This chapter is divided into four sections. First, we briefly describe the European electricity market liberalisation policy and its transposition in the member states. We note that, despite the existence of common European rules concerning both economic and social and political regulation, the liberalisation processes are not similar in the member states. In addition, for environmental reasons, the electricity produced from renewable energy sources (RES-E) can derogate from some EU competition rules (state aid rules, public service obligations). Indeed, under certain conditions, state aid can be granted to RES-E in order to compensate for the absence of the internalisation of environmental externalities in the liberalised electricity market. Moreover, the liberalisation directive of 1996 allows the member states to impose public service obligations on market operators in order to promote RES-E in the liberalised electricity market.

The second section presents the community guidelines on state aid for environmental protection, according to which support to RES-E may, under certain circumstances, be legitimated as state aid.

The third section introduces a very important ruling of the ECJ for the RES-E policy: the PreussenElektra case against Schleswag. The impact of this case went well beyond the German utilities' interests, as it questioned the legality of the German feed-in tariff system and was perceived to be a threat to all the other similar RES-E policy instruments in Europe.

The fourth section summarises the European renewable energy policy and stresses the impact of this policy on the electricity sector (Green Paper, White Paper, RES-E directive). Even more than in the case of the liberalisation policy, there is an absence of a specific institutional model prescribed by EU policy. Therefore, a plurality of RES-E policies still coexists in Europe with very limited harmonisation between member states.

Figure 1: European liberalisation and RES-E policies: the milestones

EU policy							
Green paper SER	White paper SER		Working paper E-SER	Draft directive E-SER	Directive E-SER + ECJ judgement		
1996	1997	1998	1999	2000	2001	2002	2003
Directive liberalisation					Guideline state aid environment		Directive liberalisation

II. European Liberalisation Policy

After a long decision-making process[1], the EU Council of Ministers and Parliament adopted a directive concerning common rules for the internal market in electricity (Directive 96/92/EC[2]) in 1996. At that time, only a handful of EU countries had already liberalised their electricity markets: the UK, Norway, Finland and Sweden. The 1996 directive resulted from a long and highly conflicting negotiation process between the EU Commission, the EU Parliament and the member states. It represents a compromise that imposes general guidelines for the reform of the sector, but member states are free to choose between several different implementation modes on key provisions of the reform (new generation capacity, third party access to the network, unbundling,

---

[1] See Matlary, 1997, for an in-depth analysis of the European energy policy.
[2] Directive 96/92/EC of the European Parliament and of the Council of 19 December 1996 concerning common rules for the internal market in electricity.

degree and speed of market opening, market regulator). Therefore, the potential for diversity in the EU internal market is very high. Nonetheless, as J.M. Glachant notes, "in spite of the predictions, national implementations have generally converged towards the same options, those considered most competitive" (2003, p. 9). However, the EU internal electricity market nowadays is still neither unified nor uniform, especially because of the diversity of the liberalisation reforms in the EU member states. In addition, according to J.M. Glachant, "it is not enough to declare the electricity sector 'open' to competition for desirable competitive mechanisms to appear in all dimensions of the reform" (*Ibid.*, p. 8).

Because important shortcomings and possibilities for improving the functioning of the internal electricity market remained, Parliament and Council adopted a new directive in 2003 repealing the 1996 directive (Directive 2003/54/EC[3]). The main areas in which the 2003 directive addresses shortcomings are: the creation of a level playing field in generation (authorisation and tender procedures), access to the network (regulated tariffs with obligatory introduction of a regulator), unbundling rules (legal unbundling), protection of small and vulnerable customers (public service obligations and labelling), and different degrees of market opening between the member states (speed-up).

Table 1 shows convergences and divergences between the five member states analysed in this research on the key provisions of the reform.

First, the 1996 directive introduces competition in the internal electricity market for new electricity generation capacity. The member states were free to choose between two procedures: the authorisation procedure and the tendering procedure. In the authorisation procedure, generators make the basic decision to build, with member states only setting up criteria for the grant of the authorisation. These criteria must be objective, transparent and non-discriminatory. They may relate to safety and security of the electricity system, protection of the environment, land use, energy efficiency, the nature of the primary source, characteristics particular to the applicant, etc. The tendering procedure allows the member states to plan the construction of new capacity and set up an inventory of the need for future generating capacity. The tendering procedure must also follow objective, transparent and non-discriminatory criteria. In the EU member states, the authorisation procedure has been the most popular one (see table 1). Convergence towards authorisation was then strengthened by the 2003 directive, which imposes the authori-

---

[3] Directive 2003/54/EC of the European Parliament and of the Council of 26 June 2003 concerning common rules for the internal market in electricity and repealing Directive 96/92/EC.

sation procedure as the regular one for new capacity building and authorises the tendering procedure when the security of supply is challenged.

Table 1: Electricity sector liberalisation reforms, status in 2003

	Generation	Unbundling[a]	TPA	Opening[b]	Regulator[c]
Belgium	authorisation	– legal	regulated	80%[d]	– A: strong
		– legal			– DS: yes
Denmark	authorisation	– legal	regulated	100%	– A: weak
		– legal			– DS: yes
Germany	authorisation	– legal	negotiated	100%	No regulator[e]
		– accounts			
Netherlands	authorisation	– ownership	regulated	63%	– A: weak
		– legal			– DS: no
UK	authorisation	– ownership	regulated	100%	– A: strong
		– legal			– DS: yes

[a] Distinction between unbundling of transmission system operator (1st line) and unbundling of distribution system operators (2nd line).

[b] Declared market opening in percentage (data collected by DG TREN during the second half of 2003).

[c] Distinction between the autonomy of the regulator v/v Ministry (A, 1st line) and its intervention in dispute settlement (DS, 2nd line).

[d] Full market opening in Flanders, but not yet in Wallonia and Brussels.

[e] But a regulator was finally installed in Germany in July 2005.

Source: DG TREN, Third Benchmarking report on the implementation of the internal electricity and gas market, 2004.

To avoid discrimination, cross-subsidisation and distortion of competition, unbundling rules were imposed in the 1996 directive. Vertically integrated utilities must therefore keep separate accounts for their generation, supply, transmission and distribution activities in their internal accounting (unbundling of accounts). Access to these unbundled accounts must be granted to the competent authority (regulator or ministry). While only accounting unbundling was required in the 1996 directive, the 2003 directive considerably strengthens the unbundling rule, since it imposes the legal and functional unbundling of the transmission system operators (TSOs) and distribution system operators (DSOs). Legal unbundling requires that separate legal entities are responsible for respectively transmission and distribution activities ("network companies"). Functional unbundling only imposes the separation and independence of the network activities (transport and distribution) within the vertically integrated utilities. However, the possibility of exemptions from the requirement of legal and functional unbundling for DSOs is provided for in the 2003 directive. Finally, the EU does not impose any

European Policy

ownership unbundling, but modifications in the ownership of the TSOs can be observed in several member states (the Netherlands and the UK, see table 1).

A third key provision of the EU directives is third party access (TPA), namely the non-discriminatory access to the networks (transmission and distribution). The 1996 directive gives the opportunity to the member states to choose between three models: negotiated TPA, regulated TPA and single buyer. Under negotiated TPA, the users have to negotiate access to the network with the TSO or DSO. Tariffs and access conditions can therefore vary from one contract to the other. In case of regulated TPA, the tariffs and conditions for network access are set and published by a supervisory public body (public regulator). The single buyer model applies when a legal person is responsible for centralised electricity purchasing and selling. Often, the single buyer is also the transmission system operator. This model can be combined with either negotiated or regulated TPA. Regulated TPA has been the most popular in the EU member states, while the negotiated TPA was only adopted in a few countries (e.g. Germany, see table 1) and the single buyer model has not been implemented significantly anywhere. However, the 2003 directive now imposes the regulated TPA model on all member states, which means that Germany was forced to review its approach (see the chapter on Germany in this study).

The 1996 directive provided a gradual electricity market opening in three steps: February 1997 (about 26% of the market), February 2000 (28%), February 2003 (33%). All member states are required to open their market according to those minimum thresholds. But most countries chose to exceed the minimum targets (e.g. Germany, Denmark, Flanders, the Netherlands, see table 1). With a view to achieving a fully operational internal electricity market, the 2003 directive provides a new market-opening timetable that speed up the process. All non-domestic consumers must now be eligible from July 2004 and, in July 2007 at the latest, all domestic consumers will finally be eligible.

The 1996 directive remained very vague about the regulatory authority. Therefore, diverging institutional models applied in the member states. Genoud (2004, p. 196) identifies three different models of regulatory institutions for electricity in Europe: the "politico-administrative" model (e.g. the Netherlands), the "quasi-judicial" model (e.g. Germany), and the independent regulatory authority (e.g. the UK, Belgium, Denmark). The first model is characterised by a strong dependence of the electricity sector regulation on the political authorities (the government). In those cases, the ministry is often in charge of the regulation. In the second model, for which Germany is the only case, the regulatory authority intervenes only ex-post in case of conflicts on the market. The regulation of the sector is therefore based on the general competition regulation. Finally, the existence of independent regulatory institutions is the most common model. The institutions differ in terms of power, resources and competence between the EU countries, but the basics are the same for all: independence of the regulator vis-à-vis political authorities (strong or weak, see table 1) and combination of ex-ante and ex-post interventions on the market (intervention in dispute settlement, see table 1). In the light of this diversity in the national regulatory authorities, the 2003 directive introduces more prescriptive provisions on the subject. It requires member states to give responsibility for a number of decisions (especially supervision of network access) to designated regulatory authorities wholly independent from the interests of the electricity industry. These regulatory authorities are required to ensure non-discrimination, effective competition and the efficient functioning of the market. This does not necessarily require the regulator to be separated from existing government structures, but a separate regulator appears as the most common and desirable model (e.g. the new energy regulator due in Germany in 2004, see the chapter on Germany in this book).

In order to reconcile the liberalisation of the electricity sector and the pursuit of services of general interest in the electricity market, the 1996 directive gives member states the option of imposing public service obligations (PSOs) on electricity undertakings in their market. Member states define these obligations individually, but they must notify them to the Commission and ensure that PSOs neither distort competition nor slow down the opening of the market. Therefore, the obligations must be clearly defined, transparent, non-discriminatory, verifiable and published. Five categories of obligations are recognised in the directive: security of supply, regularity, quality and price of supplies and environmental protection. The new electricity directive of 2003 strengthens the existing provisions in force regarding PSOs by describing more precisely the measures that member states can or must impose based on considerations of: universal services for households and small consumers (e.g. default supplier predetermined or designated by the regulator, uniform tariff), consumer protection (e.g. pre-payment meters and/or "free" supply amount for vulnerable customers), and labelling (electricity suppliers must specify in or with the bills the fuel mix and its related environmental impact). Other considerations also justify the imposition of PSOs: environmental protection (promotion of RES, energy efficiency, climate change mitigation) and security of supply (e.g. ex-ante monitoring of supply/demand balance, construction and maintenance of network infrastructures). Table 2 gives an overview of several key PSOs concerning universal services and consumer protection in the five member states compared in this study. Since the principle of subsidiarity is predominant in the definition of the PSOs, one expects to observe significant differences between the member states, which is not always the case (e.g. the default supplier).

Table 2: Public service obligations in the electricity sector, status in 2003

	Default supplier	Uniform tariff	Pre-payment meters	"Free" supply amount
Belgium	Predetermined	Yes	Yes	Yes
Denmark	Predetermined	No	No	No
Germany	Predetermined	No	Yes	No
Netherlands	Predetermined	No	No	Yes
UK	Predetermined	No	Yes	-

Source: DG TREN, Third Benchmarking report on the implementation of the internal electricity and gas market, 2004.

The member states can give compensations (direct aid or exclusive rights) to the undertaking which carries out the PSO, but this must be done in a non-discriminatory and transparent way. Compensation relating to the accomplishment of a PSO is not a state aid if four conditions are met (Altmark case[4]): the company is responsible for the implementation of the obligation, the calculation of the cost is objective and transparent, the compensation does not exceed what is necessary, and when the selection of the company to carry out the obligation is not done by a public tendering procedure, the compensation level must orient itself on the cost of a well-managed company.

III. State Aid for Environmental Protection

In 1994, the European Commission first promulgated Community guidelines on state aid for environmental protection[5]. The purpose was to clarify when and under which conditions financial aid administered by member states for environmental protection is or is not compatible with the internal market. In 2001[6], these guidelines were revised as a consequence of the increasing and evolving interventions of the member states in the field of the environment (in particular following the adoption of the Kyoto Protocol). Accordingly, competition and environmental policy are not considered mutually antagonistic; rather the requirements of environmental protection need to be integrated into the

---

[4] Judgement of the court of 24 July 2003, in Case C-280/00, Almark Trans GmbH and Regierungspräsidium Magdeburg against Nahverkehrsgesellschaft Altmark GmbH.

[5] Community guidelines on State aid for environmental protection, JO C 72 of 10 March 1994.

[6] Community guidelines on State aid for environmental protection, 2001/C 37/03.

competition policy rules. This does not mean that all state aid must be authorised, but that consideration has to be given to the proportionality between the effects of state aid on environmental protection and on competition and economic growth. It is within the discretion of the Commission to determine on a case-by-case basis whether or not the state aid measures are accepted (Jansen, 2003, p. 13).

In the guidelines, support to RES-E is explicitly recognised as acceptable state aid, subject to certain conditions (e.g. as long as it does not distort the internal electricity market disproportionately). The definition of RES-E covers all renewable energy sources, but with a restriction for hydropower plants (only of a <10MW capacity). Three cases of state aid for promoting the use of RES-E are identified in the guidelines: tax exemptions or reductions for a period limited to 10 years, investment grants (from 40% to 100% of investment) covering extra costs compared to a conventional power plant, and operating aid provided that it covers the difference between the cost of RES-E production and the electricity market price.

IV. The PreussenElektra Case against Schleswag[7]

In 1998, a German court referred three questions regarding the interpretation of the EU Treaty to the European Court of Justice (ECJ)[8]. These questions were raised in proceedings between PreussenElektra AG and Schleswag AG, two German electricity supply companies, concerning the German RES-E feed-in tariff policy (purchase obligation of RES-E by supply undertakings with compensation[9]). The ECJ had to decide whether the German law was in line with the EC Treaty, especially in terms of quantitative restrictions on imports (A. 28) and state aid (A.87). In March 2001, the ECJ ruled that, under current conditions, the German 1998 feed-in law neither constituted state aid, nor was incompatible with the quantitative restrictions on imports provision. This decision had an important impact on the RES-E policy under construction both at the EU level (directive 2001) and in the other member states (e.g. Belgium, Denmark, the Netherlands), since it legitimised the German policy and recognised that the member states have a remarkable margin of manoeuvre in choosing and implementing particular RES-E support schemes. With the adoption of the directive on RES-E in September 2001, the European Parliament and Council con-

---

[7] See Poli, 2002, pp. 209-232, for an in-depth analysis of the PreussenElektra v. Schleswag case.

[8] Judgment of the court of 13 March 2001, in Case C-379/98, PreussenElektra AG against Schleswag AG.

[9] Feed-in Law of 1990 as amended in 1998.

firmed this position and at last introduced legally binding legislation with respect to RES-E.

V. European RES-E Policy

Since 1978, the European Union has taken an interest in the development of renewable energy in Europe as part of its objective of security of energy supplies. However, initially its actions remained very limited. Then, from the end of the 1980s onwards, environmental issues gained more importance on the European agenda, especially following the 1992 Rio Conference on climate change. The actual integration of environmental and energy policies gave rise to only few policy proposals (e.g. the carbon tax proposal), but none of them were very successful (Matlary, 1996). At the same time, the economic importance of developing new renewable energy technologies grew in Europe. The development of renewable energies in Europe was then perceived as central, not only from an environmental point of view, but also from an economic perspective (Dehousse and Iotsova, 2000). Nevertheless, it was not until 1996 that the Commission produced a Green Paper on the RES, and that a true European renewable energy policy emerged.

The 1996 Green Paper from the Commission[10] was the first phase of a two-step process that defined a European strategy aiming at increasing the use of RES in the European Union. It describes the current situation of RES in Europe and proposes several guidelines for the development of a European strategy with the objective of doubling the contribution of RES to the European energy balance by 2010 (12% by 2010). The purpose of this paper was to stimulate discussions with both the actors within the sector and the member states. Therefore, it contains many ideas (e.g. an early formulation of the idea of tradable green certificates) (Lauber, 2002), but no detailed policy proposals.

In 1997, the Commission submitted a White Paper on RES[11] on the basis of the results of the Green Paper consultation. The need for a Union-wide strategy on RES is justified by several European policy objectives: the security of energy supplies and the reduction of dependency on energy imports, the protection of the environment (e.g. the reduction of greenhouse gas emissions), business and employment opportunities of the RES industry, and the achievement of greater social and economic cohesion within the Community through regional development. The White Paper endorses a 12% (indicative target) contribu-

---

[10] European Commission, Green Paper for a Community Strategy. Energy for the Future: Renewable Sources of Energy, COM(96)576.

[11] European Commission, White Paper for a Community Strategy and Action Plan. Energy for the Future: Renewable Sources of Energy, COM(97)599.

tion of renewable energy to the EU gross domestic energy consumption by 2010. In addition, it requires member states to set their own targets (to 2005 and to 2010) and to propose their own action plans in relation to the Union's strategy. The subsidiarity principle is prevalent in the paper and the member states are recognised as the main actors in this policy area. Therefore, the European action plan proposes different categories of measures that it considers to be priorities for the development of renewable energies, but it does not prescribe any of them (no European institutional model). Due to the construction of the internal energy market during the same period, the issue of the competitiveness of RES is emphasised (market access, internalisation of external costs) as well as the need for market-oriented support mechanisms.

The strategic importance of the electricity market in the development of RES had been stressed in the 1997 White Paper, but no specific proposition was made at this stage. In 1999, the Commission published a working paper on RES-E and the internal electricity market[12]. It represents a significant milestone in the preparation of a directive on the promotion of electricity produced from RES in the internal electricity market[13]. The purpose of this working paper was to launch a debate on the necessity for a European directive on the RES-E and on the most appropriate way to support RES-E in the internal electricity market. The need for common rules for RES-E support in the EU was justified by the increasing risk of trade and competition distortion, linked with the existence of different RES-E support schemes in the member states. However, due to considerable disagreement between the member states, the Commission did not pick any "best" support mechanism, but analysed the merits and disadvantages of the existing approaches in the member states in the light of the constraints of the internal electricity market, though the competition-based mechanisms (e.g. tradable green certificates, tenders) were perceived at that time as preferable in a competitive market. This position provoked a controversial debate among the member states and the actors of the sector, both having a strong influence on the preparation of the directive (Lauber, 2002).

Finally, due to conflicts between the member states about common rules for RES-E support mechanisms, the RES-E directive of 2001 abstains from proposing a harmonised Union-wide support system. In

---

[12] European Commission, Commission working document. Electricity from renewable energy sources and the internal electricity market, SEC(1999)470.

[13] Directive 2001/77/EC of the European Parliament and of the Council of 27 September 2001 on the promotion of electricity produced from renewable energy sources in the internal electricity market.

the first directive proposal of 2000[14], Commissioner Loyola de Palacio had already avoided this issue and proposed that the EU remain neutral vis-à-vis national support schemes as long as they comply with European law (state aid guidelines, fair competition on the internal electricity market). However, the directive indicates that the Commission must present a summary report on the implementation of the directive by the end of 2005, and that, if appropriate, it could submit with the report a proposal for such a harmonised support scheme, taking into account the experiences gained in the member states with the operation of the different national support systems. But if a harmonisation of the support schemes is proposed, a transitional seven-year period is guaranteed to the existing rules-in-use in order to ensure legal security for investors. One of the chief obligations that the directive prescribes to all member states is the establishment of guarantees of origin (GoO) for electricity produced from RES. The GoO shall ensure that the origin of RES-E is guaranteed according to objective, transparent and non-discriminatory criteria laid down by each member state. In case of later harmonisation of the support schemes, the GoO would facilitate the recognition of RES-E on the internal electricity market, but nowadays they are not necessarily part of the national support schemes relying on tradable green certificates.

The final directive also abstains from imposing binding national targets for future production or consumption of RES-E, despite efforts to this effect by the European Parliament. Since the White Paper, the objective of 12% of European energy consumption to be supplied by RES by 2010 was translated into the electricity sector, where the share of RES-E in total electricity consumption was expected to rise to 22.1% by 2010. Accordingly, the directive proposes non-binding national indicative targets (table 3) and requires the member states to set their own targets that must be consistent with on the one hand, the European objectives of 12% of energy consumption and 22.1% of electricity consumption, and, on the other hand, the Kyoto national commitments to reduce their greenhouse gas emissions (see table 3). Therefore, the Kyoto national commitments set by the burden-sharing agreement of 1998 and the national indicative targets detailed in the directive are not directly related, but member states are required to set national targets consistent with both objectives. In fact, the indicative targets are based upon firstly, a projection exercise with an energy policy simulation model (SAFIRE) that takes into account the economic and technological potential of each member state, and secondly, negotiations in the Coun-

---

[14] European Commission, Proposal for a directive of the European Parliament and of the Council on the promotion of electricity from renewable energy sources in the European Union's internal electricity market, COM/2000/279.

cil on the 2000 Commission proposal (the targets of Portugal, Finland and the Netherlands were cut compared to the initial proposal) (Jansen, 2003, p. 20).

Table 3: GHG emission reduction targets by 2008-2012 and indicative targets for the contribution of RES-E to gross electricity consumption by 2010 (including large hydro)[15]

	GHG reduction targets (%)	RES-E (TWh) in 1997	RES-E (%) in 1997	RES-E (%) in 2010
Austria	-13	39.05	70	78.1
Belgium	*-7.5*	*0.86*	*1.1*	*6*
Denmark	*-21*	*3.21*	*8.7*	*29*
Finland	0	19.03	24.7	31.5
France	0	66	15	21
Germany	*-21*	*24.91*	*4.5*	*12.5*
Greece	25	3.94	8.6	20.1
Ireland	13	0.84	3.6	13.2
Italy	-6.5	46.46	16	25
Luxembourg	-28	0.14	2.1	5.7
Netherlands	*-6*	*3.45*	*3.5*	*9*
Portugal	27	14.3	38.5	39
Spain	15	37.15	19.9	29.4
Sweden	4	72.03	49.1	60
UK	*-12.5*	*7.04*	*1.7*	*10*
EU	**-8**	**338.41**	**13.9**	**22.1**

Source: Eurostat and directive 2001/77/EC.

Two other controversial issues in the preparation of the directive were a) the definition of RES-E and b) the conditions of network access. Firstly, conflicts between member states and between the European institutions (Council, Commission and Parliament) arose in regard to the question of the recognition of electricity produced from large hydropower plants (>10 MW) and industrial and municipal waste incineration facilities as RES-E[16]. The 2000 directive proposal of limiting hydropower plants to 10 MW in the definition of RES-E was finally eliminated and, due to pressures from some member states (e.g. the Netherlands, UK), the biodegradable fraction of industrial and municipal waste was recognised as renewable. Therefore, the final consensus on the European definition of RES-E was very broad as it had to take into account the preferences of all member states. Secondly, the 2000 direc-

---

[15] The percentage contributions of RES-E in 1997 and 2010 are based on the national production of RES-E divided by the gross national electricity consumption.
[16] Lauber, 2002, "Renewable energy at the EU level", p. 25.

tive proposal proposed to give "priority access" to the grid for RES-E. This was not endorsed by the Council, which instead defined a "guaranteed access" to the distribution and trans-mission grids. However, the idea of priority dispatching for RES-E generating installations (insofar as the operation of the national electricity system permits) was kept.

In conclusion, the preparation of the 2001 directive on RES-E revealed several conflicting issues among the member states and resulted in a final document that is far less ambitious in terms of policy harmonisation than it was supposed to be.

Today in the context of the increased climate change and energy security challenges, the European Commission and the European Council have decided to develop a far more ambitious European energy policy. The Commission presented in March 2006 a new Green Paper called "A European strategy for sustainable, competitive and secure energy". Then, during the 2007 spring European Council, the European leaders agreed to set a binding overall goal of 20% of RES by 2020 as well as a binding minimum target of 10% for the share of biofuels in the transport sector by 2020. These ambitious targets must now be translated in action plans and public consultations have been launched on different issues (biofuels, energy efficiency, market-based instruments, etc.).

CHAPTER 3

RES-E Policy in Belgium[1]

I. Introduction

Belgium has been a federal state since 1993[2]. As a result of a slow process of federalisation, which started in the 1970s, federated bodies were progressively given great measures of autonomy in the management of social, economic and environmental matters. In addition, it is important to point out that there is no hierarchy in edicts, whether they come from the federal government or from the Regions and Communities: a regional or community decree, providing it concerns matters over which the latter have authority, is equivalent to a federal law. In case of conflict, the court of arbitrations delimits which edict prevails according to the division of responsibilities.

The institutional system is very complex, with two different kinds of federated entities: three linguistic Communities (French, Flemish and German-speaking), which are responsible for cultural, social and education matters, as well as three Regions (Wallonia, Flanders and Brussels-Capital), governing matters concerning economic and regional development, environmental protection, public transport, housing and energy. As presented in figure 1, these two kinds of federated entities do not overlap perfectly, mainly because the Brussels-Capital Region is bilingual and because the small German speaking part of Belgium has some autonomy concerning cultural competences but depends on Wallonia for other issues such as economic affairs or energy, for instance.

---

[1] The sections 4.1 to 4.5 of this chapter have already been published in a different version in: de Lovinfosse I. And Varone, F. (2004), "From private self-regulation to public renewable electricity policy: a paradigmatic change in Belgium". In de Lovinfosse I. And Varone F. (eds.), *Renewable Electricity Policies in Europe. Tradable Green Certificates in Competitive Markets*, Louvain-la-Neuve: Presses universitaires de Louvain, pp. 73-120.

[2] The federal status of Belgium was formally recognised in the Belgian Constitution in 1993.

Figure 1: The institutional configuration of the Belgian public authorities after 1989[3]

Federal State	Communities	Regions
Belgium	Flemish Community	Flanders
	French-speaking Community	Brussels-Capital
	German-speaking Community	Wallonia

Source: Free adaptation from Lagasse, 1999, p. 37.

According to the special law of institutional reform of 8 August 1980[4], reviewed in 1988[5] and 1993[6], the Regions and the Federal State share competences on energy issues (see table 1). In fact, the regional authorities have major responsibilities for designing and implementing energy policies, while the federal government is responsible for nuclear power, production infrastructure and tariffs issues. In addition, the 1993 special law empowered the Regions with residual competences, which means that all issues that are not formally attributed to the federal authorities fall under the competence of the Regions in case of conflict. The responsibility for renewable energy was transferred to the Regions in 1988, along with most other energy issues. Nevertheless, some renewable energy issues remain under federal responsibility, for example: offshore wind turbines, electricity transport and fiscal measures.

---

[3] The institutions of the Flemish Community and Region merged in 1980 see grey cells in figure 1.

[4] Special Law of 8 August 1980 related to the institutional reforms, Moniteur Belge (MB), 15 August 1980. In Belgium, special laws (outright majority necessary) are required for sensitive and important issues, especially for constitutional matters.

[5] Law of 8 August 1988 that modifies the special law of 8 August 1980 on institutional reforms, MB, 13 August 1988.

[6] Special Law of 16 July 1993 aiming at completing the federal structure of the state, MB, 20 July 1993.

Table 1: Division of responsibilities for energy policy between federal and regional governments

Federal government	Regional governments
• Large infrastructure for storage, transport (>70kV) and production of energy; • Indicative program for the electricity sector; • Nuclear fuel cycle and related R&D programs as well as research in the field of nuclear fusion; • Setting tariffs.	• Distribution and transport of electricity through networks with a maximum voltage of 70 kV; • Public distribution of gas; • New and renewable sources of energy, excluding nuclear energy; • Rational use of energy; • Use of methane and blast furnace gas; • District-heating equipment and networks; • Use of waste products reclaimed from coal tips; • Recovery of waste energy from industry or other uses.

Source: Adapted from IEA, 2001, p. 20.

It is important to note that when one talks about the Belgian RES-E policies, it actually includes both the federal and the regional policies, which might in fact refer to very different policies. Therefore, we always make the distinction in this chapter between the federal, Flemish, Walloon and Brussels policies and we stress the main differences between them. However, we have decided to keep the Belgian RES-E policies as one case and not to split them into three or four cases, and this for two reasons. Firstly, the federal government remains competent for major horizontal policy issues like large production infrastructure, electricity transport, tariffs (only for captive segments of the electricity market), fiscal measures and energy policy programs in general. Secondly, as far as RES-E policy changes are concerned, the main changes occurred in the same period in all jurisdictions (between 1999 and 2001) and they relied on similar policy models (e.g. tradable green certificates). Therefore, we clearly identify a similar pattern of change in the federal and regional RES-E policies between 1999 and 2001, and those changes can be analytically aggregated into one single case in this research.

II. Overview of the Electricity Sector

In Belgium, total electricity production has almost tripled over the last thirty years, increasing from nearly 30,000 GWh/year in 1970 to more than 80,000 GWh/year in 2000 (see figure 2). Nevertheless, there were some brief periods of slight recession in the progression of electricity production: in 1975 first, then in 1981-1982 and finally in 1993-1994. Each case was a consequence of important disruptions in oil prices.

Figure 2: Gross electricity generation by fuel (GWh), 1970-2000

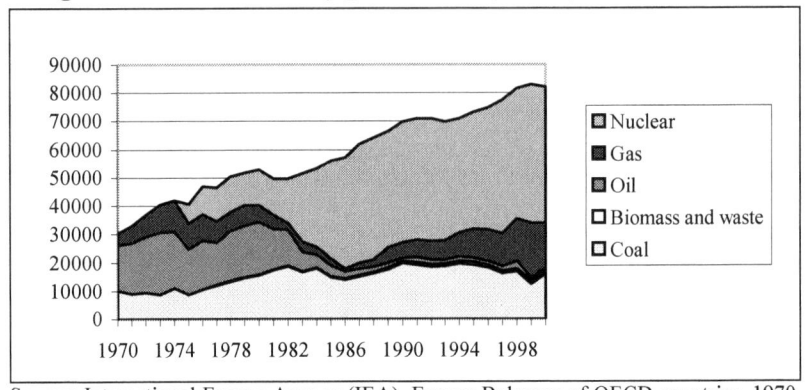

Source: International Energy Agency (IEA), Energy Balances of OECD countries, 1970-2000.

Nuclear power appeared significantly on the Belgian market in 1975, when the first installations in Tihange and Doel started, then it increased strongly from 1983 and 1984, with additional power stations on the same sites (see figure 2). Since 1984, nuclear power has covered more than 50% of total electricity production (with a peak of 68% in 1986). From the end of the 1980s, following the moratorium on new investments decided in 1988, this proportion has progressively declined, even though about 50% of the total electricity production today still comes from nuclear power.

With the emergence of nuclear power, there has been a significant reduction in oil and gas, starting from the mid-1970s (see figure 2). However, if oil seemed to gradually disappear from electricity production installations, gas reappeared at the end of the 1980s. Indeed, the moratorium on nuclear power and the development of new very efficient combined-cycle gas turbines contributed to the development of gas as privileged fuel for new investments.

Coal has remained a relatively significant fuel for the last 30 years (see figure 2). Indeed, it still covered almost 20% of the total electricity production in 2000.

Concerning renewable energy sources, we notice that electricity produced from hydro, solar or geothermal installations is insignificant (less than 1% of the total production in 2000), while biomass and waste are increasingly used in the production of electricity, yet remaining marginal, compared to other fuels (approximately 1% of the total production in 2000).

In Belgium, electricity production gradually became a monopolistic activity, benefiting a small number of electricity companies: Electrabel

(95% of market shares) and SPE (Société cooperative de Production d'Électricité, 5% of market shares). During the period 1970-2000, the share of the auto-producers in the total electricity production suffered a continuous decline (30% of the total production in 1970, compared with approximately 3% in 2000[7]). We thus observe a gradual centralisation of the means of production in the hands of a private electricity company, Electrabel, and a public company, SPE. However, at the beginning of the 1990s, both companies merged their production and transport activities into a new company, CPTE (Company for the Co-ordination of the Production and Transport of Electricity), which then exerted a monopoly on both the production and the transport sectors. Nevertheless, SPE only had a marginal control over CPTE since it owned less than 10% of its shares against more than 90% to Electrabel. With the liberalisation of the electricity sector, CPTE was dissolved in 2003 and the transport activities were transferred to a new transmission system operator (Elia), while the production plants were divided between Electrabel and SPE.

In addition, Electrabel until very recently also managed to exert a monopoly on the distribution and supply activities in Belgium (in fact it still does in Brussels and Wallonia, where the market is not yet fully liberalised). The law of 1925[8] (which was only revised in 1999) granted a monopoly on the electricity supply to the Belgian municipalities. In practice, most municipalities were grouped into inter-municipal companies that delegated the electricity distribution and supply activities to Electrabel, which had the most expertise in the field. Therefore, a distinction is to be made between the inter-municipal companies that are "pure public" (covering 20% of the electricity supply) and the ones that have "mixed" ownership (covering 80% of the electricity supply). In the end, Electrabel controlled over 80% of the electricity supply market in 2000.

As we already noticed, the share of renewable energy sources is marginal in the electricity production in Belgium (less than 2% in 2000). However, the electricity produced with renewable energy sources grew significantly at the end of the 1990s and the recent policies adopted both at the federal level and in the Regions should contribute to increasing it in the coming years. In Flanders, the current target is to reach 25% of electricity supply, to come from either CHP or RES-E by 2010, and more specifically 6% of electricity supply from RES-E by 2010[9]. In Wallonia, 8% of electricity generation is expected to originate from

---

[7] Source: Ministry of Economic Affairs, Energy division.
[8] Law of 10 March 1925 on the electricity energy distributions, MB, 25 April 1925.
[9] Beleidsnota 2004-2009 Energie en Natuurlijke Rijkdommen.

RES-E by 2010[10]. In Brussels, the promotion of RES-E is involved in the regional climate policy[11], but no specific target has been adopted so far.

Currently, the main sources of RES-E are biomass[12] (more than 60% of RES-E production) and hydropower (almost 35% of production), both growing in 1999-2000 (see table 2). Wind power has been developing lately, but it is still marginal. On the other hand, solar and geothermal energy do not exist in significant quantity in Belgium.

Table 2: Gross RES-E generation by fuel (GWh), 1990-2000

	Gross Electricity Generation (excl. pumped storage)	Gross RES-E generation	% Gross RES-E generation	Biomass	Hydro	Wind	Solar PV	Geothermal
1990	70,215	766	1.09	492	267	7	0	0
1991	71,195	772	1.08	535	229	8	0	0
1992	71,444	876	1.23	527	341	8	0	0
1993	70,079	784	1.12	522	254	8	0	0
1994	71,398	873	1.22	518	346	9	0	0
1995	73,567	948	1.29	601	338	9	0	0
1996	75,186	857	1.14	610	239	8	0	0
1997	77,920	862	1.11	549	305	8	0	0
1998	82,133	929	1.13	529	389	11	0	0
1999	83,373	1,205	1.45	851	341	13	0	0
2000	82,654	1,324	1.60	850	459	15	0	0

Source: Eurostat.

As far as RES-E producers are concerned, the big electricity companies are the main actors (Electrabel, SPE). Nevertheless, more and more small producers have recently appeared on the market, mainly in wind power production (production of 2 GWh in 2000[13]).

---

[10] Plan pour la maîtrise durable de l'énergie, December 2003.
[11] Plan Air Climat 2002-2010, November 2002.
[12] Biomass consists of wood, wood waste, municipal waste (only biodegradable fraction) and biogas.
[13] Source: Ministry of Economic Affairs, Energy division.

III. Overview of the RES-E Policy History in Belgium (1970-1999)

Before the law of 1999 on the reorganisation of the electricity market[14], the Belgian electricity sector was regulated by a law of 1925[15] that granted a legal monopoly on the electricity supply to the municipalities. The activities of electricity production and transport were characterised by a *de facto* monopoly, in the hands of a private electricity company (Electrabel) and about 80% of the distribution network also ended up in the hands of Electrabel, though partnerships with the municipalities. Until very recently, the regulation of the electricity sector has therefore been monopolised at all levels by a private company (Electrabel) under the lenient supervision of the public authorities.

During the period 1970-1999, we can identify two main stages in the development of the Belgian electricity policy. Firstly, the period from 1970 to 1988 was characterised by the switch from oil to nuclear energy for the production of electricity in a context of political instability, capture of the electricity policy by the sector and international oil crisis. Secondly, from 1988 to 1999, several changes occurred: the regionalisation of most energy competences, the moratorium on nuclear energy, the increased burden of environmental issues and the emergence of renewable energy on the political agenda. Renewable energy policy was not yet a priority on the political agenda in that period. However, several policy programs or private initiatives were adopted to foster the development of renewable energies in the electricity sector, but they remained very limited in scope and result.

A. *The Nuclear Period (1970-1988)*

From the 1950s, several nuclear research programs were carried out in Belgium, in particular within the Centre for Nuclear Energy (CEN), one of the main research centres into atomic energy, founded in Mol in 1952. At the end of the 1960s, these experimental research programs were transformed into industrial programs carried out by electricity production companies (Doel I and II, Tihange I[16]). This movement was reinforced and accelerated by the 1973 oil crisis, with the plannification of several additional installations (Doel III and IV, Tihange II and III[17]). It seems that the rise of oil prices and the subsequent supply crisis

---

[14] Law of 29 April 1999 on the organisation of the electricity market, MB, 11 May 1999.
[15] Law of 10 March 1925 on the electricity energy distributions, MB, 25 April 1925.
[16] The first three reactors were built in Doel (Flemish Region) and Tihange (Walloon Region) and they started to be operational in 1975.
[17] The next four reactors were operational in 1982 and 1985.

served the expansion of the nuclear industry. The transition of the experimental nuclear energy production to large-scale industrial production (in 1990, 60% of the total electricity production came from nuclear power) was implemented without an actual political debate. The decision to invest in nuclear energy was mainly managed by the private electricity companies with few public regulations. For instance, most of the nuclear power plants received their planning and operations permissions when they were already under construction. In addition, the law on public liability in the field of nuclear energy was only adopted in 1985[18].

Starting in the years 1973-74 (the first oil crisis), successive governments tried to elaborate an energy policy program. The projects had varying success, mainly because of a lack of government stability on the one hand, and of a lack of political mobilisation on the energy issues on the other. For the first time, the government declaration of April 1974 stated the principle of the development of an energy policy program to be adopted by government and parliament. Such a program was put on the agenda several times, but without ever being adopted. In fact, in the energy sector, the decisions were taken in very narrow actors networks (and thus without a real political debate) and the public mechanisms of control were extremely weak.

As far as the electricity sector is concerned, the Electricity and Gas Control Committee (CCEG, 1955-2003) organised a very specific system of regulation and control for Belgium, which was quite unique in Europe. It was based on a strictly contractual regime between the actors of the sector. Indeed, the CCEG was constituted by so-called "controlling organisations" (trade unions and employers' federations), which had voting rights, and on the other side of the table were the so-called "controlled organisations", which represented the electricity and the gas sectors (electricity and gas companies and representatives of the municipal utilities). The public authorities (both federal and regional) were also represented but without voting rights. Therefore, until 1999, the electricity sector had been quite independent from the public authorities since this system of self-regulation through the CCEG prevailed.

In 1975, a commission for the evaluation of nuclear energy, called a "Commission of Sages", was set up by the government. It was in charge of advising the government about the opportunity, the methods and the effects of nuclear energy. Its report was released in April 1976. Nevertheless, the nuclear program was already operational at that time, without any chance of cancelling it. In fact, this commission was an opportunity to initiate a political debate on the energy policy, but it started a

---

[18] Law of 22 July 1985 on the public liability in the field of nuclear energy, MB, 31 August 1985.

little too late to allow in-depth changes in political decision-making. As far as alternative energy was concerned, the final report of the Commission was relatively concise: "new alternative energies are not likely to play more than a marginal role in the Belgian case [but] research on solar energy deserves to be encouraged"[19].

At that time, other initiatives were taken to launch a debate on energy. But, very few led to a mobilisation of the public opinion or the politicians. A parliamentary debate started in 1979 on a White Paper prepared by the minister of energy (at the time, Willy Claes): "Elements for a new energy policy". The Regions were also consulted on this report, but too late to be able to influence its content. Finally, no decision was adopted on this report, partly because of governmental instability.

After the first oil crisis, in 1975, a national research and development pro-gram on energy started. It was managed by the Services for Scientific Policy Planning (SPPS). This program aimed at giving a new impulse to the research fields that were not targeted by the existing research programs: the analysis of the energy system (approximately 15% of the budget), the rational use of energy (more than 25% of the budget) and new and renewable energies (approximately 35% of the budget)[20]. The budget of the program (approximately 42 million euros during the period 1982-1987[21]) was marginal compared to the global effort of the Belgian government in favour of R&D in energy. It represented approximately 10% of the entire R&D energy budget, while research in nuclear energy represented more than 60% of the budget[22]. Unfortunately, the program declined from 1985 because of cuts in the budget and it was definitively dropped in 1987. The end of this program can partially explain why alternatives to nuclear energy did not emerge in Belgium at that time, unlike in other European countries.

At the beginning of the 1980s, the second oil crisis occurred just after the first institutional reform[23]. The political agenda was not mobilised a lot by energy issues since it was dominated by the development of the new regional institutions. In addition, the devolution of the energy competences to the Regions partly explains this policy paralysis.

---

[19] Final report, Evaluation Committee on nuclear energy, March 1976.
[20] Services for Scientific Policy Planning, Final Report national R&D energy program, 1988, pp. 1-4, 13.
[21] *Ibid.*, pp. 12-13.
[22] *Ibid.*, pp. 14-15.
[23] Special Law of 8 August 1980 on the institutional reforms, MB, 15 August 1980.

At the federal level, there was an attempt to increase the control of the public authorities over the electricity sector in 1980[24] (nationalisation of the CCEG and other provisions to restrain the energy companies' monopoly). But eventually, it has not been implemented.

The parliamentary debate that had been launched at the end of the 1970s with the White Paper, "Elements for a new energy policy", finally resulted in the adoption of nine parliamentary resolutions in 1983. These resolutions did not have any legal value, but they aimed at defining the main orientations of the Belgian energy policy for the future. It represented an improvement in terms of the political debate on energy issues and resulted in various specific programs, especially in the promotion of the rational use of energy (RUE)[25]. Nevertheless, the RUE programmes that emerged at the beginning of the 1980s were put aside in the middle of the 1980s, when the oil prices decreased due to the counter oil crisis.

In conclusion, we observe that support to the renewable energies was not on the Belgian political agenda in the 1970s and 1980s. Except for the R&D program from 1975 to 1987, no specific programmes in favour of renewable energies were adopted.

B. Institutional Reform, Emergence of Environmental Issues and RES-E Policy (1988-1999)

In 1988[26], the competence on renewable energy policy was transferred to the Regions. The federal government remained responsible for general environmental policy objectives, but the Regions were in charge of the implementation of these objectives in the renewable energy sector. This division of competences made it even more difficult to conduct an ambitious Belgian RES-E policy.

In Belgium, the impact of the Chernobyl disaster (1986) on the nuclear energy policy was not immediate. In 1988, the new government declared for the first time that priority needed to be given to non-nuclear scenarios in the future[27]. A new national equipment program was adopted by the electricity companies and new investments in nuclear energy were frozen. Therefore, a *de facto* moratorium on nuclear power has been observed in Belgium since 1988. This moratorium finally led

---

[24] Law of 8 August 1980 related to the financial propositions 1979-1980, MB, 15 August 1980.

[25] Royal Decision of 10 February 1983 related to the promotion of the rational use of energy, MB, 4 March 1983.

[26] Law of 8 August 1988 that modifies the special law of 8 August 1980 on institutional reforms, MB, 13 August 1988.

[27] Governmental declaration of 10 May 1988.

to the adoption, in January 2003, of a law on the progressive phasing out of nuclear power[28].

In the same period, environmental issues gradually gained more and more importance in Belgium, especially after the Rio Conference of 1992. The first environmental program focusing on the energy sector was adopted by the federal government in 1994: the Belgian national program for reducing CO_2 emissions[29]. It suggested different measures to be implemented in order to achieve the CO_2 reduction target that was adopted in 1991 by the federal government (a 5% reduction between 1990 and 2000). The Regions were competent for implementing this program, but it was not very successful in the end, mainly because it relied on the introduction of a carbon tax at the European level, which was eventually not decided.

As far as RES-E was concerned, the issue appeared on the political agenda in the middle of the 1990s due a lobby of non-profit associations[30], but it did not reach the expected results. In fact, two interrelated initiatives in favour of RES-E were introduced at that time: a recommendation of the CCEG in 1995[31] and a bill in the federal parliament in 1996[32]. Since the sector did not want to be constrained by new legislation, it decided to adopt its own program and convinced the government to withdraw the bill. Therefore, the electricity sector (CCEG) adopted a recommendation in 1995 aiming at enhancing the development of RES-E through financial support to RES-E: the so-called "Green frank system". The CCEG recommendation stated that a minimum price of 0.025€/kWh had to be paid to RES-E delivered to the grid. The premium was guaranteed for 10 years according to a contract that had to be signed between the RES-E producer and purchaser (either an electricity supplier or producer). If the purchase contract was signed with the local electricity supplier (inter-municipal utilities), the cost was borne directly by that company. If it was signed with an electricity producer, it was financed by a fund, the RUE-production fund[33], managed by the CCEG. This RUE-production fund was financed by the electricity companies

---

[28] Law of 31 January 2003 on the phasing out of nuclear energy used for industrial electricity production, MB, 28 February 2003.

[29] Belgian national program for reducing CO_2 emissions, 1 July 1994.

[30] Especially APERe (Association pour la Promotion des Énergies Renouvelables) that has been created in 1991, see Interview with J.-M. Van Nypelseer, ex-APERe, Brussels, 30 May 2003.

[31] CCEG recommendation of 25 January 1995.

[32] Bill on the supply of electricity produced from renewable energy sources to the public electricity grid, 8 January 1996.

[33] CCEG recommendation 96/15 of 27 March 1996 called "RUE and control of CO_2 emissions in the electricity generation sector".

(about 5 to 7.5 million €/year) who then passed the cost on to the price of the kWh. It was dedicated to RUE programs to be implemented by the electricity producers and to the premiums to be paid to RES-E producers. Unfortunately, no reports on the use of this fund are available since it was administrated on a decentralised and *ad hoc* basis[34].

Due to the European White Paper on renewable energies (1997[35]) and sustained pressures in Parliament for additional support to RES-E, the CCEG adopted a new recommendation in 1998[36]. This recommendation doubled the minimum price to be paid to hydro and wind electricity (with maximum 10MW capacity), from 0.025€/kWh to 0.05€/kWh, while the premium of 0.025€/kWh remained valid for all other installations (biogas, municipal waste except incinerators). In addition, it ensured favourable financial conditions for solar installations based on the difference between the amounts of electricity they supplied and purchased on the grid.

Nevertheless, the actual effects of these CCEG recommendations were very limited, and this for several reasons (de Radigues, Huart 2000). Firstly, there was a lot of long-term insecurity, as the premium was not based on a legal provision but on a recommendation of an independent organisation to its members (electricity suppliers and producers). Secondly, the number of eligible RES-E producers was very limited since several categories of producers were excluded: the auto-producers[37], the dominant public and private electricity producers (Electrabel and SPE), and the RES-E producers connected to the transmission grid. Thirdly, the discrimination among the renewable energy sources as far as the premium was concerned (0.025€/kWh or 0.05€/kWh) was not justified and no indexation of the premium applied. Finally, the financial and technical conditions for the connection to the grid were neither homogenous nor transparent (every local distributor had its own rules).

In addition to the CCEG "Green frank system", several investment grants were adopted at the regional level. A first category are the investment grants awarded to companies (around 15%), which were

---

[34] Interview with F. Delourme, CCEG, Brussels, 28 March 2003.
[35] White Paper for a Community Strategy and Action Plan "Energy for the future: renewable sources of energy", COM (97) 599, 26 November 1997.
[36] CCEG recommendation of 8 July 1998.
[37] The RES-E producers not connected to the grid.

adopted from 1992 to 1994 in Wallonia[38], Flanders[39] and Brussels[40], as part of the economic expansion or reorientation programs. However, the outcomes of these subsidies were not very significant in terms of renewable investments, and this for several reasons. Firstly, because a rate of 15% for the renewable energy investments is not more favourable than any other general investment grants, and they cannot be cumulated. Secondly, the electricity producers were not eligible for the grant. Finally, this measure was integrated in a more general policy aiming at regional economic expansion and not specifically at renewable energy development, which reduced the visibility of the measure. In addition, the regional authorities implemented other investment grants directed at households and non-commercial organisations, but most of them were dedicated to the rational use of energy purposes rather than renewable energy purposes.

At the federal level, a preferential deduction rate for investments in energy efficiency and renewable energy was included in the company tax code of 1992[41]. But it was rather directed at fiscal purposes than at the development of renewable energies.

IV. Liberalisation of the Electricity Sector (1999)

In 1999, a new law on the organisation of the electricity market[42] was adopted by the federal government in order to implement the 1996 European directive concerning common rules for the internal electricity market[43]. It marked the beginning of the liberalisation process. The transposition of the European directive required a large number of new legal provisions in Belgium, and this for two reasons. On the one hand,

---

[38] Decision of the Walloon Government of 16 September 1993 aiming at implementing a specific policy on renewable energies within the framework of article 32.13 of the law of 4 August 1978 on economic reorientation as modified by the Decree of 25 June 1992, MB, 18 January 1994.

Decision of the Walloon Government of 16 September 1993 aiming at implementing a specific policy on renewable energies within the framework of articles 5 and 5bis of the law of 30 December 1970 on economic expansion as modified by the decree of 25 June 1992, MB, January 18, 1994.

[39] Decree of the Flemish Council of 15 December 1993 on the support of economic expansion in the Flemish Region, MB, 4 March 1994.

[40] Ordinance of the Council of the Brussels-Capital Region of 1 July 1993 related to the promotion of economic expansion in the Brussels-Capital Region, MB, 31 July 1993.

[41] Law of 12 June 1992 on the confirmation of the income tax code 1992, MB, 30 July 1992.

[42] Law of 29 April 1999 on the organisation of the electricity market, MB, 11 May 1999.

[43] Directive 96/92/EC of the European Parliament and of the Council of 19 December 1996 concerning Common Rules for the Internal Electricity Market.

the electricity sector was under-regulated in the past. On the other hand, the competences in the electricity sector are shared between the Regions and the Federal State. The federal government is responsible for the access to and the operation of the transmission network (>70 kV), the licenses for the new generation plants and pricing (non-eligible consumers). The Regions are competent for the access to and the operation of the distribution network (<70 kV) as well as for the electricity produced from renewable energy sources and combined heat and power plants.

With the liberalisation of the market, the entire organisation of the electricity sector has been reformed in order to create the conditions for an actual competitive market (unbundling of the vertically integrated companies and third party access to the grids) and in order to ensure the proper regulation of the sector (end of CCEG and new independent public regulators).

Both at the transmission and at the distribution level, network and supply activities have been unbundled (different legal entities) and regulated third party access to the grids has been granted to licensed operators. In 2001, the Elia System Operator took over the transmission grid from CPTE (a joint venture of Electrabel and SPE). Electrabel remained the majority shareholder at the beginning, but it then sold most of its shares to Belgian public (Publi-T, 30%) and private investors (stock market, 40%). At the distribution level, most inter-municipal utilities kept the distribution network activities (distribution grid operators) and left their supply activities behind.

Since 2000, the prevailing sector regulator, CCEG, has been replaced by an independent public regulator: CREG (Commission of Regulation of Electricity and Gas). Three other regional regulators have been institutionalised in Flanders, Wallonia and Brussels: VREG (the Vlaamse Reguleringsinstantie voor de Electriciteits-en Gasmarkt) in Flanders, CWAPE (the Walloon Commission for Energy) in Wallonia, and IBGE-BIM in Brussels (Brussels Environment Institute). The four regulators are in charge of regulating the tariffs and conditions of grid access, whether at the transmission level (CREG) or at the distribution level (VREG, CWAPE and IBGE-BIM). They are expected to operate as independent regulators of the electricity markets and administrative jurisdiction in case of conflicts between operators in the market.

As far as electricity production in concerned, in 2002 the CREG elaborated an indicative program of the means of production that follows the major options of the energy policy and sets investment priorities[44]. This program covers a ten-year period, but is adapted every three

---

[44] CREG, Programme indicatif des moyens de production d'électricité 2002-2011, December 2002.

years (last program adopted in 2005[45]). It is only an indicative program and not a constraining investment plan, unlike previously. New production plants are subject to an authorisation to be delivered by the federal energy ministry.

In Belgium, the liberalisation of the electricity supply market followed different speeds depending on regional legislation (see table 3). The electricity consumers became eligible (they can freely choose their supplier) according to their annual electricity consumption or their connected load, but depending on the Region where they are located, the timeframe differed. In Flanders, for instance, all electricity consumers have been eligible since July 2003, while in Wallonia and Brussels households had to wait until January 2007 to freely choose their supplier. However, the consumers who decided to buy exclusively RES-E and/or electricity from combined heat and power plants (CHP) were immediately eligible in the tree regions.

Table 3: Schedule of the electricity market opening in Belgium

Deadlines	Federal State	Flanders	Wallonia	Brussels-Capital	Percentage of total market opening[46]
24-10-2000	>100 GWh/year				33%
31-12-2000	>20 GWh/year				
01-07-2001		>20 GWh/year			45%
01-01-2002		>1 GWh/year			
01-01-2003	>10 GWh/year	>56 kVA	>10 GWh/year	>10 GWh/year	59%
01-07-2003		All			79%
01-12-2003	All customers connected to national transmission grid (>70 kV)				
01-07-2004[47]			Non residential customers	Non residential customers	87%
01-01-2007			All	All	100%

Due to the liberalisation, new electricity suppliers have been operating on the Belgian market. Electrabel still remains the main supplier in

---

[45] CREG, Programme indicatif des moyens de production d'électricité 2005-2014, January 2005.
[46] According to estimates from Electrabel: Direct, No.117, 28 August 2002.
[47] Directive 2003/54/CE of the European Parliament and of the Council of 26 June 2003 concerning common rules for the internal market in electricity and repealing Directive 96/92/EC.

all Regions, but is progressively losing market shares, especially in Flanders, where the market is now fully open. For instance, Nuon Belgium, Luminus and RWE Solutions together took over some 20% of the supply market from Electrabel in Flanders in 2004. On the federal gross electricity market, RWE Solutions controlled some 14.7% of the market in 2004[48].

V. RES-E Policy Change in Belgium (1999-2004)

In Belgium, the 1999 law on the organisation of the electricity market did not only set the foundations for the liberalisation of the market, but it also introduced new policies like an RES-E policy. Indeed, since the introduction of competition on the electricity market was expected to threaten the development of less competitive technologies, such as RES-E technologies, the federal government decided to adopt new instruments to support the latter. In addition, the pressure to adopt an actual RES-E policy intensified with the failure of the "Green frank system" of the CCEG and the preparation of a new European directive on RES-E (adopted in 2001[49]). Therefore, the liberalisation of the electricity market and the introduction of a new RES-E policy are inseparable in Belgium, they both participate in the same evolution towards profound reform of the electricity sector.

Before 1999, sporadic programs had been adopted in favour of RES-E at the federal (the CCEG green frank system) and regional levels (R&D and investment subsidies), but without short or long term policy objectives having been formulated. Therefore, one can hardly speak about an actual Belgian RES-E policy before 1999.

The first signs of an actual RES-E policy developed in Belgium can be found in the government declaration of July 1999 and after that in the federal plan for sustainable development, adopted by the federal government in July 2000[50]. It sees the development of renewable energies as a strategic objective for Belgium in its commitment to reduce CO_2 emissions. A target of 3% of RES-E by 2004 (about the double of the 2000 level) and 6% by 2010 was adopted by the federal government in 2001, which corresponds to the indicative target of the European directive of 2001 on RES-E. Since the federal state is not responsible for the RES-E policy, the RES-E target adopted by the federal government was

---

[48] Press release from CREG, CWAPE, IBGE-BIM and VREG about the development of the electricity and gas markets in Belgium, 2004.

[49] Directive 2001/77/EC of the European Parliament and the Council of 27 September 2001 on the promotion of the electricity produced from renewable energy sources in the international electricity market.

[50] Federal plan for sustainable development 2000-2004, September 2000.

then proposed to the Regions as a minimal indicative target. Thereafter, several programmes of the federal government reaffirmed this objective: for instance the national climate plan adopted in 2001[51] or the revised federal sustainable development plan of 2004[52].

But since the competence on RES-E devolves on the regions, each region adopts its own policy objectives autonomously. In Flanders, the energy programme of 1999[53] (government agreement) set a target of 3% of RES-E in 2004 and 5% in 2010. In 2004, the target was increased so as to reach 25% of electricity supply to come from either CHP or RES-E by 2010, and more specifically 6% of electricity supply from RES-E by 2010[54]. In Wallonia, the government energy program states that 8% of electricity supply is expected to originate from RES-E by 2010, and 15% from CHP[55]. In Brussels, the promotion of RES-E is mentioned in the regional climate policy program[56], but no specific target has been adopted so far. So, new RES-E policy objectives were adopted in Belgium from 1999, with different but convergent RES-E targets formulated by the federal government, as well as the Flemish and Walloon governments. These changes were political decisions adopted by the governments, and not administrative decisions, with the dominant policy actors being the energy ministers and the green parties in government (see table 1).

The RES-E policy, as it has been adopted in the 1999 law and the subsequent regional decrees, relies on a new policy instrument: the tradable green certificates (see figure 3). The tradable green certificate system is based on two traditional policy instruments: a quota (on the electricity suppliers) and a certification mechanism (for RES-E production). The purpose is to create, on the one hand, a demand for the RES-E (quota) and, on the other hand, a supply of guaranteed RES-E (certificates). Therefore, the certificates serve both as an additional source of income for the RES-E producers and as proof for fulfilling the quota for the electricity suppliers. Indeed, on the one hand, the RES-E producers obtain green certificates from the regulators (VREG, CWAPE, IBGE-BIM or CREG) according to the amount of electricity they generate[57]. On the other hand, the law imposes quotas of green certificates to all

---

[51] National climate plan 2002-2012.
[52] Federal plan for sustainable development 2004-2008, September 2004.
[53] Beleidsnota 2000-2004.
[54] Beleidsnota 2004-2009.
[55] Plan pour la maîtrise durable de l'énergie, December 2003.
[56] Plan Air Climat 2002-2010, November 2002.
[57] In Wallonia and Brussels, other criteria than the electricity production condition the allocation of certificates. We will come back to this later.

electricity suppliers, according to the amount of electricity they sell to final consumers. In order to fulfil their quota, the suppliers can choose between producing RES-E on their own, purchasing green certificates to RES-E producers or paying a penalty.

Figure 3: Tradable green certificates market in Belgium

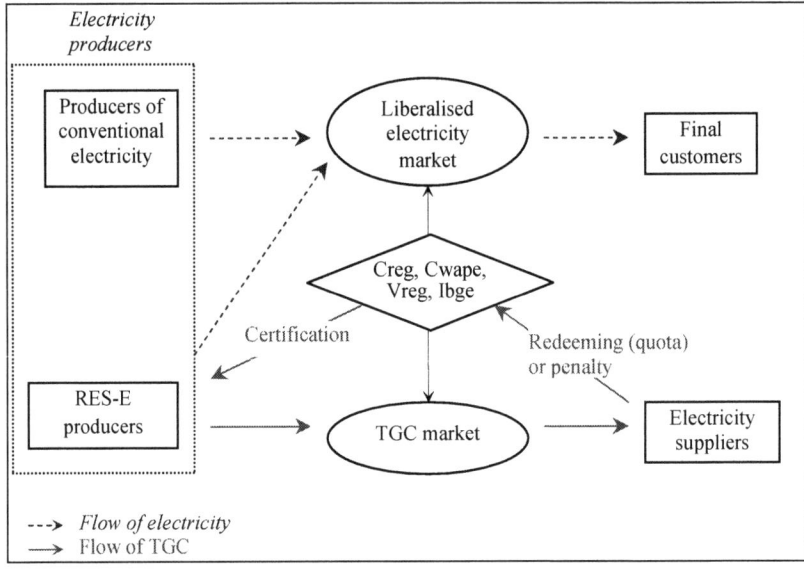

The Flemish Region[58] was the first to choose the green certificate market system because it was perceived to be the most efficient instrument to promote RES-E in a liberalised market[59]. This instrument was then adopted in Wallonia[60] and Brussels[61], albeit with some modifications. In Flanders, the idea of a green certificate market had been con-

---

[58] Flemish Decree of 17 July 2000. Decision of the Flemish Government of 28 September 2001 concerning the promotion of the electricity generated from renewable energy sources, MB, 23 October 2001.

[59] Interview with Géry Vanlommel, Cabinet of the Flemish minister for mobility, public works and energy, 29 January 2003. And interview with Wilfried Bieseman, Flemish administration of economic affairs, Natural resources and energy division (ANRE), 15 April 2003.

[60] Walloon Decree of 12 April 2001. Decision of the Walloon Government of 4 July 2002 concerning the promotion of green electricity, MB, 18 August 2002.

[61] Brussels-Capital Ordinance of 19 July 2001 on the organisation of the electricity market in the Brussels-Capital Region, MB, 17 November 2001.Decision of the Brussels Government of 6 May 2004 concerning the promotion of green electricity and CHP, MB, 28 June 2004.

sidered from 1998[62]. It stems from several external influences: the experience of the Dutch green certificate system, reports elaborated by a Dutch research institute (ECN), similar projects in other European countries (Denmark in particular) and the creation of RECS in which Flanders was among the first members[63].

Due to the regionalisation of the RES-E policy, three distinct tradable green certificate markets are currently operating in Belgium. Even if they are based on the same model (see figure 3), the rules vary according to the Regions (see table 4). Therefore, the TCG cannot be exchanged between the Flemish, Walloon and Brussels markets, unless agreements are signed between the Regions. Walloon TGCs have been eligible for Brussels quotas since 2005[64], and other reciprocity agreements are being discussed between the regional regulators: VREG, CWAPE and IBGE-BIM. The main differences between the three regions concern the eligible RES-E, the criteria for TCG allocation, quotas and penalties.

Firstly, the electricity produced from "quality"[65] CHP installations is eligible for green electricity TGC in Wallonia and Brussels, but not in Flanders[66]. However, quality CHP also benefits from a separated TGC system in Flanders[67]. In all three regions, electricity produced by quality CHP receives TGC because it helps reducing CO_2 emissions, just like RES-E, but different modalities have been adopted in Flanders, Wallonia and Brussels. In addition, hydropower plants of more than 10 MW are not eligible in Flanders and Brussels, but they are in Wallonia, until 20 MW.

Secondly, in Wallonia and Brussels, TGCs are granted not only according to the quantity of electricity produced (1 certificate for 1MWh produced), as is the case in Flanders and at the federal level (for offshore

---

[62] Interview with Wim Buelens, Cabinet Stevaert, 12 June 2003.

[63] RECS (renewable energy certification system) is an organisation that was initiated by several European electricity companies in 1999 in order to promote the development of a tradable green certificates systems in Europe. They organise their own green certificate market on a voluntary basis with their own certification mechanism and they lobby to make it recognised formally by the EU member states and the European Commission.

[64] Decision of the Brussels Government of 3 May 2005, MB, 17 May 2005.

[65] Meaning that the CHP installation must allow significant energy savings compared to standard electricity and heating installations.

[66] In Flanders, CHP benefits from its own TGC system separated from the RES-E system.

[67] Flemish Decree of 10 July 2003 modifying the Flemish electricity decree of 17 July 2000 concerning the introduction of a system of CHP certificates, MB, 11 August 2003.

plants), but also according to the CO_2 emissions avoided. In Wallonia, the number of TGCs a RES-E producer receives depends on its production and on the quantity of CO_2 emissions it saves compared to reference installations (1TGC per 456 kg CO_2 avoided). In Brussels, the same logic as in Wallonia prevails, since RES-E or CHP producers receive an amount of TGCs that corresponds to the CO_2 emissions savings it has made compared to a reference installation (1TGC per 217 kg CO_2 avoided). Wallonia and Brussels look at the CO_2 emissions savings and not only at RES-E production, because it allows them to include CHP in the system (with fewer certificates for 1 MWh than wind, for instance) and it can also prevent them from supporting less ecological energy sources like waste incinerators (if they do not comply with the CO_2 savings criteria).

Thirdly, the quota of TGC that the electricity suppliers must redeem annually differs between the regions. In Flanders, the quota has been increasing yearly, from 2% in 2005 until 6% in 2011[68]. In Brussels, similar quotas apply since the electricity suppliers must redeem TGC for 2.25% of their electricity supplies in 2005 and 2.5% in 2006[69]. However, the quota is slightly higher in Wallonia: 6% in 2006 and 7% in 2007[70]. Note that the quota in Brussels has been put at a higher level than the regional potential (which is quite low due to the urban area) in order to support the development of RES-E in the other regions. Therefore, exchange agreements with Wallonia and Flanders are vital for the Brussels market.

Fourthly, the penalty that a supplier must pay if it fails to reach its quota of TGC also differs between the regions, but they tend to converge around 100€ per missing certificate in the short term. In Brussels, the penalty per missing certificate is 75€ today, but it should increase to 100€ by 2007. In Wallonia, the penalty is 100€ and in Flanders 125€.

As shown in table 4, the TGC market is the main instrument of the RES-E policies in Belgium, but not the only one. Other instruments have been adopted by the three regions and the federal government to complement the TGC market: guaranteed minimum prices for TGC (federal, Walloon and Flemish governments); free transport on the distribution network for RES-E (Flanders); investment grants (in the three regions); and fiscal incentive for RES-E investments (at the federal level).

---

[68] Decree of the Flemish Council of 20 December 2002, MB, 31 December 2002.
[69] Ordinance of Brussels Capital of 1 April 2001, MB, 26 April 2004.
[70] Decision of the Walloon government of 15 May 2003, MB, 26 May 2003.

Table 4: The TGC systems and additional instruments in Belgium

	Flanders	Wallonia	Brussels	Federal
Eligible electricity	RES-E (hydro<10MW)	RES-E (hydro<20MW) + CHP	RES-E (hydro<10MW) + CHP	Offshore RES-E
TGC allocation	1 MWh	1 MWH/CO_2 savings	CO_2 savings	1 MWh
Quota	2.5% in 2006; 6% in 2011	6% in 2006; 7% in 2007	2.5% in 2006	n/a
Penalty	125€/TGC	100€/TGC	75€/TGC; 100€ in 2007	n/a
Minimum prices	Yes	Yes	No	Yes
Free distribution	Yes (until end 2004 only)	No	No	n/a
Invest grants	Yes: solar PV	Yes: biomass, solar boilers	Yes: solar boilers	n/a
Fiscal incentives	n/a	n/a	n/a	Yes

Legend: n/a = non applicable.

In order to limit the uncertainty of the TGC price on the market and therefore increase the security for the RES-E investors, the federal and subsequently the Walloon and the Flemish governments decided to offer a guaranteed minimum price for the TGC. Therefore, RES-E producers will have to choose between selling their TGC on the TGC market or receive the minimum price set by the government. Due to regionalisation, three minimum price systems operate simultaneously in Belgium, with different prices and terms (10 to 20 years).

First at the federal level, the transmission system operator (Elia) has the obligation to purchase TGC (federal, Flemish, Walloon or Brussels TGC) for 10 years at a price that varies according to the energy source and/or technology: 20€/MWh for biomass, 50€/MWh for onshore wind and hydro, 90€/MWh for offshore wind and 150€/MWh for solar power[71]. The transmission system operator is then allowed to sell the TGCs on the TGC market to cover its costs. Extra costs are covered by the federal energy fund, financed by an extra charge on the connection costs. A new proposition is currently being discussed to increase the support for offshore wind power plants (increased price and terms), but no decision has thus far been adopted yet.

The Walloon government then decided to adopt a similar measure in order to guarantee a minimum price for the green certificates in Wallonia[72]. Unlike the federal prices, only one generic price of 65€ per

---

[71] Royal Decree of 16 July 2002 concerning the mechanisms aiming at promoting electricity generated from renewable energy sources, MB, 23 August 2002.

[72] Decision from the Walloon Government of 6 November 2003 concerning the production subsidy granted to green electricity and modifying the Decision of the

certificate, whatever RES-E is concerned, is awarded in Wallonia by the energy ministry for a 10-year period. The price can be adapted according to the given technology to a maximum of 100€ (the penalty price). This means that there is room for negotiation between the minister and the RES-E producer when concluding the convention. However, this system has been adopted mainly as a safeguard provision in order to increase the security for the investors. The price of the TGC is expected to be higher than 65€ on the market, so that it will not be activated unless major TGC prices drop. Therefore, it has been decided that the TGC given to the ministry will be immediately redeemed by the CWAPE, without compensation for the ministry, which will otherwise have to cover the charges linked with the minimum prices.

Finally, a similar provision has been adopted recently in Flanders, with a significantly longer term (for solar power only, 20 years) and higher prices (especially for solar power)[73]. Flemish distribution operators must buy TGC from the RES-E producers connected to their grids (only if they ask for it), at a price that varies between the energy source and technology: onshore wind power and biomass, 80€ per certificate; hydro, geothermal and wave power, 95€; and solar power, 450€. Charges for this obligation are covered by the suppliers subsequently selling the TGCs on the market.

In addition, no distribution costs can be charged for RES-E in Flanders. This helps reduce the price of RES-E on the electricity market. However, since this measure only applies to RES-E generated in Flanders, the European Commission has been arguing against it. So far, the Flemish government has stuck to its position and maintains its decision, but the discussion is not over yet.

Moreover, in order to stimulate investments in RES-E technologies, the three regions have adopted several programs, including investment grants covering up to 50% of the investments, but with very different objectives in each region. In Flanders, investments of solar PV in households get the largest investment grants: 50% until 2005, and only 10% from 2006, but with a minimum price of 450€ per TGC when in operation. Other subsidies relate to more general energy savings and the rational use of energy investments. In Wallonia, a large number of investment grants are available, but mostly for heating systems (solar boilers, heat pumps, wood boilers) or energy savings investments. In

---

Walloon Government of 4 July 2002 concerning the promotion of green electricity, MB, 11 February 2004.

[73] Decision from the Flemish Council of 7 May 2004 modifying the electricity decree of 17 July 2000 concerning the TGC system and article 37§2 of the decree, MB, 8 June 2004.

Brussels, heating systems or energy savings are the main beneficiaries of investment grants as well: 50% for households investing in solar boilers and 30% for companies or public sector investment in energy saving solutions for buildings. In addition to these investment grants, tax breaks are available at the federal level for RES-E investments made by households (based on 40% of the investment for most RES-E technologies[74]) or companies (based on 13.5% of the investment).

Finally, the increased promotion of the RES-E represents an opportunity for new actors to emerge on the Belgian market. With regard to RES-E generation, the new investors include not only the incumbent electricity producers (Electrabel and SPE), but also local utilities or private national or international eco-investors. With the new legal incentives (green certificates, minimum prices, investment grants) that have been implemented lately, one can expect more investors entering the market. In addition, the renewable energy industry is not very developed in Belgium, but recently some industrial initiatives have emerged, particularly in the solar PV (in Flanders mainly), biomass, wind and CHP sectors. Some of these businesses focus more on assembling imported materials (wind turbines), while others tend to develop their own technologies (solar PV, CHP). Not all are new actors on the Belgian market, and traditional industries may take advantage of the emergence of this new market to diversify their equipment and services. A powerful professional federation operates in the solar sector: Belsolar (Belgian Solar Industry Association). And a federation of the renewable and alternative electricity producers was created in 2003 (EDORA). It aims at representing the RES-E sector in financial and political arenas and, more recently, started to act as a TGC exchange platform for Wallonia and Brussels.

VI. Interpretation of RES-E Policy Change in Belgium

In Belgium, before 1999, the development of RES-E was not a political priority. The predominance of a domestic, cheap and abundant source of electricity (nuclear power covers about 50% of electricity production in Belgium) and the lack of political and social mobilisation in favour of RES-E partly explain this situation. Several isolated RES-E programs were adopted in Flanders and Wallonia during the 1990s, but without actual long-term policy objectives for the development of RES-E. Under pressure of the federal government, the electricity companies

---

[74] Law of 10 August 2001 related to the reform of income taxes, MB, 20 September 2001. Royal Decision of 20 December 2002 modifying the AR/CIRC 92 related to the reduction of income taxes for energy savings investments in houses, MB, 28 December 2002.

adopted financial support to RES-E in 1995 (CCEG green frank system), but without commitment to any quantitative targets (see table 5). Therefore, this period was characterised by an RES-E policy based on no policy objective, but with a specific policy instrument, the CCEG green frank system. The target group of the green frank system were the distribution companies (demand actors), which were obliged to purchase the RES-E fed into their grid for a premium price per kWh (see table 5). The incentive used by the CCEG to support RES-E was thus a premium price (see table 5). Finally, this system was financed, either directly by the distribution companies or by the RUE production fund managed by the CCEG, but in both cases it was ultimately the electricity consumers who endorsed the cost through increasing their electricity bills (private budget, see table 5).

In 1999, the RES-E emerged on the political agenda and the process of RES-E policy change started (the turning point of RES-E policy change, see table 5). Due to the federal structure of Belgium, the Federal State and the Regions (Flanders, Wallonia and Brussels-Capital) share the competence on the RES-E policy. However, since the RES-E policy changes at the different levels pretty much share the same characteristics, we interpret the RES-E policy change in Belgium as a whole. This way, it also makes this case easier to compare with the other cases in this research.

Starting in 1999, the reform of the electricity sector (liberalisation of the electricity market and institutional reform with new regulators), the preparation of the RES-E directive at the European level and the arrival of green parties in the federal and regional governments, opened a window of opportunity for RES-E policy change in Belgium. New ambitious policy objectives with short and long-term quantitative targets were adopted at the federal level[75] (3% of electricity supplies from RES-E by 2004 and 6% by 2006) and at the regional level (Wallonia[76]: 8% by 2010; Flanders[77]: 5% by 2010) (see table 5[78]). In addition, new policy instruments were adopted with increased and differentiated financial support to RES-E on the electricity market (TGC, guaranteed minimum prices differentiated by technology, new investment subsidies). The central instrument of the Belgian RES-E policy is the TGC system, which imposes annual quotas of RES-E (quantity incentive) to the

---

[75] Federal plan for sustainable development 2000-2004, September 2000.
[76] Beleidsnota 2000-2004.
[77] Plan pour la maîtrise durable de l'énergie, December 2003.
[78] Since the RES-E targets by 2010 differ in Wallonia and Flanders, we decided to mention the federal targets in table 5, as they represent the national targets at the European and international level and they are very close to the regional targets anyway.

electricity suppliers (target group demand actors) and is funded by the electricity consumer bills (private budget) (see table 5).

Table 5: RES-E policy change in Belgium

	Before 1999	After 1999	Policy change
Policy objective	None	3% by 2004 6% by 2010	Yes
Policy instrument	– Target group: demand (distribution companies) – Incentive: price (green frank) – Resource: private budget (electricity bills)	– Target group: demand (suppliers) – Incentive: quantity (quota) – Resource: private budget (electricity bills)	Yes

In conclusion, Belgium is still far from being a pioneer in the renewable energy field, but the policy change that has been going on from 1999 (see table 5) create the conditions (necessary but still not sufficient) for a future development of RE-E in Belgium. The question now is, what explains the shift in RES-E policy that happened around 1999. In fact, several factors combined to introduce this policy change: new policy actors, failure of past policy, European policies, and change in governing coalitions.

VII. Explaining RES-E Policy Change in Belgium

A. Policy Actors

The progressive transfer of energy competences to the Regions and the final overlapping division of responsibilities for RES-E explain why in Belgium some RES-E policy actors are specific to one Region (political actors, RES-E associations, environmental associations), while other actors participate in the policy making process at all levels (electricity companies, consumer associations). However, we observe that the patterns of how the policy actors explain RES-E policy change are similar in Flanders, Wallonia and at the federal level. Therefore, like we did for policy change, we consider the policy actors in Belgium as a whole, without disaggregating the interpretation at the regional level. However, when significant differences are observed (especially between Flanders and Wallonia) we will mention them.

For decades (1955-1999), Belgian electricity policy has been dominated by the electricity companies through the CCEG[79]. While the

---

[79] CCEG (Electricity and Gas Control Committee): the regulator of the electricity market from 1955 to 2000.

CCEG was supposed to be a parity organ, grouping all actors of the electricity sector and controlled by the public authorities, it was in fact managed by the main electricity company, Electrabel. Since Electrabel was strongly vertically integrated (production, transport, distribution, supply) and had strong political connections (at the municipal, regional and federal levels), it was able to exert its monopoly not only on the electricity market, but also on the electricity policies. For instance, there was an adequacy between the electricity prices defined by the CCEG and the revenue the electricity companies wanted to earn. In addition, the electricity consumers were left aside during that period, with no opportunity to choose their electricity suppliers and with weak consumer associations with no interest in RES-E (Test-Achats and CRIOC). Finally, by the end of the 1990s, the promotion of RES-E was still in its infancy, with very limited RES-E generation (see table 2) and no powerful policy actors to support it: no significant RES-E industry, weak RES-E associations (APERe, ODE-Vlaanderen), and limited priority to RES-E from the environmental associations (Inter Environnement Wallonie, Bond Beter Leefmilieu).

The year 1999 marked the beginning of a change in the RES-E policy actors. The general elections in June 1999 saw the arrival of the Green Parties in the governing coalitions in Flanders (Agalev), Wallonia (Ecolo) as well as at the federal level (Agalev and Ecolo). At all levels the new government programs included the Green Parties' priorities on the Belgian energy policy: dissolution of the CCEG and replacement by new public regulators, the phasing out of nuclear power, and finally, increased political priority to RES-E (new RES-E policy objective). So, new RES-E policy objectives were adopted in Belgium from 1999, with different but convergent RES-E targets formulated by the federal (3% by 2004 and 6% by 2010), the Flemish (5% by 2010) and the Walloon (8% by 2010) governments. In Belgium, the politico-administrative culture is characterised by a dominance of the political actors over the administrative actors in the policy-making process. Most of the policies are designed by the Ministers and their cabinets (sort of personal secretariat of the Minister highly politicised) and the role of the administrations is often limited to the implementation (De Visscher C., le Bussy G. & Eymeri J.M., 2004). The RES-E policy is no exception. The RES-E policy changes that we presented were political decisions adopted by the ministers, and not administrative decisions, with the dominant policy actors being the energy ministers and their cabinets (see table 6).

In addition, in 1999 the energy portfolios were given to the green parties (Ecolo at the Federal level and in Wallonia) or to "green" ministers (the Socialist Steve Stevaert in Flanders) in the different governments (except for Brussels), which explains why the RES-E policy became a political priority in Belgium. At the federal level, the green parties negotiated the phasing out of nuclear power and the development of RES-E as a primary condition for their participation in the governing coalition and consequently Ecolo received the portfolio of State Secretary for Energy (Olivier Deleuze)[80]. In Flanders, the energy portfolio did not go to Agalev but to the Socialist Party (Steve Stevaert). However, Agalev and the Socialist party were both traditionally pro-renewable energy and against nuclear energy, and moreover, the new socialist energy Minister (Steve Stevaert) had a personal background and interest in RES-E (he was dubbed a "green" socialist minister). In Wallonia, an Ecolo MP was appointed as the energy minister (José Daras). In Brussels, the green parties were not in the new governmental coalition. The energy competences were mixed, with many other ones (employment, economy, housing) in the hands of the Socialist Party (Éric Tomas). So, unlike the other governments, no political priority for energy policy or RES-E policy could be observed in the new Brussels government in 1999, which probably explains why the RES-E policy change was less significant in that Region compared to the others.

In conclusion, as from 1999, the political actors (politicians) took over the direction of the Belgian RES-E policy from the hands of the CCEG (private electricity companies) and the presence of the green parties in the governments significantly increased the political priority to RES-E (see table 6). The civil servants of the energy administrations were involved in the discussions, especially those on the design of the new instrument (TGC) in Flanders and Wallonia, but the lead was without a doubt in the hands of the Ministers and their "cabinets" (see table 6). Finally, in 1999, the existing regulator of the electricity sector (CCEG) was dismantled, while the new ones (CREG, CWAPE and VREG) were not yet in operation, so the regulators did not play a role in the RES-E policy change in Belgium.

Table 6: Dominant policy actor in RES-E policy change (H1/1)

	H1/1
Dominant policy actor?	Politicians: – Minister of energy (Wallonia and Flanders) – State Secretary for Energy (Federal)

As far as the other policy actors are concerned (electricity companies, RES-E associations, environmental associations and consumer groups), the actors of the sector were more involved in the RES-E policy

---

[80] Note that there had not been a State Secretary for Energy for years in Belgium. This appointment shows the priority of the new government in energy policy.

process in 1999 than in the past (except the electricity companies, which had dominated the policy before). However, it is important to distinguish between the change of the RES-E policy objective that resulted primarily from political agreements within the government, with no formal participation of the actors of the sector; and the change of policy instruments (TGC) that were piloted by the energy ministers, but with the participation of the actors of the sector (especially in Wallonia). So, we focus our analysis of the influence of the actors of the sector on the formulation of the TGC systems in Flanders and Wallonia.

Table 7: Influence of the actors of the sector (H1/2)

Policy actors (H1/2)	Early / late	Preferences	Influence
Electricity companies			
– Electrabel	early	+	+
– Intermixt	early	-	+
– Inter-Régies	early	-	+
– SPE	early	+	-
RES-E associations			
– Apere	early	+	+
– Ode-Vlaanderen	early	+	+
Environmental associations			
– IEW	early	+	0
– BBL	early	+	0
Electricity consumers			
– Test-Achats	early	0	0
– Agoria	early	-	-
– FEB	early	-	-

Legend: Preference (+) favourable, (-) not favourable, (0) indifferent; Influence (+) strong, (-) weak, (0) no significant influence.

Even if the general trends of the actor's influence were similar in Flanders and Wallonia (see table 7), there were some differences. In Flanders, the consultation of the actors of the sector turned out to be less frequent than in Wallonia, the cabinet of the Minister doing most of the job by himself or with the Flemish administration of energy (ANRE)[81]. In Wallonia, the consultation of the stakeholders of the sector was more frequent, as was the consultation with the other Ministers of the Walloon government[82]. This is probably due to the "new way of doing politics",

---

[81] Interview with Wim Buelens, Cabinet Stevaert, 12 June 2003. Interview with Wilfried Bieseman, Flemish administration of economic affairs, Natural resources and energy (ANRE), 15 April 2003.

[82] Interview with Cécile Barbeaux, Cabinet Daras, 5 February 2002. Interview with Francesca Stockman, Walloon administration of energy (DGTRE), 20 May 2003.

which characterised the Ecolo party in Wallonia at that time. In addition, the links between the electricity companies (Electrabel, Intermixt, Inter-Régies) and the political actors appeared to be stronger in Wallonia than in Flanders. For instance, in Wallonia, the other ministers of the government were often represented by representatives from the electricity companies in the inter-ministerial meetings on electricity policy[83]. So, the influence of the electricity companies on the policy making process was less significant in Flanders than it was in Wallonia.

Table 7 summarises the characteristics of the influence of the actors of the sector on RES-E policy change in 1999 in Belgium (aggregated picture). At first, we observe that all the actors were consulted early in the policy-making process in Flanders and Wallonia. As we already mentioned before, the consultation of the actors of the sector was more frequent in Wallonia than it was in Flanders, but it was still significant in both regions. Secondly, few actors seem to have had a strong influence on the policy change: the electricity companies (Electrabel, Intermixt and Inter-Régies) and the RES-E associations (Apere in Wallonia and Ode-Vlaanderen in Flanders). This is explained by the strong leadership of the energy ministers and their cabinets on this issue. The other actors were either less influential (SPE, Agoria and FEB) or not influential enough (IEW, BBL, Test-Achats). For instance, the RES-E issue was not a priority of the environmental associations (even if they were rather in favour of the change), especially because specific RES-E associations were already taking care of it. The consumer watchdog association Test-Achats did not show any significant interest in the issue at the time. The two other associations representing the large electricity consumers (Agoria and FEB) opposed the policy change, but without significant impact.

As far as the preferences of the most influential actors are concerned, except for Intermixt and Inter-Régies, all the other actors who had a strong influence on the policy change were in favour (Electrabel, Apere, Ode-Vlaanderen) (see table 7). The preference of the RES-E associations seems obvious; the energy ministers were finally going in the direction they had been advocating for years: more support for the RES-E sector. Concerning Electrabel, the preference is less straightforward. As an electricity generator (and RES-E generator), Electrabel was naturally in favour of the policy change; but as an electricity distributor (through Intermixt) it was more critical of the new instrument, which imposed a quota of RES-E (target group). Inter-Régies, as an electricity distributor, did not support the policy change at first either. However, the RES-E policy change occurred at the same time as the reform of the

---

[83]   *Idem.*

electricity sector in Belgium, which represented a far more critical issue for the electricity companies and especially the inter-municipal utilities (Intermixt and Inter-Régies). So the opposition of Intermixt and Inter-Régies to the RES-E policy change was not their priority and, since they were being reformed themselves, their impact on the policy-making process was not as significant as it had been in the past.

In conclusion, the dominant policy actors of the RES-E policy change in Belgium in 1999 were clearly the politicians (ministers of energy and their cabinets) and not the administrations (H1/1 validated). Besides, the actors of the sector participated early in the policy-making process (especially for the formulation of the TGC systems in Flanders and Wallonia), but with more or less influence, depending on the actors. We observe that the influence of the actors did not depend on the stage of the policy process in which they were consulted, as all of them were consulted early, which means that we cannot either validate or invalidate the hypothesis (first proposition of H1/2 non-applicable). In addition, the most influential actors were not all in favour of the policy change, but the ones who supported the change (Electrabel, RES-E associations) had ultimately more impact than the others (Intermixt and Inter-Régies) (second proposition of H1/2 validated). So, ultimately, H1/2 is only partly validated in the Belgian case.

B. Past Policy

In 1999, the past RES-E policy (the CCEG green frank system) was largely reckoned to have failed to increase the RES-E share on the market. In addition, it also did not fit with the new context of the liberalisation of the electricity market anymore (CCEG being replaced by CREG).

The CCEG green frank system, adopted in 1995 and then revised in 1998, did not specify a quantitative target to be reached (no policy objective), which makes it difficult to evaluate its effectiveness, but also shows how un-ambitious it was ("failure"[84], see table 8). In 1999, the new Secretary of State for energy (Olivier Deleuze, Ecolo) commissioned a report about the evaluation of the past RES-E policy and future policy prospects from APERe (association for the promotion of RES). In this report, APERe stresses the weaknesses and failures of the CCEG green frank system (de Radigues, Huart, 2000). While the system of feed-in tariffs has proved to be very effective in several European

---

[84] The fact that there were no quantitative target does not represent a failure of the policy *stricto sensu* (because no point of reference to assess the failure), but yet a strong loophole that was perceived as a "failure" of the policy by the policy actors and needed to be remedied.

countries (e.g. Germany and Spain), the way it was implemented by the CCEG in Belgium happens to be very unsuccessful, and this for several reasons: prices for RES-E generators were too low (even with an increase in 1998), lack of long-term security (no legal enforcement), discrimination among eligible RES-E producers (utilities, autoproducers, producers connected to the transmission grid were not eligible), and connection costs were not transparent. In the end, only few RES-E installations have been built under this system (mainly in wind, hydro and biomass), but the overall percentage of RES-E production in the overall electricity production increased only very slightly: from 1.29% in 1995 to 1.60% in 2000 (see table 2). For instance, in the 2001 directive on RES-E, an indicative target of 6% by 2010 has been adopted for Belgium, which shows that the potential to be reached (on e condition that proper policies are adopted) is far higher than what was achieved in the past. So, in 1999, the past RES-E policy appeared to be very unsuccessful: first because it did not set any policy target[85], and second because if proved to be ineffective in significantly increasing the share of RES-E on the electricity market (see table 8). The question of its affordability could not be questioned because it was not transparent enough to be evaluated. The green frank system was financed by the electricity suppliers or the CCEG, based on a general increase of the electricity bills for RUE purposes and no transparent account of how this money was used is available, which makes the affordability of the RES-E instrument impossible to evaluate (see table 8).

Table 8: Past policy success/failure (H2)

	Past policy success/failure (H2)	
Objective	No quantitative target	"Failure"
Instrument	Green frank system: – not effective – affordability?	Failure

In addition, in 1999 the reform of the electricity sector sounded the death knell for the CCEG, which did not fit with the new model of independent sector regulation and policy making based on transparency and non-discrimination. A new public regulator was appointed to regulate the market (CREG). And the energy ministers (federal and regional) took over the formulation of electricity policies. Therefore, the green

---

[85] The past policy objective is impossible to evaluate *stricto sensu* (no point of reference), but it was still clearly perceived as a failure by the policy actors in 1999. This is the reason why the failure is in quotation marks in table 8.

frank system was to be cancelled as the CCEG was dismantled, which called for the adoption of a new policy instrument.

In conclusion, before 1999, the RES-E policy in Belgium appeared to be un-ambitious (no quantitative target), un-transparent (no control over the RUE fund) and ineffective (no significant increase of RES-E); in sum unsuccessful (H2 validated).

C. *Europeanisation*

In Belgium, we observe that in 1999 at the time the Belgian governments were changing the RES-E policy, this policy at the EU was still in a very early stage (see figure 4). Figure 4 shows that the European and Belgian RES-E policy-making processes occurred pretty much at the same time, which means that the European policy was still very uncertain. However, the European liberalisation policy was already settled (see figure 4) and the Belgian policy actors were very careful to adopt a policy instrument that would be adapted to a competitive market. They were preoccupied to fit the European policy as much as possible (even if uncertain) and what was going on at the European level was regarded as important because potentially constraining in the short and medium term. In addition, the European policy was used by the Belgian policy actors to legitimise the policy change, as they were expected to go in the same direction.

Figure 4: The European policy and Belgian RES-E policy change

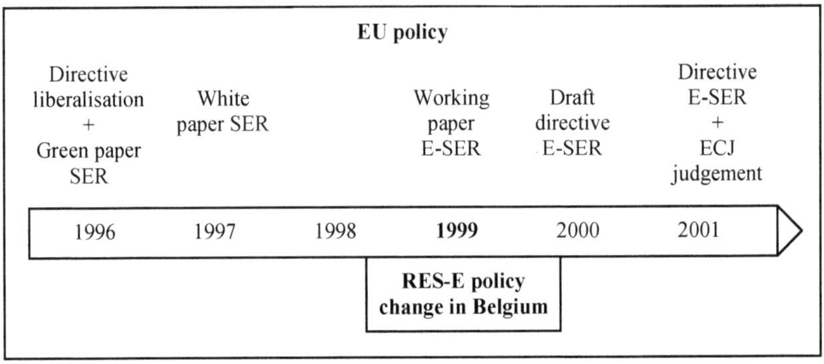

EU policy					
Directive liberalisation + Green paper SER	White paper SER		Working paper E-SER	Draft directive E-SER	Directive E-SER + ECJ judgement
1996	1997	1998	**1999**	2000	2001
			RES-E policy change in Belgium		

In 1999 when the Belgian policy actors started to formulate a new RES-E policy, the European RES-E policy was still not clear. Some propositions in terms of policy objective (12% of RES in the European gross energy consumption in the 1997 White Paper) and policy instrument (the Commission looked favourably at TGC systems in line with the liberalised electricity market) were discussed in the European Commission (e.g. in the 1999 working paper), but it was still very uncertain.

However, despite the uncertainties, those propositions were looked at seriously in Belgium at the time of formulating the new RES-E policy. In addition, the European directive on the liberalisation of the electricity market also influenced the policy change, as the Belgian actors believed that the new RES-E instrument was to be based on market mechanisms to be more adapted to the competitive electricity market. So, the Belgian policy actors perceived a misfit between the past Belgian RES-E policy and the European policy, which influenced the direction of the RES-E policy changes in Belgium. Firstly, the white paper on RES of 1997 called for ambitious RES-E targets to be set in all member states, which was not the case in Belgium (see table 9). Secondly, the green frank system operated by the CCEG did not fit in the context of the liberalised electricity market anymore, and this for different reasons: (1) it was managed by the by the monopolistic electricity companies (CCEG to be dismantled upon liberalisation), (2) without much transparency about how it operated (issue of indiscriminate access to it), (3) and how it was funded (issue of state aid) (see table 9). In a competitive electricity market, this instrument was not adapted anymore and needed to be replaced. Thirdly, in 1999, the European commission seemed to favour market-based instruments like the TGC rather than guaranteed price systems such as the FIT (1999 Working Paper).

Table 9: Europeanisation of RES-E policy (H3)

	Fit/Misfit (H3)	
Objective	No RES-E target	Misfit
Instrument	Green frank system from CCEG: managed by monopolistic actors, not transparent, access discriminated	Misfit

In addition, we observe that the Belgian policy actors used the European policy to legitimise their RES-E policy change (policy objective and instrument). The new RES-E targets adopted by the federal and regional governments were often hailed for its congruence with the European target of the 1997 White Paper, or later with the indicative targets of the 2001 directive. Moreover, in 1998 and 1999, the TGC system was expected to be more adapted in the context of a competitive electricity market because it was based on a market mechanism. The European liberalisation policy was then used by the Belgian actors (especially in Flanders[86]) to legitimise the choice of the TGC system over other potential instruments (e.g. feed-in tariffs).

---

[86] Interview with Wim Buelens, Cabinet Stevaert, 12 June 2003.

In conclusion, the misfit (or perceived misfit) between the European and the past RES-E policy in Belgium explains both the change of policy objective and instrument (H3 validated). Besides, the European policy proved to be useful to the Belgian policy actors to legitimise the policy change, but it does not directly explain the policy change.

D. *Political and Institutional Context*

In 1999, the Socialist-Christian Democrat coalition was replaced by the so-called "rainbow coalition" (Socialist, Liberal and Green Parties) in the federal, Flemish and Walloon governments. For the first time, the Green Parties (Ecolo and Agalev) participated in government and both in the federal and Walloon governments they were decisive partners. They had blackmail power over the other parties and were able to impose their priorities (e.g. phasing-out of nuclear power, new RES-E policy). In Flanders, Agalev was not critical to the coalition, but it was still able to bring a number of environmental issues on the agenda (e.g. RES-E policy change). In addition, in the federal and Walloon governments, the energy portfolios were assigned to the green party Ecolo. Therefore, the conditions for a change of the RES-E policy were optimal with the green parties in government (significant ideological distance with past government), their powerful position in the coalition, and their appointment at the energy portfolio at the federal and Walloon level (see table 10).

Belgium is a consensus democracy with multi-party governments due to the proportional electoral system. Compared to other consensus democracies, it is characterised by a large number of governing parties at the federal level due to the separation of the party system between Flemish and French-speaking parties. For instance, the 1999 federal coalition included six different political parties from three very different ideological streams (socialists, liberals and greens) and two linguistic communities (Flemish and French-speaking), which means an ideologically diversified coalition. In addition, political parties are very powerful actors in the Belgian political system (particracy). Therefore, Belgium is characterised by a large number of partisan veto players, which makes policy changes more difficult to achieve in theory (see table 10). However, as far as RES-E policy is concerned, a government consensus about the need for policy change was achieved in 1999, both at the federal and regional levels (small ideological distance over RES-E). The government agreement of the beginning of a term has a very strong value of policy commitment in Belgium, so the negotiations of this agreement is of particular importance for the governing parties and it gives a rather good idea of the policy changes that are expected to be adopted during the term of office. In 1999, the issue of RES-E policy was included by the green parties (Ecolo and Agalev) in the government agreements of both the federal and regional governments. In the Belgian political context, this means that RES-E policy change was very likely to be adopted in the end, because the partisan veto players reached a consensus on this issue.

In addition, Belgium is characterised by a small number of institutional veto players. The Belgian parliament is composed of two Chambers (bicameral system), but they share the same political majority and, since party discipline as well as government loyalty are very strong, parliament rarely vetoes government. In addition, under the Belgian constitution, the King represents another institutional veto player because he must countersign all legislation. But in reality, the King's veto power is not effective and the countersignature is just a formality in the legislative process. Finally, Belgium is a federal state, but with a strict division of responsibilities between the federal state and the Regions and no veto power from one to the other (unlike in Germany with the Bundesrat), except in very specific cases (modification of the constitution and sensitive linguistic legislations). So, in reality, the power of government is usually unchallenged by institutional veto players in Belgium, which makes policy change rather easy to achieve if the necessary consensus is achieved between the governing parties (see table 10).

Table 10: Government coalition (H4/1) and veto players (H4/2) in Belgium

Governing coalition (H4/1)	Veto players (H4/2)
– New political parties: Green Parties – Large ideological distance with past government – Significant power in coalition (energy portfolio and blackmail power)	– Few institutional veto players and internally coherent with small ideological distance: same majority in two Chambers of Parliament with strong party discipline, monarchy but no power, and strictly separated federalism – Many partisan veto players but small ideological distance over RES-E and strong internal coherence: multi-party coalition, but binding coalition agreement and strong party discipline

In conclusion, the participation of the Green Parties in the new governing coalition in 1999 opened a major window of opportunity for RES-E policy change to occur, so H4/1 is validated in this case. In addition, the presence of a large number of partisan veto players did not prevent the RES-E policy change from happening because of the coalition agreement on this issue (consensus over RES-E) and the strong party discipline (H4/2 validated).

VIII. Conclusion

We observed that the RES-E policy changed in Belgium at the turn of 1999 (at the Federal level, in Flanders, Wallonia and Brussels). Changes both at the policy objective and the policy instrument level occurred (see table 5). Those changes can be explained by a conjunction of different factors, including the policy actors involved in the policy making process (H1), the success or failure of the past RES-E policy (H2), the influence of the European policy (H3), and finally the political and institutional context (H4).

Table 11 summarises the main findings of this chapter. As far as the test of the hypotheses is concerned, table 11 shows the hypotheses are validated in most cases. Firstly, the relation between the type of dominant policy actor and the policy change (H1/1) is not fully validated, as it is confirmed regarding the change of policy objective (political actors dominant), but not regarding the change of policy instrument (civil servants not dominant), which is not a surprise in the Belgian politico-administrative system. Then, the hypothesis about the influence of the actors of the sector (H1/2) is not applicable regarding the change of policy objective, since the actors of the sector were not involved, and it is only partly validated in the case of the change of policy instrument in Flanders and Wallonia. Indeed, the relation between the earliness of the consultation and the influence of the actors (first proposition of H1/2) could not be tested in this case, but the preferences of the most influential actors and the policy change (second proposition of H1/2) are coherent with our assumption, even if it is not a highly significant explanation in those cases. Secondly, the past policy hypothesis is fully validated for the policy instrument change, but it is less straightforward for the policy objective change, as we should talk not so much about a "failure" of the past policy objective but an absence of RES-E policy objective in the past (H2). However, since the absence of policy objective was perceived by the policy actors in 1999 to be a "failure", we interpret this as the validation of the hypothesis (H2 validated). Thirdly, the hypothesis on Europeanisation (fit versus misfit, H3) is validated in the Belgian case, as the misfit between the Belgian and the European RES-E policies influenced the policy change in Belgium (objective and instrument). Fourthly, the change in governing coalition proves to be decisive in explaining the changes in the policy objective and instrument in Belgium, which confirms what we expected (H4/1). Finally, despite the presence of a large number of potential partisan veto players, the RES-E policy change actually faced no vetoes thanks to the coalition agreement that endorsed a political consensus about the RES-E policy change (no ideological distance), as well as the party discipline that characterises the Belgian party system, which validates our hypothesis (H4/2).

Table 11: Validation and relevance of the hypotheses in Belgium

	H1		H2	H3	H4	
	H1/1	H1/2			H4/1	H4/2
Validation hypotheses						
Policy objective	V	n.a.	V	V	V	V
Policy instrument	I	"V"	V	V	V	V
Explanation policy change						
Policy objective	++	0	+	++	++	+
Policy instrument	0	+	++	++	++	+

Legend: ++ (major explanation of policy change), + (minor explanation), 0 (does not explain policy change). V (hypothesis validated), I (hypothesis invalidated), "V" (hypothesis partly validated), n.a. (hypothesis not applicable).

As far as the explanation of the policy changes is concerned, we observe very similar, but still slightly different patterns regarding which hypothesis is more relevant to explain the change of policy objective or policy instrument (see table 11).

Firstly, the policy objective change seems to be explained primarily by three factors in Belgium: the dominance of political actors in the policy making process (H1/1), the participation of the green parties in the government coalitions (H4/1), and the influence of the European policy (H3). The first two explanations are closely linked, since the political actors who dominate the RES-E policy change process are the green parties in government, especially at the Federal level and in Wallonia, but also in Flanders. Besides, the misfit between the indicative RES-E targets at the European level and the absence of quantitative targets in Belgium has significantly influenced the adoption of the new targets in Belgium (H3). In addition, two other factors prove to be useful to explain the change of RES-E policy objective in Belgium, but to a lesser extent: the past policy (H2) and the veto players (H4/2). On the one hand, the lack of an RES-E target in the past ("failure") explains why it became a political priority in 1999, but it was not the main factor that explained the change of policy (H2). On the other hand, the change of policy objective was driven by the Green Parties in government, but was supported by the other parties of the coalition according to the coalition agreement (no partisan veto players) and there were no significant institutional veto players opposing, so the absence of veto players facilitated the policy change (H4/2). Finally, we observe that one hypothesis is not relevant to explain the change of the RES-E policy objective in Belgium: the influence of the actors of the sector (H1/2). The actors of the sector were not involved in the change of policy objective in Belgium, unlike the change of policy instrument, so H1/2 is not relevant in this case.

Secondly, the change of policy instrument appears to be explained primarily by three factors: the participation of the green parties in government (H4/1), the failure of the past RES-E policy instrument (H2), and the misfit of the past instrument with the European policy (H3). In addition, the influence of the actors of the sector in the formulation of the TGC system also contributed to the change of instrument, but not to a significant extent (H1/2). Besides, the adoption of a political consensus between the parties of the governing coalitions over the RES-E policy change, which was reinforced by party discipline and the absence of significant institutional veto players, has facilitated the adoption of the policy change (H4/2). Finally, one hypothesis is not relevant to explain the change of policy instruments in Belgium: the dominant policy actors (H1/1). The policy actors who managed the change of the policy instrument were primarily the Ministers of Energy and their staff (political actors) and not the administration for energy (civil servants), which we would have expected.

In conclusion, our theoretical framework is largely validated in the Belgian case. Only the hypothesis about the dominant policy actors in the case of the change of policy instrument should be revised, but otherwise all hypotheses were fully confirmed. In addition, all hypotheses prove to be relevant to explaining the change of policy objective and the change of policy instrument in Belgium, with minor differences between both changes. Therefore, the Belgian case confirms the validity and the relevance of our theoretical framework.

CHAPTER 4

RES-E Policy in Denmark

I. Introduction

The beginning of the Danish electricity policy can be traced back to 1976, after the first oil crisis. Before, the Danish electricity sector had been self-organised, with very limited government control over its activities. The Association of Danish Electricity Utilities (DEF) acted as the sector's regulator and resisted most attempts at state intervention. However, at the time of the first oil crisis, more than 90% of the Danish primary energy consumption came from imported oil. The intensity of the oil shock on the Danish energy sector demonstrated the limits of the self-regulation model and the necessity of a long-term energy public policy based on the security of energy supplies. However, according to the Danish corporatist and consensual tradition, the electricity utilities remained closely associated with the future electricity policy, especially in the RES-E sector (central role of agreements with the utilities to implement the RES-E policy during the 1980s and 1990s).

In Denmark, the definition of the electricity policy is centralised at the state level under the supervision of the Energy Minister and the Danish Parliament (Folketing). However, the municipalities traditionally enjoy considerable autonomy and power in Denmark, especially in the energy sector, through the ownership of utilities and generation installations (especially in urban areas) and land-use planning regulation (in association with the counties). In addition, electricity consumers play a significant role in the Danish electricity sector due to the tradition of consumer co-operatives in the rural areas. The successive reforms of the electricity sector (concentration and then liberalisation) did not fundamentally alter those two models of utilities ownership (Olsen and Skytte, 2001). Indeed, a great deal of the concentration process took place in the form of co-operation among the municipal and consumer co-operative utilities rather than in the form of aggressive takeovers. Moreover, the reform of the electricity sector in 1999 enforced obligatory consumer ownership in the distribution network operators and the supply obligation companies. Therefore, the main stakeholders of the Danish electricity companies today remain the municipalities and the consumers, even if the modes of management have fundamentally changed.

Denmark is often perceived as a pioneer in environmental policy (Andersen, 1997) and this is especially the case for the RES-E policy, with outstanding outcomes in terms of RES-E installed capacity and electricity production between 1980 and 2004 (RES-E covered 0% of total electricity consumption in 1980 as against 24%[1] in 2004)[2]. The definition of RES-E in Denmark includes electricity produced by wind energy, biogas, biomass, solar energy, wave energy and hydropower plants (<10 MW)[3]. Electricity produced from waste is not considered to be RES-E and is therefore not eligible for the production subsidy, even its biodegradable fraction. Wind power represents the main RES-E sector in Denmark (wind turbines covered 0% of total electricity capacity in 1980 as against 23% in 2004), thanks to substantial financial support and a competitive domestic wind turbine industry since the 1980s. The Danish RES-E policy has also been specifically targeting biomass energy (especially straw, wood chips and biogas) during the last two decades, but with less good results. The other RES-E sectors (solar, small hydro, wave energy) have been eligible for some financial support during the last decade, but to a significantly lesser extent. In sum, the Danish RES-E policy has proved to be very effective in terms of installed capacity and electricity production during the last decades, but also very selective in terms of target groups (wind power and biomass).

II. Overview of the Electricity Sector

The Danish electricity supply industry has its origin in decentralised municipal or co-operative utilities in urban and rural areas. In the second half of the 20th century, the industry became much more concentrated, but the concentration process originated from below (Grohnheit and Olsen, 2001), namely by the utilities themselves, through the Association of Danish Electricity Utilities (DEF). Concentration occurred at the generation and distribution level, but especially among generators. At the end of the 1990s, there were about 103 distribution utilities in Denmark and eight generating companies (Grohnheit and Olsen, 2001). As a result of the local historical development of the electricity industry, the electricity generators are usually owned by the distribution companies operating in their areas. Because the Danish electricity industry is divided between the eastern and western part of Denmark (at either side

---

[1] 24% excluding waste, and 29% if one includes electricity produced from waste as RES-E.
[2] Danish Energy Authority, Energy in Denmark 2004, December 2005.
[3] Electricity Supply Act, No.375, 2 June 1999.

of the Great Belt), without any connection between them, the utilities traditionally co-operate in two regional associations: Elsam (west) and Elkraft (east). In addition, the two regional associations are grouped within the DEF, which handles the structural discussions within the industry and the negotiations with the Danish government.

The reform of the electricity sector in 1999 led to several changes in the organisation of the sector, except in the ownership tradition (municipal and consumer co-operatives), which persisted. After further mergers, only two large generators were left in 2001 (Elsam in western Denmark and Energi E2 in the east) and about 88 local distribution companies[4]. The transmission grids in both regions were transferred to independent companies according to the legal unbundling requirement: Eltra in the west and Elkraft System along with Elkraft Transmission in the east. In 2005, a single energy grid operator was established from the merger of the electricity transmission companies (Eltra, Elkraft Systems and Elkraft Transmission) and the gas transmission company (Gastra): Energienet.dk. However, the gas and electricity sectors remain managed by separate subsidiary companies: Eltrasmission.dk for the electricity transmission and Gastransmission.dk for the transport of gas.

The structure of the electricity generating capacity has evolved during the last three decades towards a steady increase of the CHP plants. By the end of the 1990s, almost 50% of Danish electricity production originated from CHP plants (Grohnheit and Olsen, 2001; Grohnheit, 2002). This results from an intentional strategy of the Danish government to make the energy sector more efficient. The increase of CHP in Denmark is also closely linked with the development of district heating systems in both urban and rural areas during the 1980s. For instance, from the early 1980s, every new electricity generation plant had to be provided with the ability to perform CHP and to supply heat to the district heating networks. In addition, the Heat Supply Act of 1979 was determining for the spread of CHP and district heating systems in Denmark. In 1999, almost 80% of the district heat was produced from CHP plants (either large scale or small scales ones)[5].

---

[4] IEA, Energy Policies of IEA Countries. Denmark 2002 Review, 2002.
[5] *Idem*.

Figure 1: Electricity generation by fuel, 1972-2002 (GWh)

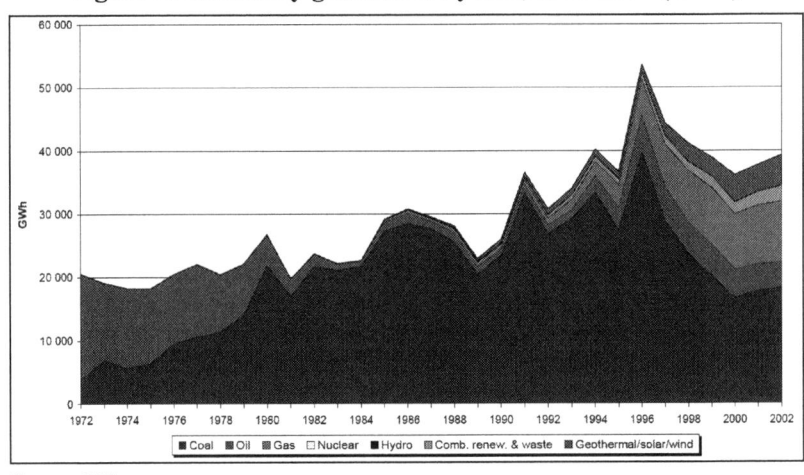

Source: IEA.

As can be seen from figure 1, the Danish power generation is highly variable due to the volatility of imports and exports with the Northern hydropower countries (Norway and Sweden) (see figure 2). In the Nordel market, Denmark acts as a swing producer thanks to its flexible fossil-fuel power stations that can compensate for the lack of hydropower on the market in dry seasons. For instance, imports of cheap hydropower from the Nordic country reached an unusual level in 1989, while 1996 set a record for Danish electricity exports due to the lack of hydro-production (see figure 2) (Asano, 2004). So, despite a relatively smooth electricity demand growth, Danish electricity generation is very volatile due to its interconnection with the Nordic hydropower market.

At the beginning of the 1970s, the Danish electricity sector was largely dependent upon imported oil (about 70% of the electricity generation in 1972). But from the first oil crisis, the Danish electricity sector switched back from oil to coal in a very short time (see figure 1). Indeed, by 1980, coal replaced oil as the main electricity fuel and in 1990, coal amounted to 90% of total electricity generation in Denmark. This extraordinary conversion from oil to coal was made possible by the multi-fuel facilities at most of the existing Danish power stations. However, two main energy issues result from Danish increasing dependence on coal. Firstly, all coal used in Danish electricity generation plants is imported, which only shifts the energy security issue from oil but does not solve it. Secondly, coal is a strong CO_2 emissions contributor, which turned out to be very problematic in the 1990s, with the increased political priority on climate change policy. The Danish government reduced the impact of those side-effects of coal by, on the one hand, promoting the use of CHP, both in centralised and decentralised installations in order to use energy more efficiently, and on the other hand, by diversifying the energy supplies and developing domestic energy sources (e.g. gas, oil, wind and biomass).

Figure 2: Danish electricity imports and exports, 1977-2001 (GWh)

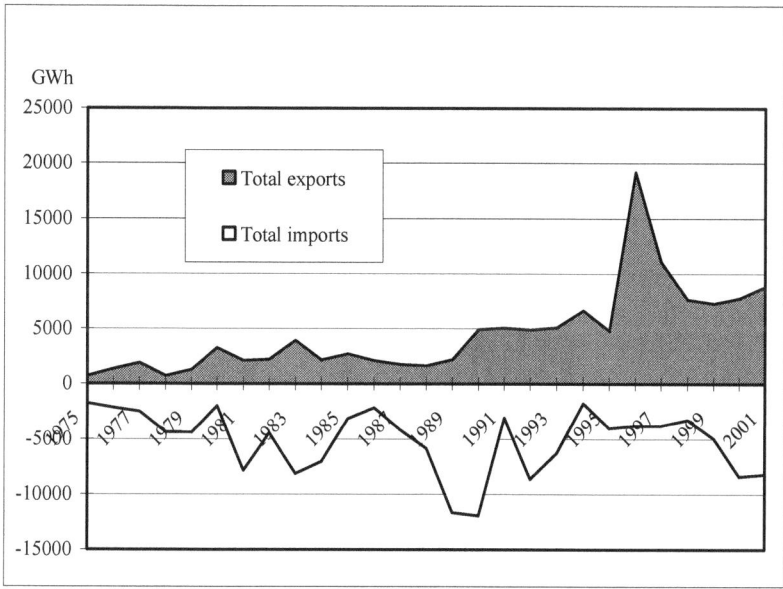

Source: Asano (2004), Association of Danish Energy Companies (2003) Elforsyningen Ti-års statistik.

During the same period, explorations in the North See for domestic gas and oil reserves were accelerated and new facilities for gas distribution were built. This led to a steady substitution from coal to gas during the 1990s. In 2000, the share of coal in the domestic electricity production was reduced to about 50%, while the share of gas increased to about 25% (see figure 1). During the same period, the use of oil in electricity production increased slightly, due to the exploitation of domestic reserves in the North Sea.

Despite political and economic interest in nuclear power during the 1970s and the first half of the 1980s, as stated in the Energy Reports of 1976[6] and 1981[7], the Danish nuclear program has never been implemented, mainly due to a strong opposition movement from Danish

---

[6] The Ministry of Trade, Danish Energy Policy, 1976.
[7] The Ministry of Energy, Energy Plan 81, 1981.

grass-roots associations (especially OVE and OOA). In 1985[8], the nuclear project was finally cancelled by the Danish parliament.

During the 1990s, additional energy sources increasingly contributed to the diversification of the Danish electricity market: wind power, biomass and waste (see figure 1). While RES-E had been insignificant in 1990, its share in the domestic electricity supply increased quickly in the following years to reach almost 29% in 2004 (see table 1).

The increase in RES-E during the last decade is to a large extent due to the extraordinary development of wind turbines (see figure 3). Wind power alone accounted for 18.5% of the total electricity supply in 2004 (see table 1). The first investments in wind power started with decentralised auto-producers building small-scale wind turbines in the rural areas. This phenomenon of wind turbine co-operatives owned by citizens is very specific to Denmark and explains why the NIMBY problems were less prominent in this country than elsewhere (e.g. the UK, Belgium, and the Netherlands). It is one of the success factors of the development of wind turbines in Denmark. Therefore, the Danish Wind Turbine Owner's Association was a powerful actor in the development of the Danish wind power policy from the early beginning. The investments in large-scale onshore and offshore wind turbines by the electricity utilities followed later, under pressure from the Danish government with whom the utilities made several agreements (1985-1998). The development of wind power in Denmark can be attributed to three main factors: the existence of a long-lasting competitive domestic wind turbine industry, significant financial support from the Danish government, and the commitment of the Danish citizens in favour of this technology.

---

[8] Parliamentary decision on public energy planning without nuclear power, 29 March 1985.

Figure 3: Wind power capacity and percentage share of electricity supply, 1980-2004 (MW and %)

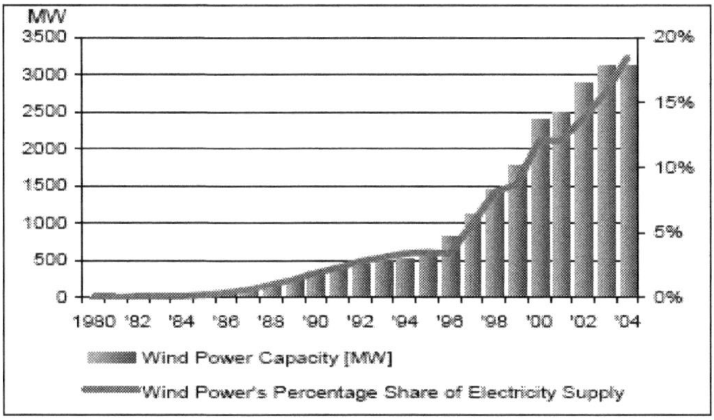

Source: Danish Energy Authority (2005), Energy Statistics 2004.

The next most important RES-E sector in Denmark is biomass, especially wood, straw and biogas. Even if their increase is less spectacular than wind, they still accounted for more than 5% of the electricity supplies in 2004 (see table 1), and further increase is to be expected in the future. The increasing use of biomass cannot be covered by the domestic reserves anymore, but originates in imports from wood pellets and wood chips. Finally, electricity produced from waste has also been increasing (4% of total electricity supply in 2004). Waste is primarily used for district heating, more than for electricity generation in Denmark, but, due to the increase in CHP installations, its impact on the electricity market has been growing in the last decade.

Table 1: Share of RES-E in domestic electricity supply, 1994-2004 (%)

	1994	1996	1998	2000	2001	2002	2003	2004
Solar Energy	0.0	0.0	0.0	0.0	0.0	0.0	0.0	0.0
Wind Power	3.4	3.4	8.1	12.1	12.1	13.8	15.8	18.5
Hydropower	0.1	0.1	0.1	0.1	0.1	0.1	0.1	0.1
Biomass	0.6	0.8	1.2	1.2	1.6	2.5	4.0	5.2
Biogas	0.3	0.3	0.5	0.6	0.6	0.7	0.8	0.7
Wastes	1.56	2.13	2.50	3.52	3.69	3.93	4.18	4.07
Total	**6.0**	**6.8**	**12.4**	**17.5**	**18.1**	**21.0**	**24.8**	**28.6**

Source: Danish Energy Authority (2005), Energy Statistics 2004.

III. Overview of the RES-E Policy History in Denmark (1976-1999)[9]

The main objectives of the Danish energy policy have been the same for the last three decades: security of energy supplies, economic efficiency and environmental protection. However, the relative priority of these three objectives has changed over time (Grohnheit, 2002). In the first two energy plans of 1976[10] and 1981[11], the priority was clearly on security of energy supplies, with emphasis on the substitutes for imported oil (coal, domestic oil and gas, wind power). Then, in the next two energy plans in 1990[12] and 1996[13], political priority shifted to environmental protection (especially CO_2 emissions reduction), with RES-E as one of the main solutions to promote the development of a sustainable energy sector in Denmark. The economic efficiency objective has been a constant priority of the energy policy since the oil crisis, but with a different focus over time. Economic efficiency was first enforced for old and new energy production installations (e.g. larger plants, fuel savings, district heating, combined heat and power production). Then the Danish government relied on the liberalisation of the sector and increasing competition to ensure reduced energy costs and prices.

A. 1976-1990

By 1973, when the first oil crisis occurred, more than 90% of energy supplies in Denmark were oil imports. This was the result of a lack of a provisional energy policy from the Danish government and cumulative decisions of the Danish electricity utilities after the Second World War (Lucas, 1985). Therefore, the oil crisis was a great shock in Denmark, since the entire energy system was threatened, but also because the Danish government realised that it was time to exert more control on the energy sector and develop a real short- and long-term energy policy.

In 1976, the first energy plan was presented by the Ministry of Trade (who was in charge of energy at that time) and the first energy supply act[14] was adopted by the Danish parliament. On the one hand, the energy plan submits the government strategies to increase the security of Danish energy supplies in the future and to increase the overall economic

---

[9] This section draws partially from Asano (2004).
[10] The Ministry of Trade, Danish Energy Policy, 1976.
[11] The Ministry of Energy, Energy Plan 81, 1981.
[12] The Ministry of Energy, Energy 2000, 1990.
[13] The Ministry of Energy and Environment, Energy 21, 1996.
[14] Act on electricity supply, 25 February 1976, No.54.

efficiency of the energy industry. On the other hand, the energy supply act reinforces the control of the state on the electricity utilities (authorisation for new generation capacity, price control) and re-organises the sector (non-profit principle).

The 1976 Energy Plan presented the three goals of the Danish energy policy: security of energy supplies, economic efficiency and environmental protection. Security of energy supplies was the main goal at that time and the main actions to be undertaken by the government and the industry were: the exploration of domestic oil and gas resources in the North Sea, the introduction of nuclear power, the conversion of the electricity power plants from oil to coal and the reform of the heating system (district heating and CHP). The development of RES-E (especially wind turbines) was also considered, but it was not a priority in this plan. In a short time, this plan proved to be a success in terms of the reduction of oil consumption and the increased security of energy supplies. This was the result of the increasing exploitation of domestic gas resources (combined with the construction of an extensive gas network in Denmark), the rapid and massive shift from oil to coal in the power generators (thank to flexible installations) and the development of district heating in the cities, combined with the increase of CHP plants (replacing traditional heat plants). The only failure of the plan relies in the non-implementation of the nuclear programme. In fact, a strong opposition against nuclear power arose from the very beginning, led by environmental and anti-nuclear associations (OOA and OVE) and a group of energy experts from Danish universities. They published two alternative energy plans excluding nuclear power in 1976[15] and 1983[16], as a response to the government plans of 1976 and 1981, and managed to attract the attention of a growing number of policy experts and politicians. In addition, the Three Mile Island nuclear accident of 1979 and the nuclear debate in Sweden reinforced the position of the Danish anti-nuclear movement. Finally, in 1985 (before the Chernobyl accident), the Danish parliament cancelled the nuclear program and decided that nuclear power should not be considered an option for the future Danish energy supply anymore[17].

The electricity supply act of 1976 filled a gap in government control over the electricity sector, but the electricity utilities succeeded in receiving generous commercial concessions in return (electricity prices negotiated and generous terms for investment depreciation, capital for

---

[15] Blegaa et al., Outline of Alternative Danish Energy Plan, Copenhagen: OVE, 1976.

[16] Hvelplund et al., Energy for the Future, Copenhagen: Borgen, 1983.

[17] Parliamentary decision on public energy planning without nuclear power, 29 March 1985.

new investments to be levied on consumer prices). The main contributions concerned the establishment of a Price Committee in charge of approving the electricity prices, the regulation of the power plants (>25MW) for which a government authorisation was henceforth required, and the pursuit of the consumer ownership tradition through the non-profit principle for the electricity utilities (redistribution of profits to the consumers) (Asano, 2004).

Due to the increased political priority on energy issues after the first oil crisis, the Danish Energy Agency was established in 1976 to control the energy sector and implement energy policies. Soon afterwards, a new Ministry of Energy was created in 1979 as an offshoot of the Ministry of Trade. The supervision of the Danish Energy Agency was transferred from the Ministry of Trade to the Ministry of Energy. So, institutional changes and policy changes converged at the end of the 1970s to create the conditions for an improved future energy policy.

In 1981, a second energy plan was submitted by the Ministry of Energy. It built further on the priorities of the previous plan (security of energy supplies), but increased the focus on economic efficiency and socio-economic issues (e.g. energy price reduction), given the drastic energy price hikes in 1979 and 1980 (Ronne, 2001). The environmental considerations also gained importance in this plan, but still marginally compared to the other two priorities (target of 6% RES in primary energy consumption by 2000 and 1.3 TWh wind power production in 1995) (Lucas, 1985; Meyer, 2004).

Denmark has developed a tradition of implementing the energy policies with broad political support (large political consensus) and involving a broad range of actors (e.g. electricity utilities, local associations, the energy industry, municipalities, research institutions, consumers) (Ronne, 2001). For instance, the electricity utilities remained central actors in the RES-E policies, especially until the end of the 1990s, and several RES-E support schemes were actually implemented by agreements with the utilities rather than by laws. Therefore, two categories of support schemes in favour of RES-E were implemented between 1976 and 1990: government programs (RD&D programs and investment grants) and utilities programs through agreements with the government (voluntary agreement for RES-E investments), the Danish Wind Industry Association and the Danish Wind Turbine Owners' Association (wind power purchase agreement).

In 1976, the Danish government launched a new energy RD&D program in accordance with the 1976 Energy Plan: the Energy Research Program (ERP). This programme aimed at stimulating the development of new energy technologies, especially in the field of oil, natural gas, coal, wind turbines, etc. During the 1980s, wind turbines and the other

RES-E technologies were not the first priority of the Danish energy RD&D program, but they still received about 25% of the total budget (about 5 M€/year on average)[18]. At that time, almost 80% of the RES-E RD&D budget was invested in the wind sector to improve the quality and efficiency of large and small wind turbines.

In addition, in 1979[19], the Danish government started to provide investment subsidies for RES-E installations. At the beginning of the program, the subsidies were especially granted to wind turbines. The grants amounted 30% of the investment costs, but due to the rapid improvement of the economics of wind turbines, they were gradually reduced to 10% by the end of the wind program in 1989. During the 1980s, the program started to support biomass and biogas with grants of 20-40%. Then, it became more diversified during the 1990s (Asano, 2004).

In addition to those programs, the electricity utilities negotiated power purchase agreements for wind power with the Danish Wind Industry Association and the Danish Wind Turbine Owners' Association, under the supervision of the Danish government. The first purchase agreement for wind power was adopted in 1979 and then revised in 1984. It provides that the utilities must purchase the wind power generated and fed into the grid in their distribution areas at a rate of 85% of the consumer electricity price in the same area. This corresponds to a feed-in tariff system organised at the local level providing different rates according the areas. The average rate during the 1980s amounted to 0.15 to 0.23 DKK/kWh (Asano, 2004).

Finally, in 1985, the Danish government and the electricity utilities adopted a "voluntary" agreement providing that 100 MW of wind turbines were to be erected by the utilities by 1990. It was then revised in 1990, 1996 and 1998, with additional targets. These agreements were relatively compliant for the utilities and less voluntary than they appeared.

In conclusion, the first developments of the RES-E policies in Denmark relied to a large extent on two types of support schemes: government RD&D and investment programs, combined with utilities agreements to invest in wind turbines and to offer FIT to wind power fed into their grid. Most of the programs especially targeted wind power and the other RES-E were only marginally considered. The electricity utilities remained powerful but they had to comply with increasing RES-E requirements from the Danish government. During that period, the

---

[18] IEA, Danish RD&D bugdet, 2005.
[19] Act on the utilisation of renewable energy, No.2, 1979. Several times amended until 1999.

environmental and anti-nuclear grass-roots associations as well as the wind turbine industry succeeded in influencing the Danish policy in favour of RES-E, especially wind power.

B. 1990-1999

During the 1990s, the priority of the Danish energy policy shifted to the protection of the environment. The main issue of the energy sector became the reduction of CO_2 emissions, and the development of environmentally-friendly energy technologies, especially RES-E, was seen as one of the main solutions. The objective of the Danish government, besides security of energy supplies, was to support the implementation of a sustainable energy industry in Denmark by the end of the century.

In 1990, a third Energy Plan[20] was submitted by the Ministry of Energy. It introduced the goal of sustainable development of the energy sector, which involved the reduction of GHG emissions by explicit targets: 20% by 2005 compared to the level of 1988. The main actions to be taken to reach its goal were, on the one hand, the reduction of energy consumption by increased energy savings and energy efficiency and, on the other hand, the conversion to cleaner sources of energy, especially RES-E. The share of RES in the total energy consumption was expected to double by 2005 (about 12-14% of total primary energy supplies), and this involved a wind turbine target of 1,500 MW installed capacity by 2005 (corresponding to wind power covering about 10% of electricity supplies by 2005). Use of biomass was expected to increase to 85 PJ by 2005 (especially from straw, wood chips and waste), and a solar PV installed capacity of 300 GW was targeted by 2005.

At the institutional level the increased priority on environmental protection in the energy sector also induced changes. In 1994, the Ministry of the Environment and the Ministry of Energy merged to establish a new Ministry of the Environment and Energy. It became a very powerful Ministry in the Danish government due to the extent and importance of its competences. The increased interaction and interdependence between the environmental and energy policies in the Danish strategy towards CO_2 emissions reduction explains to a large extent this institutional change. However, we will see that in 2001, the new government returned to the pre-1979 situation, since it took the energy competence away from the Ministry of the Environment and gave it to the Ministry of the Economy, which means that no independent Ministry of Energy

---

[20] Ministry of Environment, Energy 2000. A Plan of Action for Sustainable Development, April 1990.

remained. Institutional changes often demonstrate political changes; this is a typical example.

Due to the development of new market conditions at the European level, with the liberalisation of the electricity sector as well as increased political priority to RES-E, a fourth Energy Plan was prepared in 1996 by the new Ministry of Environment and Energy[21], laying down the strategy of the Danish Government for the coming years. It contains stronger policy measures to reach the goals of the 1990 Energy Plan, and also outlines longer-term goals for 2010 and 2030. As far as CO_2 emissions are concerned, the Danish government declared to be willing to accept a reduction target of 50% by 2030 compared to the 1990 level, but only if the international and technological conditions support its efforts. In the shorter term, the national 20% reduction target by 2005 was reaffirmed. This is a more ambitious target than the European Burden Sharing Agreement (reduction of GHG by 21% in relation to 1990 during the period 2008-2012) according to the Kyoto Protocol, which demonstrates the commitment of the Danish government to the climate change issue. The plan establishes a long-term target for the development of the RES by 2030 (35% of RES in total primary energy supplies by 2030), which means an increase of about 1% per year. In addition, the contribution of the electricity sector to this target is expected to be substantial. The plan states that RES-E (excluding municipal solid waste) is expected to cover 20% of total electricity consumption by 2005, 29% by 2010 and about 68% by 2030. So, the RES-E policy objectives of the Danish government were very ambitious in 1996 and they have remained the reference under the subsequent governments so far.

Many different actions were taken by the electricity utilities, the Danish government and the municipalities to implement those Energy Plans in the field of RES-E. The tradition of implementing electricity policies through agreements with the electricity utilities persisted until the liberalisation of the sector in 1999, but it grew weaker during the 1990s, especially when the new Ministry of the Environment and Energy was created, because of the growing government influence on the sector compared to the utilities. In addition, while wind power had for a long time been the main target of the RES-E policy, it changed during the 1990s, with an increased priority on biomass (biogas, wood, straw, waste) and the other RES-Es (e.g. solar PV).

Important RES-E agreements were adopted between the government and the utilities during the 1990s: the renewal of the wind power agree-

---

[21] Ministry of Environment and Energy, Energy 21, April 1996.

ment in 1990 and 1996 and its extension to offshore plants in 1998, and the new biomass agreement of 1993.

The wind power agreements of 1990[22] and 1996[23] increased the target of installed wind power capacity to an additional 100MW by 1996 and 200MW by 2000 respectively. Those targets were easily reached and even exceeded as the wind power market exploded beyond expectations during the 1990s (see figure 3). Wind offshore demonstration programs started at the end of the 1980s and proved to be very successful (Meyer et al., 2003). Therefore, an agreement was signed between the Danish government and the utilities in 1998[24] to establish a 750MW capacity offshore wind power (including 5 farms of 150MW capacity each) by the end of 2008.

In 1993[25], the biomass agreement was signed between the Danish government and the electricity utilities in order to increase the use of biomass in power production. According to this agreement, the electricity generators agreed to use 1.4 million tonnes of biomass for power production by 2000. The target was divided between straw (1.2 million tons) and woodchips (0.2 million tons). Biomass was intended to replace coal in the large power and CHP installations. However, limited market competition on the biomass market resulted in price increases for straw and wood, which prevented the utilities from switching to biomass (Odgaard, 2000). Therefore, the agreement was revised in 1997[26] in order to introduce more flexibility in the straw/wood chips ratio as adopted in 1993. In 2000, the target fell short of about half of the expected biomass use, primarily due to high biomass prices compared to the other fuels (Meyer et al., 2003).

In 1991[27], the Danish parliament adopted a new act on the subsidies for electricity production. It guarantees a production subsidy to all RES-E generators in the form of two grants: 0.17 DKK/kWh as a compensation for the electricity tax and 0.10 DKK/kWh as compensation for the CO_2 tax. Only the private and decentralised electricity generators are eligible for the 0.17 DKK/kWh grant. This means that the RES-E plants owned by the electricity utilities received only 0.10 DKK/kWh. For the first time, all RES-E was able to benefit from this support, but most of the budget was still allocated to wind or biomass power plants due to their supremacy in Denmark. No long-term guarantee was linked with

---

[22] Wind Power Agreement of 20 March 1990.
[23] Wind Power Agreement of 14 February 1996.
[24] Wind Power Agreement of 27 September 1997.
[25] Biomass Agreement of 14 June 1993.
[26] Adjustment of Biomass Agreement of 1 July 1997.
[27] Act on subsidies for electricity production, No.944, of 27 December 1991.

these grants, but the legal basis of the system makes it rather stable, unlike the utilities agreement schemes. This production subsidy scheme to a large extent contributed to the success of RES-E in Denmark during the 1990s (see table 1). However, due to rapid technological developments in wind turbines, the economics of wind power improved and the production subsidy proved to be too high by 1999 (Orgaard, 2000).

In addition to the production subsidies that we described above, wind power and electricity generated from decentralised CHP biomass plants benefited from feed-in tariffs from the electricity utilities. As far as wind power is concerned, the same rate as that provided by the purchase agreement of 1984 prevailed, namely 85% of the consumer price (correspond to 0.33 DKK/kWh in average). The rate of the feed-in tariff for electricity produced from decentralised biomass CHP plants varied depending on the load period[28], but it was 0.33 DKK/kWh on average. In conclusion, from 1992, private decentralised wind power generators and decentralised biomass CHP plants received a total subsidy of 0.6 DKK/kWh: 0.27 DKK/kWh production subsidy + 0.33 DKK/kWh feed-in tariff.

During the 1990s, the budget of the RD&D programs of the Danish government in favour of RES-E increased, according to the new objectives of the "Energy 2000" and "Energy 21" plans[29]. Overall, the RES share in the energy RD&D budget increased from about 25% during the 1980s to an average of 40% during the 1990s[30]. The main energy RD&D programme, the "Energy Research Program", put more priority on and earmarked a higher budget for new RES-E technologies projects (especially wind turbines) from 1991. In 1986, 56% of the budget was given to energy savings and renewable energy. In 1998, it increased to 85% of the budget[31]. In addition, a new programme for RES started in 1992[32], the "Development and demonstration program for renewable energy". It aims primarily at financing investments in commercially available technologies (see below), but a percentage of its budget (about 15%) is also dedicated to demonstration projects of non-commercial technologies (20-40% investment grant)[33]. In 1997, an additional program was

---

[28] See Asano, 2004, p. 203 for a detailed account of the load periods and the related rates.

[29] Danish Energy Authority, Follow-up on Energy 21 Status of Energy Planning, June 1999.

[30] IEA, Danish RD&D bugdet, 2005.

[31] Danish Energy Authority, Follow-up on Energy 21 Status of Energy Planning. Appendices, June 1999.

[32] Consolidated Act on the Utilisation of Renewable Energies, No.837, 7 October 1992.

[33] Danish Energy Authority, Follow-up on Energy 21 Status of Energy Planning, June 1999.

established in the field of RES-E, the so-called "Fund for the Development of New Renewable Energy Technologies". It aims at contributing to the development and testing of new RES technologies and supplementing the ongoing efforts to reach the long-term targets of the "Energy 21" plan (Odgaard, 2000). It was administered in co-operation with the "Development and demonstration program for renewable energy".

Several new investment programs for RES-E were adopted during the 1990s. The main program, the "Development and demonstration program for renewable energy" of 1992[34] seeks to promote the commercialisation of close-to-market RES technologies with a grant that covers 15-30% of the investment costs (e.g. solar heating, heat pumps, biofuel systems, biomass boilers). It also supports less commercially available technologies with grants of 20-40% of investment costs, but to a smaller extent (e.g. solar PV, household wind turbines). In addition, a specific program for decentralised CHP biomass started in 1992[35], the so-called CHP fund, to support the conversion from biomass-based district heating to CHP (Odgaard, 2000). It provides investment grants for 10-25% of the investment costs. This program was first supposed to last until 1997, but was then extended to 2002. In addition, a new RES subprogram was launched in 1997, the "Renewable Energy Island Program" (Odgaard, 2000). The objective of the program was to demonstrate the possibilities of a small community to base its entire energy supply on RES. The island of Samsoe was selected for this program. Finally, a specific demonstration program for solar PV was launched in 1998, "Solar 300". It financed the installation of 300 roof-top PVs of 20-25 m^2 each on private households by 2000. In the 1990s, the priority of the investment programs of the Danish government moved from wind power to other less competitive, but still close-to-market, technologies.

In order to encourage the traditional model of consumer ownership in the electricity sector, fiscal incentives for wind turbine owners were adopted in the 1990s. It provides that any individuals who participate in wind turbine co-operatives benefit from specific income tax breaks (Asano, 2004).

Finally, the proliferation of onshore wind turbines raised land-planning issues in Denmark in the early 1990s. Therefore, the Danish government established the Wind Turbine Sitting Committee in 1991 to take care of the wind turbines planning issues and to look into how to co-ordinate the various interests involved in land-planning at the local level. When it tried to locate the sitting possibilities for the future wind power developments according to the "Energy 2000" and "Energy 21"

---

[34] Consolidated Act on the Utilisation of Renewable Energies, No.837, 7 October 1992.
[35] Act on Subsidies for Decentralised CHP from Biomass, No.3, 3 January 1992.

plans, the Committee discovered that most regional plans included specifications for wind turbine sittings, but that almost all municipal plans did not (Nielsen, 2002). Consequently, in January 1994[36], the government sent a circular on wind power planning to all municipalities, requesting that they draw up municipal plans proposals with possible turbine sittings in their areas, looking both at new wind turbines sittings and the replacement of older ones (Nielsen, 2002). The purpose of the circular was to make wind turbine planning a permanent part of regional and municipal planning. The high acceptance of wind power in the population and the land-use planning co-ordination between the state, the counties and municipalities, were key factors to the success of wind power in Denmark during the 1990s.

IV. RES-E Policy Change in Denmark (1999-2004)[37]

The first steps towards the liberalisation of the Danish electricity market date back to 1996[38]. With this act, Denmark already fulfilled some of the obligations of the EU directive 96/92/EC. The main provisions of the acts concern: the partial opening of the electricity market (consumption >100 GWh open from 1 January 1998), a better regulation of public service obligations, the prioritisation of CHP and RES-E electricity on the grid, and the obligation to all electricity consumers to purchase electricity from prioritised plants. Therefore, the prerequisites for the future liberalisation of the market and RES-E policy were already laid down, but the main provisions of the electricity reform were still missing.

On 3 March 1999 a large majority of the Danish parliament[39] entered into an agreement on comprehensive reform of the electricity sector, and part of this agreement concerns RES-E[40]. In case of important policy reforms in Denmark, a large consensus is necessary between the minority government and the other parties in parliament. This also concerned the agreement of 1999 on the reform of the electricity sector. This agreement was the result of several months of political debate. It outlines the framework of the new legislation concerning the future organisation of the electricity sector, the opening of the electricity market, and

---

[36] Government Circular No.21 on planning for wind turbines, January 1994.
[37] This section draws partially from Asano (2004).
[38] Act on amendment of the electricity supply act and the heat supply act, No.486, 12 June 1996.
[39] The Danish Government (namely the Social Democrats and the Social Liberal Party), the Liberal Party, the Conservative Party, the Socialist People's Party and the Christian People's Party.
[40] Political Agreement of 3 March 1999 about the electricity reform.

how to protect the environment (and develop RES-E) in the liberalised electricity market. The reform is needed to unite the fulfillment of the long-term environmental objectives (Energy 21 plan) with effective consumer protection and the introduction of increased competition in the electricity sector (Directive 96/92/EC). In addition, the reform is expected to contribute to improving the government finances by approximately DKK 2 billion annually, when fully implemented[41].

As far as RES-E is concerned, the agreement provided that the on-going production subsidies for RES-E shall be discontinued and replaced by a more competitive mechanism: a tradable green certificate system. In addition, the RES-E policies should not be financed by public budgets anymore, but by the electricity consumers. Given that a growing share of the electricity consumption is expected to be covered by RES-E in the future, the political parties agreed that it is important that the development of RES-E take place as cost-effectively as possible (hence the tradable green certificate system) and with as little public spending as possible (hence the funding by the electricity consumers). The political agreement provided that 20% of the total electricity consumption should originate from RES-E by 2003 (excluding municipal solid waste), namely at the end of the agreement period. This objective accelerated the timing of the 1996 energy plan, which aimed at 20% by 2005 only. In reality, in 1999, the percentage of RES-E in the total electricity supplies had been increasing so quickly since 1996 (see table 1), that it seemed realistic to be more ambitious. However, no longer-term targets were adopted in 1999, as the targets of the 1996 energy plan for 2010 (29%) and 2030 (68%) remained the reference. So, the political agreement of 1999 did not modify the Danish RES-E policy objectives, but proposed a new policy instrument to attain them.

The political agreement was converted into legislation in May 1999 by the adoption of five new acts by the Danish Parliament: the electricity supply act[42]; the act on CO_2 quotas for production of electricity[43]; the act on the amendment of the act on subsidies for electricity production[44]; the act on the amendment of the act on the utilisation of renewable energy sources[45]; and the act on the amendment of the act on heating supply[46]. The new legislation sought to regulate the organisation of the

---

[41] Notes on the Electricity Supply Act, No.375, 2 June 1999.
[42] Electricity Supply Act, No.375, 2 June 1999.
[43] Act on CO_2 quotas for production of electricity, No.376, 2 June 1999.
[44] Act on the amendment of the act on subsidies for electricity production, No.377, 2 June 1999.
[45] Act on the amendment of the act on the utilisation of renewable energy sources, No.378, 2 June 1999.
[46] Act on the amendment of the act on heating supply, No.379, 2 June 1999.

liberalised electricity market in Denmark in accordance with the 1996 EU directive on the liberalisation of the European electricity market (e.g. unbundling, regulated third party access to electricity networks, independent energy regulators). It ensured continued consumer protection and consumer influence in the sector, but also imposed new obligations on the consumers (quota of RES-E). In addition, specific provisions were adopted to make sure that the priorities of the plan "Energy 21" in terms of CO_2 emissions reduction and RES-E development could be fulfilled, despite the increased competition in the electricity sector (RES-E market, ceiling for CO_2 emissions).

The main elements of the new framework for the liberalised electricity market includes: the legal unbundling (separate companies) between the public service activities (network management) and the commercial activities (electricity supply, electricity generation) of the electricity utilities; the regulated third party access to the distribution and transmission networks; the establishment of an independent sector regulator (the Danish Energy Regulatory Authority DERA), the progressive opening of the electricity market to competition (see table 2); and finally the pursuit of the consumer ownership tradition at the distribution level (distribution grid operators).

Table 2: Agenda of the opening of the Danish electricity market[47]

Eligible consumers	Opening date
>= 100 GWh/year	1 January 1998[48]
>= 10 GWh/year	1 April 2000
>= 1 GWh/year	1 January 2001
All	1 January 2003

As far as RES-E is concerned, the 1999 reform reaffirms the objectives of the 1996 "Energy 21" plan and adopts a short-term target of 20% of the total electricity consumption to be covered by RES-E by 2003. However, the amendment to the act of subsidies for electricity production provides that the existing production subsidy scheme for RES-E is abolished and the cost of the future support to RES-E transferred to the electricity consumers through increases of the consumer bill[49]. So, the RES-E policy objectives remain the same, but the policy instrument adopted to reach them is modified. The new policy instrument designed to replace the previous production subsidies (the so-

---

[47] Electricity Supply Act, No.375, 2 June 1999.
[48] Act on the amendment of the electricity supply act and the heat supply act, No.486, 12 June 1996.
[49] Act on the amendment of the act on subsidies for electricity production, No.377, 2 June 1999.

called RE-market) is based on a market mechanism and financed through the consumer bill. Accordingly, the electricity consumers are obliged to purchase a certain percentage of their electricity from RES-E on the RES-E market (first provisional quota of 20% by 2003)[50]. RES-E generators receive green certificates according to the amount of RES-E they produce (1 green certificate per 1 kWh) and the certificates are used as proof of the quota by the electricity consumers. If they do not comply with their quota, the electricity consumers must pay a penalty of 0.27 DKK/kWh of RES-E missing[51]. In this system, the RES-E generators compete with each other on the RES-E market, which is expected to stimulate increased cost-effectiveness in the RES-E sector. As stated in the political agreement: "It is thus essential that the future electricity market will be in a position to utilise more competition-based mechanisms that can ensure the cost-effective development of renewable energy production"[52]. In addition, the extra cost linked to the mandatory purchase of RES-E is financed by the consumer bill and no longer by the public budget. Given the expected increase in cost-effectiveness induced by the RES-E market, the surcharge for the Danish consumers is supposed to be moderate.

During the transition period towards the RES-E market (2000-2002), fixed settlement prices were laid down to secure a similar income as under the previous system to the already established plants. In addition, until the RES-E market was fully operational, new plants were ensured fixed minimum prices for the electricity they supplied to the grid during a fixed period of time (see table 3)[53]. The RES-E market was expected to be fully operational by January 2003, so the transition scheme was only valid in the period up to 2003. The transitional scheme appears rather complicated since it must comply with both the existing and future RES-E investors' requests for price security and the political parties' objective to rely more and more on the market to settle the more cost-effective price.

During the transition period (2000-2002), the RES-E existing plants were ensured to receive the same prices as before the reform, namely the equivalent of the 85% rule (0.33 DKK/kWh) plus the compensations for the electricity and CO_2 taxes (0.27 DKK/kWh). The installations built during the transition period were granted a guaranteed settlement price of 0.33 DKK/kWh (for wind power) or 0.50 DKK/kWh (other RES-E) plus a 0.10 DKK/kWh price premium that corresponds to the minimum

---

[50] Electricity Supply Act, No.375, 2 June 1999.
[51] *Idem.*
[52] Political Agreement of 3 March 1999 about the electricity reform.
[53] Electricity Supply Act, No.375, 2 June 1999.

price for the green certificates (see table 3). In addition, a replacement scheme for old small wind turbines started in 1999 (Meyer *et al.*, 2003). It provides that turbines of <=100 kW capacity that are replaced by new turbines with three times the initial capacity are guaranteed a payment of 0.60 DKK/kWh for the first 12,000 full load hours of production (corresponding to about 5 years) (see table 3). From 2000, new wind turbines received a smaller settlement price than the other RES-E installations (except under the replacement scheme) because they are more competitive, and thus need less financial support to enter the market.

The implementation of the RES-E market was first postponed in December 1999 due to the delay of the European Commission's approval of the system (the final approval was only received in October 2000), the anticipation of the 2001 directive, the appearance of implementation issues to frame the new RES-E market[54], and the rise of disagreements about the system among the experts in charge of drafting it (Nielsen and Backer, 2003).

A new political agreement was adopted in March 2000 as a follow-up of the electricity reform[55]. It brings several amendments to the reform and some of them concern the RES-E policy. Firstly, it takes note of the postponement of the RES-E market, but reaffirms that is should be able to be fully operational by 2003. Secondly, it specifies the rules of the transitional scheme for the existing wind turbines and calls for a replacement scheme for the old wind turbines. Thirdly, the agreement presents a new biomass plan that pursues the previous agreements of 1993 and 1997, which did not reach its target (namely 1.4 million tons of biomass used for electricity production by 2000). The new agreement extends the time limit to 2005, but it also aimed to establish 2-3 new big biomass plants by the end of 2003. Moreover, in order to cut costs as much a possible, when securing construction of subsequent plants, the agreement provides that calls for tenders could be issued.

In autumn 2001, hearings were held in parliament about the last report of the Danish Energy Authority on the green certificate system. Strong opposition against and criticism of the system arose from different stakeholders of the RES-E sector (e.g. the Danish Wind Industry Association, the Association of Danish Wind Turbine Owners, academic experts). Two external events contribute to explain the sudden rise opposition to the system: the adoption of the RES-E European directive in September 2001 and the decision of the European Court of Justice in the so-called PreussenElktra case in March 2001 (Nielsen and Backer,

---

[54] Danish Energy Authority, Report on the Danish green certificate market, December 1999.
[55] Political Agreement of 22 March 2000 on the Reform follow-up.

2003). During the preparation of the directive on RES-E, the Danish government pressed for a harmonised system at the European level based on an RES-E market (similar to its own new RES-E scheme). However, the member states where a feed-in tariff system has been implemented so far (especially Germany and Spain) were very reluctant to give up this system and very skeptical about the proposed RES-E market model. In the end, the conflict between the two models prevented the European Commission from taking a decision about the harmonisation of the RES-E instruments and the directive ended up postponing the decision about the harmonisation for another four years. Given that the Danish agreement about the green certificate system presupposed that it would be coupled with a European RES-E market, the decision of the Commission to postpone such a market questioned the Danish system itself and resulted in increased claims not to implement it. In addition, the decision of the European Court of Justice in March 2001 to legitimise the German feed-in tariff system against the claims that it would contravene the European state aid rules, reinforced the arguments in favour of a return to a feed-in tariff system in Denmark. Finally, in October 2001, the Danish government decided to put the green certificate system on hold for two years (Nielsen and Backer, 2003). Of course, the European events alone do not explain the decision of the Danish government. The Danish Energy Authority faced serious issues in the implementation of the system and the growing disagreement among the experts in charge of the implementation did not facilitate its work (Nielsen and Backer, 2003).

The decision to freeze the implementation of the Danish green certificate system was then endorsed by the parliamentary coalition of the Electricity Reform Agreement in June 2002. It was agreed that the implementation of the Danish system would be postponed "until it becomes possible to establish a common market with a number of EU countries"[56]. As a consequence, a new extended transitional scheme was adopted for the old and new RES-E, given that the previous transition scheme was to expire at the end of 2002. The succession of transitional schemes and the insecurity linked with their short-term periods was a very important factor against the development of new RES-E capacities in the period 2000-2003.

Due to the freezing of the green certificate system for an unlimited period of time, the scheme presented in the June 2002 agreement actually constitutes the new Danish support scheme for RES-E rather than a transitional scheme (see table 3). After the transition period, as from

---

[56] Political agreement between the Government (the Liberal Party and the Conservative People's Party), the Social Democratic Party, the Socialist People's Party, the Social Liberal Party and the Christian People's Party of 19 June 2002.

January 2003, the new RES-E plants receive the electricity market price (no more settlement price) plus a premium price of 0.10 DKK/kWh as an environmental bonus. In addition, a ceiling price of 0.36 DKK/kWh cannot be exceeded[57]. In the end, the financial conditions for the new RES-E plants are less favourable than they were before the electricity reform, especially for wind power. It is therefore not surprising to observe that the wind turbine capacity increased at a reduced rate during the first decade of 2000, compared to what it used to be during the 1990s. Moreover, the new wind power capacity corresponds to a large extent to the construction of new offshore wind farms or the replacement of older turbines, and not to new onshore capacities as in the 1990s.

Table 3: Transitional schemes for RES-E 2000-2004 and new scheme from 2005

RES-E			Settlement price	Production subsidy	Price premium	Comments
Wind power	Existing plants (before 01/01/2000)	<=200 kW	0.33 DKK/kWh	0.17 DKK/kWh	0.10 DKK/kWh	First 25,000 full-load hours
		201-599 kW	0.33 DKK/kWh	0.17 DKK/kWh	0.10 DKK/kWh	First 15,000 full-load hours
		>= 600 kW and 12,000 full load hours	0.33 DKK/kWh	-	0.10 DKK/kWh	-
	New plants (01/01/2000 – 31/12/2002)		0.33 DKK/kWh	-	0.10 DKK/kWh	For 10 years
	Replacement scheme (1999-2003) (<=100 kW)		0.33 DKK/kWh	0.17 DKK/kWh	0.10 DKK/kWh	First 12,000 full-load hours for prod. subsidy
	Offshore (2002-2003)**		0.46 DKK/kWh	-	-	First 42,000 full-load hours
	New plants from 01/01/2003 or Existing plants after at the end of transition period		Market price	-	0.10 DKK/kWh	Max 0.36 DKK/kWh
	Replacement scheme for old wind turbines (2005-2009)		Market price		0.12 DKK/kWh	First 12,000 Full-load hours only; max 0.48 DKK/kWh
Biogas	Existing plants (before 01/01/2000)		0.33 DKK/kWh	0.17 DKK/kWh	0.10 DKK/kWh	Same as before
	New plants 01/01/2000 – 31/12/2002		0.50 DKK/kWh	-	0.10 DKK/kWh	For 10 years
	New plants from 01/01/2003		Market price	-	0.10 DKK/kWh	Max 0.36 DKK/kWh

---

[57] Political Agreement of 19 June 2002 and Executive Order No.151 of 10 March 2003.

	New plants from 01/01/2005	0.60 DKK/kWh (10 years) then 0.40 DKK/kWh (10 years)	-	-	Upper limit of 8 PJ by 2008
Biomass	Existing plants (before 01/01/2000)	0.33 DKK/kWh	0.17 DKK/kWh	0.10 DKK/kWh	Same as before
	New plants 01/01/2000 – 31/12/2002	0.50 DKK/kWh Or 0.30 DKK/kWh*	-	0.10 DKK/kWh	For 10 years
	New plants from 01/01/2003 or Existing plants after at the end of transition period	Market price	-	0.10 DKK/kWh	Max 0.36 DKK/kWh
	From 01/01/05 for all plants	Market price	-	Premium determined on individual basis	For 20 years; max 0.34 DKK/kWh
Other RES-E	Existing plants (before 01/01/2000)		0.17 DKK/kWh	0.10 DKK/kWh	-
	New plants 01/01/2000 – 31/12/2002	0.50 DKK/kWh	-	0.10 DKK/kWh	For 10 years
	New plants (from 01/01/2003) or Existing plants after at the end of transition period	Market price	-	0.10 DKK/kWh	Max 0.36 DKK/kWh
	From 01/01/2005 for all plants	Market price	-	Premium determined on individual basis	For 20 years; max 0.34 DKK/kWh

* 0.50 DKK/kWh for electricity produced from biomass in private decentralised plants and 0.30 DKK/kWh if produced in utilities' plants.
** Only for offshore wind farms built by the utilities according to the agreement with the government.

The follow-up of the agreement of June 2002 was only adopted in March 2004[58]. It aims at framing the new RES-E policy, given the freezing of the RES-E market. As far as wind power is concerned, the agreement cancels the consumer's obligation to purchase electricity produced from wind turbines and replaces it with financial support in accordance with the transitional scheme in force (see table 3). For the wind turbines that are no longer covered by the transition rules, the support includes the market price plus a price premium of up to 0.36 DKK/kWh. In order to simplify the calculation and ensure more

---

[58] Political Agreement between the government (the Liberal Party and the Danish Conservative Party) and the Social Democrats, the Socialist People's Party, the Social Liberals and the Christian People's Party, on wind energy, decentralised power and heat, etc (follow-up to the agreement of 19 June 2002), 29 March 2004.

security, the market price is fixed as a monthly average and is no longer based on the spot price. Two new offshore wind farms of 200 MW each are planned by 2007/2008. They will be assigned by a tender procedure in order to reduce the costs as much as possible. The tariff for the electricity produced by those wind farms will be fixed on an individual basis, according to the results of the bid. These tenders represent the follow-up of the 1998 agreement that has not been fulfilled yet. The new scrapping regulation for old wind turbines seeks to remove approximately 900 old wind turbines (<= 450 kW) and replace them by a new overall capacity of up to 350 MW by the end of 2009. The price premium for the new turbines is of 0.12 DKK/kWh during the first 12,000 full-load hours, with a total support ceiling of 0.48 DKK/kWh (see table 4). The consumer purchase obligation from the other decentralised RES-E plants is also cancelled and replaced by price premiums guaranteed for 20 years from the installation of the plant to a maximum of 0.34 DKK/kWh (see table 4). Finally, electricity produced from new biogas plants will receive a settlement price of 0.60 DKK/kWh for the first 10 years and 0.40 DKK/kWh for the next 10 years. This financial support aims at doubling the installed capacity of biogas installations in Denmark by 2008.

In the meantime, the general elections of November 2001 brought to power a new governing coalition composed of the Liberal and the Conservative Party. The energy policy objectives of the new government shifted towards a higher priority on cost-effectiveness, but without neglecting the other objectives, such as the security of energy supplies and environmental protection. The past RES-E policy objectives were not reconsidered and the targets of 20% of RES-E by 2003 (1999 political agreement) and 29% by 2010 (1996 Energy Paper and EU directive of 2001) were reaffirmed by the new government. Moreover, long-term objectives for 2025 were presented in 2005 in a new energy plan, "Energy Strategy 2025"[59], and the projections for RES-E varied according to the scenarios from 36% (moderate rise in oil prices and CO_2 allowance prices) to 80% (if high oil and CO_2 allowance prices) of the total electricity consumption by 2025. Therefore, those projections are congruent with the projection of the "Energy 21" plan of 1996 that about 68% of electricity consumption should be covered by RES-E by 2030.

However, there is no doubt that the energy policy was a less important priority for the new Danish government than it had been since 1979, especially the RES-E policy from 1994 to 2001. The institutional changes that followed the 2001 elections clearly show this. In the new governments of 2001 and 2005, the competence on energy policy was

---

[59] Ministry of Transport and Energy, Energy Strategy 2025, 2005.

assigned to the Ministry of the Economy in 2001 and the Ministry of Transport and Energy in 2005. The previous Ministry of Energy was dissolved and its staff reduced while being incorporated in other ministries.

From 2002, the pursuit of the energy policy objectives (especially RES-E development) became subordinated to the increased cost-effectiveness of the energy sector. This strategy was endorsed by the Social Democrats and most of the other political parties in parliament through the agreements that were adopted in 1999, 2000, 2002, 2003 and 2004. The main reasons that explain this new strategy are, on the one hand, the need to reduce public expenses, and, on the other hand, the increased competition in the energy sectors due to the liberalisation of the electricity and gas markets. As described at length above, the instrument used to support the development of RES-E on the electricity market was modified in order to gradually reduce the subsidies to RES-E and transfer the cost of the subsidies to the consumers. The other dimensions of the RES-E policy, the investment subsidies and the RD&D program, also experienced significant changes from 2002, due to drastic public spending reductions in the energy sector.

In 2002 and 2003, a significant cut in the energy RD&D budget was decided by the conservative government. About half of the budget of the previous years was withdrawn (about 46 million € in 2001 to about 23 million € in 2002 and 2003)[60]. Given that the share of the RES budget in the total energy budget did not change (about 40%), the overall budget for RD&D in the renewable energy technologies was reduced by half in 2002 and 2003 (about 10 million €). The objective of the government was to reduce public spending in the energy sector, given that it was expected to become more competitive, in order to transfer the money to other political priorities (e.g. health care, quality of life for the elderly). However, in the political agreement with the opposition parties in May 2003[61], it was agreed that RD&D in the energy sector should be re-emphasised. So, from 2004, an additional 47 million DKK have been allocated each year to research, development, and demonstration activities, with the aim of promoting the exploitation of new energy-efficient technologies. Out of this budget, 7 million DKK/year are specifically allocated to type approval and quality assurance for RES-E, but new RES-E technologies also benefit from the rest of the budget through the

---

[60] IEA, Danish RD&D bugdet, 2005.
[61] Political Agreement between the government (the Liberal Party and the Danish Conservative Party) and the Social Democrats, the Danish People's Party, the Socialist People's Party, the Social-Liberals, and the Christian People's Party, regarding the development of the Danish energy market and measures to improve the development of new technologies, 9 May 2003.

Energy Research Program or the research and development programs implemented by the system operators.

In addition, in 2000, a new three-year development program for solar PV started with a budget of 10 million DKK/year. It is the successor of the Solar 300 program that ran from 1998 to 2000. The purpose of the program is to develop and demonstrate solar PV solutions in co-operation with housing and urban-renewal associations for a target of 1,000 new roof-top systems to be installed by 2004.

So the RD&D in RES-E certainly decreased significantly in 2002 and 2003 due to budget cuts, but it restarted slightly in 2004, under the impulse of a large majority in parliament. However, the capital grants for RES-E investments were cancelled in 2002 and have not been replaced so far since the economics of RES-E are supposed to be improved (especially for wind turbines) and for the technologies that are not (e.g. solar cells, biomass plants, biogas plants) financial support focuses on the RD&D level (e.g. solar cells) or call for tenders (e.g. biomass plants, offshore wind farms) instead.

V. Interpretation of RES-E Policy Change in Denmark

In Denmark, a large coalition of political parties (government coalition plus opposition parties) agreed on a reform of the electricity sector in 1999. This reform included the introduction of a new RES-E policy instrument (tradable green certificates), but did not significantly modify the RES-E medium and long-term policy objective (29% of RES-E in total electricity consumption by 2010 and 68% by 2030) (see table 4).

As far as the Danish RES-E policy objective is concerned, the only change that the political agreement of 1999 introduced was an acceleration of the short-term target of 20% RES-E in total electricity consumption, which was no longer expected by 2005, but by 2003. This was due to the rapid increase of RES-E in Denmark at the end of the 1990s, which showed that the 20% target would easily be met by 2003 (see table 1), but it does not represent a significant change of the policy objective. In fact, the RES-E targets for 2010 and 2030 as formulated in the 1996 "Energy 21" paper of the ministry of the environment and energy were not modified in 1999. This means that, as far as the RES-E policy objectives are concerned, the main references are not the political agreements, but the Energy Papers from the Ministry of Energy (e.g. the 1996 "Energy 21" paper in 1999).

The 1999 electricity reform provides that the previous production subsidies to RES-E are to be discontinued and replaced by a new instrument (TGC). The main decision that endorsed this policy change is the political agreement of March 1999, while the adoption of the act in

June 1999 only represents the legal acknowledgement. So, as far as the RES-E policy instrument change is concerned, the main references are the political agreements (1999, 2002 and 2004) and not the law. However, it is important to point out that the agreement on the electricity reform of 1999 originates in the work of a "reform group" in the DEA (the administration) and not in political debates. The RES-E policy instrument change had been formulated by the DEA and then endorsed by the political parties, but, as said, the main actor of the process was the administration and not the politicians.

Regarding the content of the policy instrument change, the TGC system is based on a market mechanism, the green certificate market, which was expected to be more adapted to the context of the liberalised electricity market[62]. In addition, the costs of the system are transferred from the public budget to the electricity consumers (change of resources to private budget), who are obliged to purchase a minimum quantity of RES-E per year (change of target group (demand) and incentive (quantity)) (see table 4). So the new instrument is expected in future to support the development of RES-E at a lesser cost to the state budget. In addition, the cost of the system to the electricity consumers is supposed to be reduced thanks to the competition mechanism introduced among the RES-E generators. The political agreement also provides that during the transition period between the end of the production subsidies and the full implementation of the green certificate system (anticipated for January 2003), a sufficient income is guaranteed to the existing and new RES-E plants through a complex transitional scheme in which premium prices (financed via the electricity bill) mix with market prices.

Table 4: RES-E policy change in Denmark

	Before 1999	After 1999	Policy change
Policy objective	20% by 2005; 29% by 2010 and 68% by 2030	20% by 2003; 29% by 2010 and 68% by 2030	No
Policy instrument	– Target group: RES-E generators (supply) – Incentive: price (production subsidies) – Resource: public budget	– Target group: electricity consumers (demand) – Incentive: quantity (quota) – Resource: private budget (electricity bills)	Yes

The change of policy instrument was planned to be implemented smoothly. However, shortly after having adopted this new RES-E policy instrument, the same political coalition decided first to postpone (in December 1999 and March 2000) and then to freeze (in October 2001

---

[62] Political Agreement of 3 March 1999 about the electricity reform.

and June 2002) its implementation, due to the criticism and uncertainty of the actors of the sector about the TGC system and how to implement it successfully in Denmark. So, in the end, the change in policy instrument was never implemented and the transitional scheme is not "transitional" anymore, but fully operational. Some important changes introduced in 1999 remain in today's operating subsidy system compared to the pre-1999 system (partly market-based and funded by private budget). Firstly, the tariffs guaranteed to RES-E are based on the electricity market price supplemented by a price premium with a maximum limit, which keeps the idea of a market-based instrument with an incentive to cost-effectiveness. Secondly, the financial burden of the system is now endorsed by the electricity consumers through a surcharge on the electricity bills (private budget).

In conclusion, when we look at the RES-E policy in Denmark, we first observe a change of policy instrument in 1999 (turning point), then the freezing of this change (2001-2002), which makes things a little complicated from an analytical point of view. Therefore, we decided to focus our analysis on the change of 1999 (the political agreement of March 1999), even if it has not been implemented so far, and this for two reasons. Firstly, some aspects of the TGC system have been incorporated in today's operating production subsidy system (market-based subsidy and private budget), so there has been a change in the policy instrument anyway. Secondly, the freezing of the tradable green certificate does not represent a new RES-E policy change because the option of the tradable green certificate is still open for the future and the existing production subsidy system is still more "transitional" than "definitive" today, politically and legally. So it makes more sense to explain the policy change of 1999, rather than the uncertain and "transitional" change of 2001-2002.

VI. Explaining RES-E Policy Change in Denmark

A. Policy Actors

In Denmark, the adoption of politically important policy reforms is traditionally approved by a large consensus of political parties (government and opposition parties). This is due to a large extent to the fact that most Danish governments are minority governments and must therefore rely on a larger majority in parliament to introduce policy changes. Moreover, the strong democratic and pluralist culture in Denmark usually allows the stakeholders of a policy to participate in the policy making process. However, we observe that the RES-E policy change of 1999 departed from those traditions on several levels.

Firstly, we pointed that the RES-E policy change of 1999 must be traced back to the coalition agreement of March 1999, in which most of the Danish political parties participated: the coalition partners (the Social Democrats and the Social Liberal Party), and the opposition parties (the Liberal Party, the Conservative Party, the Socialist People's Party and the Christian People's Party). However, it appears that the dominant actor who formulated the policy change (namely the new tradable green certificate system), were the Danish Energy Agency (DEA) and the Minister of Environment and Energy (Svend Auken), and not the political parties or the members of parliament[63]. In 1998, a specific group of experts on the energy policy was formed within the DEA to prepare the reform of the electricity sector (the so-called reform group). One of their tasks was to revise the existing production subsidy system for RES-E, which was perceived to be too expensive for the public budget and no longer adapted to the context of the liberalised electricity market. They came up with a proposal for a new RES-E policy instrument in August 1998, based on the model of a TGC system. The proposal was then debated and largely questioned by the RES-E associations (DWIA, DV, OVE), but without the DEA revising its position about the change of instrument (no influence). In March 1999, the agreement about the energy reform was adopted by the Danish political parties, including a section about the RES-E policy instrument change in favour of a TGC system, without this decision being preceded by an intensive political debate in parliament. It appears that the DEA and the Minister of the Environment and Energy persuaded the political parties rather easily. However, since the TGC system, which the DEA proposed, had not been subject to in-depth analysis and open debate with the actors of the sector prior to the adoption of the agreement (and in June 1999 the law), the agreement states that by the end of 1999 a detailed assessment of the implementation of the TGC with a consultation of all parties interested was to be conducted by the DEA. So, the actor who dominated the formulation of the policy instrument change (DEA) appears to be different than the actor who actually adopted it (political parties) (see table 5).

Table 5: Dominant RES-E policy actor (H1/1)

	H1/1
Dominant policy actor?	Civil servants: Danish Energy Agency (reform group) with Minister of Environment and Energy (Svend Auken).

---

[63] Interview with Ole Jess Olsen, Roskilde University, June 2005. Interview with Soren Krohn, ex-Danish Wind Industry Association (DWIA), June 2005. Hvelplund, 2001, pp. 71-99.

In addition, we observe that the influence of the actors of the sector prior to the adoption of the political agreement and the law was very limited (except DEF) (see table 6). Two main reasons can explain this: the late consultation of the actors of the sector by the DEA (fall 1998 when the proposal for a TCG system had already been finalised) (see table 6) and the apparent unwillingness of the DEA to question the basic principles of the change of instrument at that time (lack of in-depth analysis of the TGC system and how to implement it in Denmark) (Hvelplund, 2001). So, there was a debate about the policy change before the adoption of the agreement in May 1999, but it proved to be too late for the criticism to be considered by the DEA (e.g. DWIA, DV, and OVE). The only actor who participated in the formulation of the TGC earlier in the process is the Association of Danish Energy Companies (DEF), which was in favour of the TGC system and seemed to have benefited from a more careful attention from the DEA (1st and 2nd propositions of H1/2 validated) (see table 6) (Hvelplund, 2001).

Table 6: Influence of the actors of the sector (H1/2)

Policy actors (H1/2)	Early/Late	Preference	Influence
Electricity companies			
Association of Danish Energy Companies (DEF)	early	+	+
RES-E associations			
Danish Wind Industry Association (DWIA)	late	-	-
Danish Wind Turbine Owners' Association (DV)	late	-	-
Organisation for Renewable Energy (OVE)	late	-	-

Legend: Preference (+) favourable, (-) not favourable, (0) indifferent; Influence (+) strong, (-) weak, (0) no significant influence.

In fact, the criticism of the other actors of the sector (e.g. DWIA, DV, OVE) were listened to and addressed only during the implementation debate in the fall of 1999. At that time, the DEA was in charge of organising the implementation of the TGC in consultation with the actors of the sector (the reference group) and the opposition of the sector proved to be more fundamental than expected (Nielsen and Backer, 2003). The uncertainties and the criticism of the sector ultimately led the political parties involved in the agreement of May 1999 to review the agreement in June 2002 and to freeze the implementation of the TGC[64].

---

[64] Interview with Ole Jess Olsen, Roskilde University, June 2005. Interview with Soren Krohn, ex-Danish Wind Industry Association (DWIA), June 2005.

So, it appears that the actors of the sector did not have a strong influence in 1999 at the time of adopting the TCG, but they did in 2002, when the policy change was frozen.

In conclusion, what we observe in the Danish case of RES-E policy change, supports what we expected (H1/1 and H1/2). Firstly, the change of policy instrument was piloted by bureaucrats and not by politicians (DEA). The then Ministry of the Environment and Energy was also a strong advocate of instrument change, but the central actor remained the DEA, under its supervision (H1/1). Secondly, we observe that the actor of the sector who had the main influence on the policy-making process is the actor that was involved at the early stage of the process (DEF), while the other actors who were consulted later on happened to have a smaller influence on the policy change (first proposition of H1/2). In addition, since the DEF was in favour of the change of instrument, it contributed to the policy change (second proposition of H1/2).

B. Past Policy

Denmark is one of the pioneering European countries in RES-E, both in terms of RES-E industry development and in terms of RES-E market increase. This is primarily due to very successful RES-E policies during the 1980s and 1990s, based on a variety of instruments, especially the feed-in tariff scheme, with production subsidies during the 1990s. As far as RES-E penetration on the electricity market was concerned (policy objective), the past RES-E policy appeared to be very successful in 1999 (about 15% of RES-E in total electricity supplies in 1999; see table 1). Accordingly, the ambitious RES-E targets set for 2010 (29%) and 2030 (68%) were expected to be within reach (success of policy objective, see table 7).

In addition, the past RES-E instrument proved to be very effective in terms of RES-E installed capacity and RES-E share in the electricity supplies (see figure 3 and table 1). For instance, the target for wind power capacity (to be installed by 2005) of the 1990 energy plan "Energy 2000" (1,500 MW) was already achieved in 1998, so seven years in advance (see figure 3). However, the development of RES-E happened to the detriment of the Danish public budget, given that during the 1990s the production subsidies took the form of compensations for the electricity and CO_2 taxes (fiscal instrument financed by the state budget). Moreover, the rapid increase of the RES-E shares on the electricity market at the end of the 1990s and the expected pursuit of this trend according to the targets of 2010 and 2030 led the Danish government to decide that it could no longer bear those costs anymore in the future[65].

---

[65] Political Agreement of 3 March 1999 about the electricity reform.

This explains why the RES-E policy changed in 1999 with a new instrument based on a market mechanism (to reduce the costs of RES-E in the future) and funded by private budgets (the electricity consumer bills) (see table 7). The change of instrument was estimated to entail a state financial improvement for the years 2000-2002 of DKK 1,040, 1,210 and 1,340 million, respectively[66]. In addition, the cost-effectiveness of the new instrument has been a very important objective of the Danish government since 1999. Indeed, since the cost of RES-E was borne by the electricity consumers, it was crucial to keep it as low as possible to avoid an exaggerated increase of the electricity prices.

Table 7: Success versus failure of the past RES-E policy in Denmark (H2)

	Past policy success/failure (H2)	
Objective	RES-E target (29% by 2010 and 68% by 2030) within reach	Success
Instrument	– Effectiveness to reach the RES-E targets (e.g. 2005 wind power target reached in 1998 already) – Too expensive for public budget	Failure

In conclusion, in 1999, the past RES-E policy appeared very successful in terms of market development and therefore no change in the ambitious policy objectives for the future was to be considered. However, a serious failure in terms of public budget loss made it impossible for the Danish government to leave the policy instrument unchanged. Therefore, while the Danish RES-E policy objectives were not challenged, the policy instrument used to attain them had to be modified to solve the failure of the past (H2).

C. *Europeanisation*

Denmark has been a pioneer country in Europe concerning the RES-E policy. At the time of developing the first proposals of the European RES-E policy (see figure 4), the RES-E already represented a significant share of the Danish electricity market (see table 1) and ambitious policy objectives for the future had been adopted (1990 "Energy 2000" plan and 1996 "Energy 21" plan). In addition, the Danish government has always been a strong advocate of ambitious environmental policies at the European level, for instance regarding the GHG emission reductions and the implementation of the Kyoto Protocol, especially between 1994 and 2001 under Svend Auken as the Minister of the Environment and

---

[66] Act on amendment of the act on subsidies for electricity production, No.377, 2 June 1999.

Energy. At the beginning of 1999, when the Danish RES-E policy changed, the European RES-E policy was still in its infancy (see figure 4). No clear policy proposals had been submitted by the European Commission yet. Therefore, it is difficult to assess whether the European and the Danish RES-E policies fitted or not. However, we can analyse the perception of the Danish dominant policy actor on the fit/misfit between the past Danish RES-E policy and the European policy (see table 8). Looking at the perceptions of the policy actors on the fit/misfit will tell us more about the Europeanisation of the Danish RES-E policy than an objective assessment of the fit/misfit, since the policy actors are at the core of the Europeanisation process.

Figure 4: The European policy and Danish RES-E policy change

EU policy					
Directive liberalisation + Green paper SER	White paper SER		Working paper E-SER	Draft directive E-SER	Directive E-SER + ECJ judgement
1996	1997	1998	**1999**	2000	2001
			RES-E policy change in Denmark		

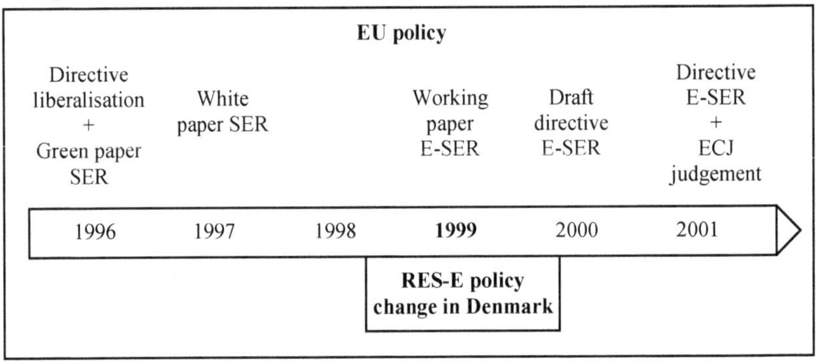

In 1998, when the DEA started to revise the Danish RES-E policy and formulate the new policy instrument, the Danish RES-E policy objective (29% by 2010) seemed congruent and even more ambitious than the European target (12% of RES in the European gross energy consumption) laid down in the 1997 White Paper (fit; see table 8). So, no change in the Danish policy objective seemed to be justified in the light of the European target. On the contrary, the existing RES-E policy instrument (production subsidies from public budget) was perceived by the DEA to be incongruent with the existing European liberalisation policy (competitive electricity market) and with the expected harmonisation of the European RES-E policy on the TGC model[67]. Therefore, this perceived misfit with the European policy contributed to the adoption of a TGC system in Denmark (see table 8), even if it was not the main reason for the policy change.

---

[67] Hvelplund, 2001. Political Agreement of 3 March 1999 about the electricity reform.

Table 8: Europeanisation of Danish RES-E policy (H3)

	Fit/Misfit (H3)	
Objective	29% by 2010	Fit
Instrument	Production subsidies: not market-based, TGC more adapted to competition	Misfit

In addition, the European policy (competition in the electricity market and the expected European TGC market) was also used by the DEA and the Minister of the Environment and Energy to legitimise the policy change when facing the criticism of the partisans of another policy instrument (feed-in tariffs) (Hvelplund, 2001).

After the adoption of the TGC in parliament in 1999, the Danish government worked towards the establishment of such an instrument based on the Danish model at the European level during the preparation of the 2001 directive. However, sufficient support could not be obtained, so the directive finally postpones the harmonisation of the RES-E instruments in Europe[68]. That decision not to harmonise the European RES-E market based on a green certificate system partially explains the freezing of the Danish green certificate system in 2002. In addition, the judgment of the ECJ in 2001 in favour of the German feed-in tariffs system increased the influence of the feed-in tariff advocates in Denmark (e.g. DWIA, DV).

In conclusion, the European policy definitely influenced the policy change in Denmark in 1999. The misfit hypothesis is validated in this case, though it is actually based not on an actual misfit but on a perceived misfit (H3).

D. Political and Institutional Context

Denmark is a consensus democracy with multi-party governments (usually two to three parties). What is specific to Denmark is that the Danish governments are most often minority governments, governing with the support of one or more opposition parties in Parliament (counter-majoritarian veto players). In other words, a government is not required to have the confirmed majority support. That is one reason why a minority government consisting of one or more parties is the usual outcome of government negotiations in Denmark. Consequently, Danish policy-making is characterised by large inter-party compromises, used to overcome the ideological distance between the political parties, which may include the parties in government and other parties in Parliament.

---

[68] Ministry of Environment and Energy, Energy Policy Review 2001, April 2001.

In this context, major policy change is more difficult to implement since it requires the approval of a large coalition of parties (many partisan veto players; see table 9). However, in 1999 the Ministry of the Environment and Energy succeeded in building a large consensus in favour of the electricity reform: the Liberal Party, the Danish Conservative Party, the Social Democrats, the Danish People's Party, the Socialist People's Party, the Social Liberal Party, and the Christian People's Party. In that case, the RES-E policy change was made possible even in the wake of many partisan veto players, because of the small ideological distance between the Danish political parties regarding the RES-E policy. This consensus in favour of the RES-E policy change was primarily due to the fact that the policy problem behind the reform was acknowledged by all parties (liberalisation of the electricity sector and reduction of public expenses for RES-E support). So, despite the presence of a large number of partisan veto players in the Danish political system, the RES-E policy change was supported by a large coalition of parties, which means that there were actually no partisan veto players who opposed the policy change.

As far as the institutional veto players are concerned, we count only few of them with real power in Denmark (see table 9). Parliament is unicameral, which means there are no double checks and balances in the legislative process. The Queen must countersign all legislation, but like in most other constitutional monarchies in Europe, this is only a formal power in reality. A lot of responsibilities and powers are delegated to the provinces and municipalities in Denmark (especially concerning electricity distribution and land-use planning), but the definition of the Danish energy policy remains the prerogative of the central government, so the provinces and municipalities can influence the formulation of the policy but they do not have the power to veto it.

Table 9: Governing coalition (H4/1), veto players (H4/1) and RES-E policy change in Denmark

Governing coalition (H4/1)	Veto players (H4/2)
– No change in the governmental coalition prior to 1999 – Minority governments, so less influence of governmental change on policy change (e.g. change of governing coalition in 2001)	– Many partisan veto players: multi-party governments and minority governments (counter-majoritarian veto players), but small ideological distance between parties over RES-E (large political consensus) and party discipline – Few institutional veto players: mono-cameral parliament (Folketinget) and Queen (no veto power), but small ideological distance over RES-E and party discipline

No change in the Danish governing coalition occurred prior to the adoption of the RES-E policy change in 1999, which means that H4/1 cannot be validated in this case (see table 9). In addition, due to the tradition of minority governments and large political consensus on policy changes, a change in the governing coalition is likely to have less impact on the policies in Denmark than in other consensus democracies with majority governments (H4/1; see table 9). Indeed, if the new governing coalition does not have the majority in Parliament, it will not be able to change the policies without making agreements with the opposition parties. This has been the case under the new Liberal-Conservative government (which was created at the end of 2001) for the amendments of the electricity reform agreement about the RES-E policy in 2002 and 2004 (freezing of the TGC). The change of government coalition did not explain the freezing of the TGC, as it was adopted by a large consensus of political parties including government and opposition parties (same parties as in 1999 except the Danish People's Party).

In conclusion, the first hypothesis (H4/1) is invalidated in the Danish case, but the second one is validated (H4/2). Firstly, there was no change in the governing coalition before the 1999 RES-E policy change (H4/1 invalidated). Besides, with the Danish tradition of minority governments, even if the governmental majority changes, it does not necessarily lead to policy changes due to the power of counter-majoritarian veto players (e.g. in 2001 with the freezing of the TGC) (H4/1 invalidated). Secondly, the Danish political system is characterised by a large number of potential partisan veto players, but thanks to the small ideological distance between them regarding the RES-E policy and the strong party discipline, a large consensus in favour of the RES-E policy change was reached in 1999 and thus the potential veto players were overcome in reality (H4/2 validated).

VII. Conclusion

In Denmark, a large coalition of political parties (government coalition plus opposition parties) agreed on a reform of the electricity sector in 1999. The Danish RES-E policy was part of this reform. The RES-E policy objective was not revised, as the medium and long-term policy objective (29% of RES-E in total electricity consumption by 2010 and 68% by 2030) of the 1996 energy paper of the Ministry of the Environment and Energy remained the reference. However, the past RES-E policy instrument (production subsidies financed by public budget) was to be replaced by a new instrument (TGC financed by electricity consumers) in order to reduce the burden of the RES-E policy on the Danish public budget (replacement of public budget by private funding) and to increase the cost-efficiency of the system to be more adapted to the

How and Why Do Policies Change?

competitive electricity market (competition between RES-E generators). In conclusion, the RES-E policy change in Denmark involved only the change of the policy instrument and not the policy objective.

Table 10: Validation and relevance of the hypotheses in Denmark[69]

	H1		H2	H3	H4	
	H1/1	H1/2			H4/1	H4/2
Validation hypotheses						
Policy objective	*V*	*n.a.*	*V*	*V*	*V*	*n.a.*
Policy instrument	V	V	V	V	I	V
Explanation policy change/no change						
Policy objective	+	0	+	+	0	0
Policy instrument	+	+	++	+	0	+

Legend: ++ (major explanation of policy change/no change), + (minor explanation), 0 (does not explain policy change/no change). V (hypothesis validated), I (hypothesis invalidated), n.a. (hypothesis not applicable).

Table 10 also shows whether each hypothesis is validated on not in the Danish case. Out of the six hypotheses, five are validated and one is not. The first two hypotheses concern the policy actors who participated in the process of policy change (H1/1 and H1/2). What we expected about the civil servants being the dominant policy actors in the case of change of the policy instrument (versus the political actors in the case of change of the policy objective) is observed in this case. Indeed, even if the policy change was adopted in a political agreement, it had been formulated by the energy administration (DEA) and then adopted by the political parties without much discussion, so the DEA is the actor that dominated the policy change process and formulated the new instrument (H1/1 validated). In addition, the influence of the actors of the sector who opposed the policy change (RES-E associations) remained limited (late consultation), while the influence of the energy companies who supported the policy change was greater (early consultation) (H1/2 validated). According to the third hypothesis (H2), the failure of the past policy instrument (unaffordable for the public budget) is one of the main reasons why the policy instrument was modified, as all the Danish public actors (administration and politicians) agreed that this failure was to be rapidly solved. On the contrary, the past policy objectives were perceived to be successful, and thus not modified, which consolidates the validation of H2. Then, the fourth hypothesis (H3) looks at the influence of the European policy on the domestic policy changes. We

---

[69] The policy objective lines are in italics because there was no change of policy objective in Denmark, so validation and relevance of the hypotheses can only be assessed by default for the policy objective.

observe that even if there was not a clear European RES-E policy yet at the time of the Danish policy change, the interpretations and expectations of the Danish policy actors about the European policy influenced the change of policy instrument (misfit) and the non-change of the policy objective (fit) (H3). Finally, among the two last hypotheses concerning the role of the governing coalition (H4/1) and the number of veto players (H4/2), one is validated (H4/2) while the other is not (H4/1) in the Danish case. On the one hand, the governing coalition did not change prior to the change of policy instrument, and even if it did, with the tradition of minority governments, it would not have had the impact it could have in other countries, which invalidates our assumption (H4/1). However, the absence of change in the policy objective validates the assumption by default; but again, we maintain that even if the governing parties had changed, it would not necessarily have resulted in a change of the policy objective because of the minority government tradition. On the other hand, even in presence of a large number of potential partisan veto players in Denmark, the Ministry of the Environment and Energy succeeded rather easily in having its reform proposal adopted by a large coalition of political parties thanks to the small ideological distance between the Danish parties regarding the RES-E policy and the strong party discipline, which validates our hypothesis about the role of the veto players (H4/2).

Finally, table 10 summarises the findings of this chapter about the relevance of the hypotheses in explaining the RES-E policy change in Denmark in 1999. We observe that out of the six hypotheses that we looked at, five prove to be relevant in explaining the change of policy instrument or the absence of policy objective in Denmark (H1/1, H1/2, H2, H3 and H4/2), and one appears to be irrelevant (H4/1).

On the one hand, we can consider the relevance of the hypotheses to explain the non-change of the RES-E policy objective in Denmark in 1999 by default. Only three hypotheses appear to be relevant to explaining why the policy objective did not change (H1/1, H2 and H3). The actors who dominated the policy change process were not the politicians but the administration (DEA), which probably partly explains why the policy objectives were not revised. In addition, the past policy targets were not questioned because they had been set for a long-term period (2010 and 2030) and they appeared within reach (H2). Finally, the past policy targets were congruent with the European indicative targets (H3).

On the other hand, those hypotheses aim at explaining policy changes, so it is more interesting to look at their relevance to explaining the change of policy instrument. In fact, one hypothesis emerges as the most significant one to explain the change of policy instrument: past policy (H2). The main factor that explains the change of policy instrument in 1999 in Denmark is the failure of the past policy instrument (production subsidies no longer affordable for the public budget) (H2). In addition, three other factors proved to be relevant to explaining the change of policy instrument, but to a less significant extent: the perception of the misfit of the past instrument with the European liberalisation policy (competition in the electricity market) along with the expectation of the harmonisation of the European RES-E policy on the TGC model (H3); the fact that the policy change was dominated by the energy administration (DEA) and not by the politicians (H1/1), and influenced by the Danish energy companies (DEF) in favour of the change and not by the RES-E lobby (which opposed the change) (H1/2); and the support of a large coalition of political parties (government majority and minority in the Folketinget) for the RES-E policy change, with no significant veto players (H4/2). Finally, one factor is not relevant in this case, the change of government coalition (H4/1), because there was no change of coalition before the policy change in 1999.

In conclusion, the hypotheses of our theoretical framework are validated in the Danish case (except H4/1) and they have proved to be relevant to explaining the RES-E policy change (especially the past policy success/failure). The only exception is the hypothesis derived from the partisan theory about the influence of governing parties on the policy change, which is not validated in this case, because Denmark is characterised by minority governments. But this is not a surprise, as the partisan theory literature has already observed that it was not applicable to non-majoritarian democracies (Schmidt 1996). So, our hypothesis is invalidated, but our findings confirm the existing literature.

CHAPTER 5

RES-E Policy in Germany

I. Introduction

Germany has been a federal country since the end of Second World War, with a division of competences between the federal government and the Länder. This encompasses three special features (Pehle, 1997): a distribution of the legislative competences that distinguish between exclusive and concurrent legislations (e.g. energy policy), the involvement of the Länder in the legislative process at the federal level through the Bundesrat (the Länder's chamber in the federal parliament), and the enforcement of the federal law through the Länder. Traditionally, the Länder have been very active in the field of RES-E, both through regional policies and through active policy co-operation or opposition in the Bundesrat (e.g. revision of the Renewable Energy Sources Act in 2004).

In the federal government, the RES-E competence was in the hands of the Ministry of Economic Affairs (BMWi) until 2002, when the Red-Green coalition decided to transfer it to the Ministry of the Environment (BMU). Since the Ministry of Economic Affairs is traditionally very close to the German electricity companies (as opposed to RES-E) and the Ministry of the Environment is more linked with the RES-E sector, the change of ministry represented an important factor in favour of RES-E. In addition, the Ministry of Research was also an important actor in the energy policy until 1998 since it managed the energy R&D programmes, and it often stood along with the Ministry of Environment in favour of RES-E. The municipalities also played an important role in the development of RES-E in Germany, especially in favour of solar cells during the second half of the 1990s (the Aachen model).

Germany is the world leader in the use of wind energy (16,629 MW of installed capacity by the end of 2004[1]) and it accounts for about half of all wind turbines built worldwide thanks to a strong wind turbine industry. In addition, Germany has installed the highest solar cell capac-

---

[1] BMU, Renewable energy sources in figures – national and international development, June 2005.

ity in Europe (708 MW of installed capacity by the end of 2004) and it accounted for about one fifth of the worldwide stock of solar cells at the end of 2003. This position of leadership in the wind and solar sectors results from the successful RES-E policies that have been implemented for decades (R&D subsidies, market introduction programmes and feed-in tariffs) and also from the industrial policy in favour of the German wind and solar manufacturers since the 1990s. It also relies on a strong and growing coalition of actors (both public and private) which have managed to advocate and defend the RES-E policies against the powerful electricity sector utilities during the last decade (Jacobsson and Lauber, 2005). Finally, the development of RES-E in Germany cannot be explained without considering the strong commitment of the German government in favour of the climate change policy (the energy sector plays an important role in the CO_2 emissions reduction strategy) and the commitment – and eventually decision – to phase-out nuclear power (with RES-E expected to contribute significantly to its replacement).

In Germany, the definition of eligible RES-E for financial support on the electricity market (mostly through feed-in tariffs) has evolved over time towards an ever larger scope of sources and technologies. Typically, electricity produced from wind, solar, biomass (max 5 MW), hydro (max 5 MW) as well as landfill and sewage treatment (max 5 MW) have been able to receive feed-in tariffs since 1991. Later, mine gas and geothermal energy have been included in the new feed-in tariff system from 2000 and the limit for biomass plants has been increased to a maximum of 20 MW. And recently, the 2004 revision of the policy enlarged the scope to include hydropower up to 150 MW.

II. Overview of the Electricity Sector

Germany is the largest European electricity market and its position in the centre of Europe makes it an important transit country. The structure of the electricity sector in Germany is decentralised, with a large number of privately or publicly owned utilities (supra-regional, regional and municipal utilities). However, the mergers and acquisitions among the German electricity companies that followed the liberalisation of the market in 1998 resulted in a less decentralised industry than there used to be (significant reduction in the number of supra-regional and regional utilities) and in an increasingly privatised and centralised ownership structure (most regional and municipal utilities being owned or controlled through contracts by the supra-regional utilities). Today, four big supra-regional utilities dominate the German electricity market (RWE AG, E.ON Energy AG, Vattenfall Europe AG and EnBW AG)[2]. They

---

[2] IEA, 2002, pp. 99-100.

operate at the generation (80% of total electricity generated), transmission (owners of the six transmission system operators) and supply levels (directly and indirectly through contracts with the regional and municipal suppliers). The regional utilities (about 56 companies) distribute electricity over a well-defined area to the municipal utilities and final consumers. They also generate electricity, but only to a small extent (10% of total electricity generated)[3]. Local utilities (about 840 companies) distribute electricity to final customers and some also have their own generating capacities (10% of total electricity generated)[4]. Finally, electricity traders have entered the German market after the liberalisation. Since few of them have generating capacity of their own (due to the centralisation and protectionism of the market), they can only engage in marketing activities (including green power and trading business) through the European Power Exchange (EEX) in Leipzig. The big supra-regional utilities usually speak for themselves on important political questions, but the other electricity companies usually lobby either through the VRE (association of the national grid operators and regional utilities), VDEW (association of electricity suppliers) or VKU (association of municipal utilities).

Due to its domestic reserves, coal still remains the main fuel for electricity generation in Germany (about 50% of total electricity generation today), though its share has been steadily declining, with the increase in the total electricity production and the development of nuclear (see figure 1). Nuclear power has represented the second most important share of the German electricity generation market since the 1980s (about 30% of total electricity generation). However, due to the absence of new investments, it has been stagnating since the end of the 1980s (see figure 1) and is expected to decrease in the future according to the decision of June 2001 to phase-out nuclear power in Germany. Electricity generation from gas only accounts for about 10% of electricity generation (see table 1), but it is expected to increase in the future, with the reduction of nuclear power. Finally, RES-E represents less than 10% of the total electricity production so far (see figure 1), but has been increasing rapidly during the last decade and is expected to further develop in the near future (political targets of 12.5% by 2010 and 20% by 2020).

---

[3] *Idem.*
[4] *Idem.*

Figure 1: Electricity generation by fuel, 1972-2002 (GWh)

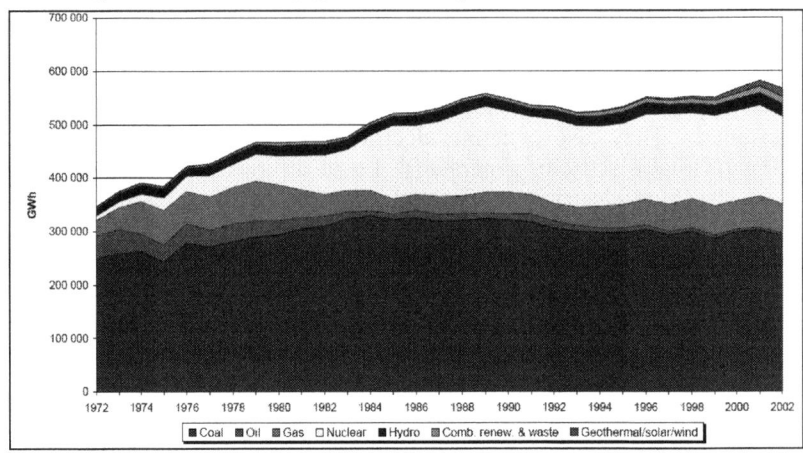

Source: IEA.

Before 1990, hydropower was the only significant RES-E in Germany (see figure 1), with still a limited share in total electricity generation (3.2% in 1990[5]). It has remained relatively constant so far, while all the other RES-E grew rapidly (see table 1).

On the contrary, wind power was non-existent in 1990 but exploded during the 1990s to become the fastest growing RES-E in Germany (see table 1). In 2004, wind power generation topped hydropower (see table 1). Since the future potential for onshore wind power sites has become limited, most of the future investments are expected to be offshore wind power plants. Research and development programs as well as pilot projects are being developed with the support of the government by the German wind industry for large offshore wind turbines.

Biomass has also increased steadily during the last decade (see table 1) and is expected to further increase in the future thanks to the increased feed-in tariffs for electricity produced from biomass installations since 2004 (especially for small plants and new technologies).

---

[5] IEA, 2002, pp. 135.

Table 1: RES-E installed capacity (MW) and electricity production (GWh), 1990-2004

	Hydro		Wind		Biomass		Solar PV		Geothermal		Total	Share of gross elec. cons.
	GWh	MW	GWh	MW	GWh	MW	GWh	MW	GWh	MW	GWh	%
1990	17,000	4,403	40	56	1,422	190	1	2	0	0	18,463	3.4
1991	15,900	4,403	140	98	1,450	N/A	2	3	0	0	17,492	3.2
1992	18,600	4,374	230	167	1,545	227	3	6	0	0	20,378	3.8
1993	19,000	4,520	670	310	1,570	N/A	6	9	0	0	21,246	4.0
1994	20,200	4,529	940	605	1,870	276	8	12	0	0	23,018	4.3
1995	21,600	4,521	1,800	1,094	2,020	N/A	11	16	0	0	25,431	4.7
1996	18,800	4,563	2,200	1,547	2,203	358	16	24	0	0	23,219	4.2
1997	19,000	4,578	3,000	2,082	2,479	400	26	36	0	0	24,505	4.5
1998	19,000	4,601	4,489	2,875	2,800	409	32	45	0	0	26,321	4.7
1999	21,300	4,547	5,528	4,444	3,020	448	42	58	0	0	29,890	5.4
2000	24,936	4,572	9,500	6,112	4,129	585	64	100	0	0	38,629	6.7
2001	23,383	4,600	10,456	8,754	5,065	825	116	178	0	0	39,020	6.7
2002	23,824	4,620	15,856	11,965	5,962	1,510	188	258	0	0	45,830	7.8
2003	20,350	4,640	18,919	14,609	7,982	1,694	333	408	0	0	47,584	8.0
2004	21,000	4,660	25,000	16,629	9,367	2,061	459	708	0.4	0.2	55,826	9.3

Source: BMU, Renewable energy sources in figures, June 2005, p. 12.

Solar PV capacity has been increasing significantly since 2000, especially in 2004, with 300 MW of newly installed capacity (see table 1). Germany has the highest installed capacity in Europe. However, solar power still remains marginal in terms of electricity generation, but has been increasing steadily lately and it is expected to contribute even further to the total electricity generation in the short and medium term.

Finally, geothermal energy is marginal in Germany (see table 1), but the increased feed-in tariffs for geothermal electricity since 2004 might well give a boost to the sector.

III. Overview of the RES-E Policy History in Germany (1974-2000)[6]

In order to give a clear overview of the RES-E policy in Germany before 1998, we draw on Jacobsson and Lauber (2005) to distinguish two phases. The 1974 to 1989 period was characterised by government

---

[6] This section draws substantially on Lauber and Pesendorfer (2004), Lauber and Mez (2004) and Jacobsson and Lauber (2005).

support to the research and development of technical innovations in the RES-E sector (mostly directed to wind power and solar cells). Then, the 1989-2000 period saw the government reorient its policy towards the market penetration of RES-E by means of investment programs (the 1,000 solar roofs program and 100-250 MW wind power) and financial support to RES-E on the electricity market (feed-in tariffs).

A. Research and Development (1974-1989)

The German electricity sector has never been heavily dependent upon oil, mostly because of its domestic coal reserves. However, the oil crisis of the 1970s still impacted the German energy policy in different ways. First, the German government wanted to increase the security of its energy supplies by promoting domestic energy sources Coal, nuclear and, to a lesser extent at that time, RES-E were targeted as domestic electricity sources to be further expanded. Coal was supported by direct subsidies and fiscal advantages to the German coal-mining industry and R&D programs for new technologies (Schmitt, 1983). The nuclear industry has benefited from considerable R&D subsidies since the 1950s, and especially during the 1970s, as a consequence of the oil crisis. However, the outcome in terms of installed capacity was less significant than first ambitioned. Indeed, public opposition against nuclear power started to constrain the government and delay the construction of the nuclear plants from the mid-1970s onwards. In the end, nuclear power accounted to some 30% of the total electricity production in the 1980s, while it could have contributed a lot more, had the social context been less hostile (Jahn, 1992). Moreover, after the Chernobyl accident in 1986, the anti-nuclear movement grew significantly stronger, with about 70% of the public opinion expressing opposition to nuclear power (*Ibid.*, p. 396). From then on, a *de facto* moratorium on further nuclear investments was applied, until the final phasing-out agreement was signed between the German government and the utilities in 2001.

Beside increased support to coal and nuclear, the German government also decided to stimulate research in new RES-E technologies, but to a lesser extent[7]. The annual budget for RES-E increased progressively from 1974 (about 20 million DM) to 1982 (with a peak of 300 million DM), and then decreased until 1986 (about 164 million DM) (Lauber and Pesendorfer, 2004, p. 133) when the Chernobyl accident put an end to the German nuclear policy and obliged the government to thoroughly revise its energy policy.

---

[7] See Lauber and Mez, 2004, for further details on the German wind and solar R&D policy.

In addition to research grants, a first (but limited in outcome) attempt to create an RES-E market started in 1979, with an amendment to the competition law that obliged the electricity distributors to purchase the RES-E produced in their area at a price based on the principle of avoided costs (*Ibid.*). This was not successful in terms of market development, because the compensations based on the avoided costs were too low. This is not a surprise, due to the fact that the policy targets (the actors who were expected to modify their behaviour) were at the same time the actors in charge of implementing the policy, namely the electric utilities.

B. Market Penetration (1989-1998)

With the Chernobyl accident and the increased concern about climate change, the second half of the 1980s marked a shift of the German energy policy to increased support to RES-E and a stronger commitment to develop an RES-E market. Two kinds of instruments were adopted for this purpose: investment grants for wind and solar power, as well as a feed-in tariff to RES-E delivered to the electricity grid.

Two significant market creation programs were adopted in 1988 and started in 1989: the 100 MW wind programme (then extended to 250 MW) and the 1,000 roofs program (*Ibid.*). The former intended to reach 100MW (and then 250 MW) of wind turbine installed capacity in Germany thanks to a guaranteed payment of 0.04€/kWh (then reduced to 0.03€/kWh) for electricity produced from wind turbines. With the adoption of the feed-in law of 1990, the target was quickly reached and even exceeded. The 1,000 roofs program aimed at stimulating investments in solar PV through granting applicants 50% of the investment costs from the federal government plus 20% from the Länder. The outcomes exceeded the expectations, since 2,250 roofs had eventually been equipped with solar PV at the end of the program (1995) (*Ibid.*, p. 134). Therefore, it is not surprising that Germany today appears to be a leader in both the wind and solar power sectors, since those RES-E technologies have been the most stimulated since the end of the 1980s.

In addition to those specific programs, a new law was adopted in 1990 to grant favourable tariffs to the decentralised RES-E supplied to the German electricity grid: the electricity feed-in law. This law represents a change in government strategy. The government would have preferred a self-commitment of the utilities to provide for increased RES-E compensations. Then, when the utilities refused, the government used the traditional legal framework to enforce its policy (Lauber and Mez, 2004). After several initiatives by Members of Parliament, especially Conservative backbenchers and Greens, against the government and the Conservative leadership, the feed-in law was finally adopted by a consensus of all parties (*Ibid.*). The objective of the law was "to level the playing field for renewable power by setting feed-in rates at levels that took account of the external costs of conventional power generation" (*Ibid.*). No quantitative target was adopted, but it was expected to play a minor role on the overall electricity market, with only a few hundred MW of small hydro plants to be installed. This explains why the Conservatives and the utilities did not oppose the law from the beginning (they tried later on, when the law unexpectedly proved to be successful) (*Ibid.*). According to the law, the electric utilities were obliged to connect the RES-E generators to their grid and buy the electricity at a favourable price, namely 65% to 90% of the tariff paid by their final customers (see table 2). Table 2 shows the annual tariffs for each RES-E, depending on the varying percentages of the average final customers' price (Staiss, 2000, II-27).

Table 2: Feed-in tariff rates under the Feed-in Law, 1991-2000 (in pfennigs/kWh[8])

	1991	1992	1993	1994	1995	1996	1997	1998	1999	2000
Wind, solar (90%)	16.6	16.5	16.5	16.9	17.2	17.2	17.1	16.7	16.5	16.1
Biomass up to 5 MW (75%, after 1994 80%)	13.8	13.7	13.8	14.1	15.3	15.3	15.2	14.9	14.6	14.3
Hydro, landfill and sewage station methane up to 0.5 MW (75%, and 80% from 1994)	13.8	13.7	13.8	14.1	15.3	15.3	15.2	14.9	14.6	14.3
Same as above but from 0.5 MW to 5 MW (65%)	11.9	11.9	11.9	12.2	12.4	12.4	12.3	12.1	11.9	11.6

Source: Lauber and Pesendorfer, 2004, p. 137.

The feed-in law was first amended in 1994, but only marginally (increased tariffs for biomass and small hydro, landfill and sewage stations). Due to the unexpected success of the law, especially for wind

---

[8] 100 pfennigs = 1DM = 0.51€.

power (installed capacity increase from 20-60 MW in 1990 to 12,000 MW in 2002) (Lauber and Pesendorfer, 2004, pp. 136-137), opposition increased on the utilities' side. By the end of the 1990s, the electricity utilities tried to defeat the law both through political lobbying (special links with the Conservative party and the Ministry of the Economy) and legal proceedings (national courts and DG Competition). But public opinion and a majority of Members of Parliaments were strongly committed to the pursuit and even the reinforcement of the law, so the utilities' attempt to weaken it by political means failed, albeit only narrowly (Lauber and Mez, 2004).

The second amendment to the feed-in law was adopted only in 1998 as part of the Energy Act (the act transposing Directive 96/92/EC on the liberalisation of the electricity market). It finally resolved a problem that appeared from the beginning of the system, but only turned out to be critical in the second half of the 1990s with the drastic increase of wind turbines in the northern part of Germany: the absence of a proper compensation mechanism to redistribute the costs of the feed-in tariff among the German utilities. The Energy Act establishes a mechanism of ceilings for the quantity of RES-E that the utilities were required to buy (5% of the total volume of electricity distributed). Above that 5% ceiling, the purchase obligation fell on the upper stream network operator (first ceiling) and one level further up if that operator also reached the 5% (second ceiling). Nevertheless, a more comprehensive revision of the feed-in law was then postponed due to a lack of political consensus.

While the feed-in law (in conjunction with the 100/250 MW programme) proved to be very successful for wind power, the tariffs for solar power were still too low to cover the costs of this technology. Therefore, an original initiative in favour of solar power appeared in the second half of the 1990s, starting in Aachen (hence the name, "the Aachen model"). Several German municipalities imposed the purchase of solar power on their municipal electricity utilities at full cost rates, even if those rates exceeded the avoided costs of those utilities. This measure resulted from an intensive lobby by citizen groups and solar power advocates to preserve and increase the market for solar PV in Germany after the 1,000 roofs programme stopped and proved to be successful.

C. Liberalisation of the Electricity Market (1998)[9]

For decades, the German electricity sector was dominated by large supra-regional utilities, which were vertically integrated so as to monopolise the entire electricity market (generation, transmission, distribu-

---

[9] This section draws on Lauber and Pesendorfer (2004), Mez (2003).

tion). In 1998, at the time of the liberalisation of the electricity market, there were eight of them operating in a limited territory, most of them owned by public bodies, HEW, VEBA, BEWAG, VEW, VEAG, RWE, VIAG and EnBW. As we mentioned earlier, following the liberalisation of the electricity market, the sector experienced mergers, acquisitions and privatisation, which resulted in a less decentralised, more privatised electricity sector, but still very much vertically integrated through ownership.

The process of liberalisation of the electricity market and the reform of the electricity industry was slow and controversial in Germany. The first proposals to reform the sector started at the end of the 1980s, but they were strongly opposed by the large electricity companies. With the preparation of the European directive on the liberalisation of the European electricity markets, the German government was forced to introduce changes. Several domestic actors also urged the German government to reform the electricity sector: the Federal Cartel Office, the ministry of the Environment, the Liberal Party (FDP) and the industrial associations (e.g. BDI). The first drafts of the Energy Industry Act were introduced by the ministry of the Economy in 1994 and 1996. But then the Bundesrat opposed the proposal of the ministry of the Economy in 1996, when the Green Party and the Social Democrat Party introduced an alternative proposition (including, for instance, improved conditions for RES-E in the electricity market). In the end, the Energy Industry Act came into force in April 1998, after having been adopted in the Bundestag but without the involvement of the Bundesrat[10]. This act transposes the directive of 1996 on the liberalisation of the European electricity markets. Firstly, it provides for the immediate opening of the electricity market as from April 1998, except in the new Länder. So, all the electricity consumers became eligible at once. However, due to the dominant position of the large German electricity companies in the German market, the competition in the supply market was not very effective. Secondly, the unbundling of the network and sales activities is enforced, but only an accounting unbundling, which allowed the large electricity companies to remain vertically integrated and maintain their domination on the entire sector. Finally, unlike in most other European countries, the access to the German electricity networks was negotiated and not regulated. Indeed, the German government decided not to introduce an independent regulator for the electricity sector and to let the electricity companies self-regulate the access to their network. The Federal and Länder Cartel Offices were in charge of controlling *ex post* the application of the competition rules. In this context, the German

---

[10] The government claimed this act did not affect the Länder, so the involvement of the Bundesrat was not necessary for the act to be adopted.

electricity companies adopted a voluntary agreement on tariffs for third party access to the transmission and distribution networks in 1998, which stated the rules for new market entrants to connect to the German networks, the so-called Association Agreement (VV in German). This agreement was subsequently revised in 1999 and 2001 to improve the conditions of network access. This procedure was supposed to ensure transparent and non-discriminatory access to the electricity networks, but it was not the case in reality, since there was no legal basis to the Association Agreements and no *ex ante* control from independent authorities. This was criticised by a number of actors in Germany, especially the Federal Cartel Office, the ministry of the Environment, and the new electricity companies; but the German government supported this system, even the new Red-Green coalition of 1998.

In 2003, however, the German government decided to revise its position and establish an independent regulator for the electricity (and gas) sector. This decision resulted from the accumulation pressure to do so, both at the domestic and European levels. Firstly, a new directive on the liberalisation of the European electricity markets was adopted in 2003 and it made it mandatory for each member state to adopt regulated third party access to the electricity networks. Therefore, an independent regulator was to be established in Germany to regulate access to the network. In addition, with the competence on energy being transferred to the hands of the ministry of the Environment in 2002, the influence of the ministry of the Economy on this subject (traditionally strong links between the latter and the electricity companies) was reduced. Finally, a court judgment against the Association Agreements, in Berlin in March 2003, stated that they violated the German Competition Act. Therefore, confronted with such pressure, the German government decided, against the large supra-regional utilities, to establish an independent regulator. Finally, the amended Energy Industry Act of July 2005, which transposed the Directive of 2003, charged the Federal Network Agency for Electricity, Gas, Telecommunications, Post and Railways to regulate the electricity market. However, whether this Agency will be able to circumvent the powerful electricity companies is still open to question, especially in the light of its numerous and diversified tasks.

IV. RES-E Policy Change in Germany (2000-2004)[11]

In 1998, a Red-Green coalition (Social Democrats and Greens) replaced the former Conservative-Liberal government. As expected (in view of their past opposition), the new coalition opted for strong com-

---

[11] This section draws on Lauber and Pesendorfer (2004), Lauber and Mez (2004) and Jacobsson and Lauber (2005).

mitment in favour of the nuclear phase-out and the promotion of RES-E in their coalition agreement, which marked a significant shift in objectives of the German electricity policy. As far as RES-E was concerned, four important decisions were adopted in the period 2000-2004: the launch of a new solar PV programme in 1999 (the 100,000 roofs programme), the replacement of the old feed-in law by a more ambitious act in 2000 (the Renewable Energy Sources Act – RESA[12], at least doubling RES-E by 2010) still based on feed-in tariffs, the "advanced" law for solar PV in 2003 (advanced amendment to RESA), and the new amended RESA, adopted in 2004[13] with stronger commitment to RES-E (20% by 2020). Therefore, while the policy instruments used by the German government to stimulate the RES-E market from 2000 did not vary, the policy objectives to be attained changed significantly, with RES-E expected to increase substantially so as to replace nuclear power in the medium term.

Since the feed-in tariffs for solar power were too low to ensure the development of the German solar PV industry and the further penetration of solar power in the electricity market, the government decided to launch a new solar program in 1999, the 100,000 roofs program. It was expected to fill the gap until the old feed-in law was revised. Such a program had been requested by several members of Parliament as well as the industry for years, but it had faced too much opposition in the previous government. This program provides subsidies for solar PV investors in the form of low-interest loans. Initially, the goal was to achieve an installed capacity of 300 MW, but it finally led to the installation of 350 MW at the end of the program in 2003, thanks to the additional support of increased solar power feed-in tariffs from 2000 onwards (Lauber and Mez, 2004). Besides its contribution to the sustainable German energy supply, this program also aimed at supporting the German solar manufacturers to sustain their leading position on the international market, hence it contributes to both German energy and industrial policies. Then, in 2003, the solar lobby again asked for a new program or increased feed-in tariffs for solar power in order to avoid the outsourcing of the industry and ensure that solar power significantly contributes to the 20% RES-E target by 2020. Since the revision of the RESA was under way but not ready yet, a special "advance" law, with increased rates for solar power, was adopted in 2003 and subsequently incorporated in 2004 in the new RESA. With those increased rates, solar

---

[12] Renewable Energy Sources Act, 1st April 2000.
[13] Amended Renewable Energy Sources Act of 21 July 2004.

power became competitive, even without additional support and resulted in a boom in solar PV installations in 2004[14].

Another larger market program started in 1999 promoting increasing investments in renewable energy sources in general (especially for heat-producing RES technologies): the so-called Market Incentive Program (MAP). It grants direct investment subsidies or soft loans to solar thermal systems, biomass heating systems, biogas, small hydropower, geothermal and solar PV installations (only at schools). However, since it is especially directed at heating systems, we will not develop it further here[15].

The 1990 feed-in law was very successful in terms of RES-E capacity increase during the 1990s (especially for wind turbines), but it needed to be revised for several reasons: increased long-term security for investors, better redistribution of the costs among the utilities, and more differentiated and "cost-related" feed-in tariffs. In addition, from 2000, RES-E was not considered to be a marginal energy option in Germany anymore, but a serious alternative to nuclear power and a decisive solution to address climate change. Therefore, the RESA of 2000 not only revised the feed-in tariffs system, but also acknowledged the commitment of German policy to promoting a sustainable energy supply to which RES-E is expected to contribute a great deal (at least the doubling of RES in total energy consumption compared to 2000, hence a minimum of 12.5% of total electricity consumption by 2010). RESA provides that the feed-in tariffs are guaranteed for 20 years, which represents a significant improvement in terms of financial security for the investors. In addition, the tariffs are not dependent anymore on the average electricity prices but fixed in advance for each RES-E category, based on the cost of each technology (see table 3), which resulted in significantly higher rates for all technologies. However, given the strong variations in revenue depending on wind quality, the rate for wind power now also depends on the quality of the location. In order to avoid overcompensation and account for the technological learning curves of each technology, the rates decline annually for the new investments in biomass, wind turbines and solar cells (see table 3). Finally, the additional cost of this system for the utilities is now redistributed equally nation-wide on the pro-rata of the electricity sales of each utility, which represents a significant improvement compared to the previous "ceiling" system. Therefore, the feed-in tariff system of RESA is based on the same instrument as the 1990 feed-in law, but the

---

[14] Jacobsson and Lauber, 2005.

[15] See Bechberger and Reiche, 2004, for further details on the MAP program and other RES heating promotion schemes in Germany.

settings were significantly improved to ensure a more effective, efficient and secure RES-E market in the future.

Table 3: Feed-in rates under RESA 2000 (in €ct)[16]

		2000	2001	2002	2003	2004	2005
Hydro, methane from landfill, coal mines, sewage tations	<500kW	7.67	7.67	7.67	7.67	7.67	7.67
	>=500kW – 5MW<	6.6	6.6	6.6	6.6	6.6	6.6
Biomass	<500kW	10.2	10.2	10.1	10.0	9.9	9.8
	>=500kW – 5MW<	9.2	9.2	9.1	9.0	8.9	8.8
	>5MW – 20MW<	8.6	8.6	8.6	8.5	8.4	8.3
Geothermal	<20MW	8.9	8.9	8.9	8.9	8.95	8.9
	>=20MW	7.1	7.1	7.1	7.1	7.16	7.1
Wind power	For at least 5 years**	9.1	9.1	9.0	8.8	8.7	8.6
	Final compensation***	6.2	6.2	6.1	6.0	5.9	5.8
Solar radiation*	Non built-up areas <100kW Buildings <5MW	50.6	50.6	48.1	45.7	43.4	41.2

* Obligation to pay this tariff ends for new photovoltaic facilities once a cumulative capacity of 350MW is exceeded. In 2002, this threshold was raised to 1,000MW. In 2003, it was abolished.

** For at least 5 years from the date of installation (9 years for offshore), afterwards decline to definitive rate depending on site quality. The 5 years apply to facilities achieving 150% of the performance of a reference facility. For others the higher rate is extended for 2 months for every 0.75% that the facility remains below the 150%. For older installations at least 4 years after entry into force of RESA.

*** Final compensation refers to the compensation received once the special rates listed above-limited in time-have run out.

Source: Lauber and Pesendorfer, 2004, p. 161.

RESA was supposed to be evaluated and revised after two years. The evaluation report of 2002 served as a basis for the amendment of RESA, but the final document was only adopted in 2004, due to a bitter conflict between the ministry of the Economy (previously in charge of this policy) and the ministry of the Environment (in charge of RES-E policy as from 2002). Indeed, the policy priorities and preferences of both ministries differed to a large extent and a compromise was more difficult to reach than had been expected at first. In addition, the bill of the

---

[16] RESA provides the tariffs until 2010, but since it was amended in 2004 our table stops in 2005. See Staiss, 2003 for further details.

amended RESA was first opposed in the Bundesrat led by conservative Länder, before agreement was reached between the red-green coalition and the opposition in the Conciliation Committee. However, the final compromise provides only for modest changes compared to the initial ministry of the Environment bill (e.g. the reduction of the onshore wind power rates).

The main contribution of the new RESA lies in its reaffirming and reinforcing the commitment of the German government in favour of RES-E. A new minimum target of 20% of RES-E by 2020 was formally adopted and a further target of 50% of RES-E by 2050 was discussed. In addition, other related policy objectives were reaffirmed, such as the internalisation of external costs of fossil fuels, the development of new technologies by the German industry, and the potential of RES to prevent involvement in international military conflicts about energy resources. The new feed-in rates were increased for most technologies (e.g. solar PV, offshore wind turbines, biomass and geothermal) and reduced for onshore wind turbines (see table 4). In addition, the rates appear to be more diversified than before, with an increased number of conditions and exceptions that primarily seek to make the system more efficient (reduction of the risk of overcompensation). However, in the end, they carry the risk of being very complex to handle. For instance, a bonus for innovation has been created for biomass technologies. Hydropower generated from installations up to 150 MW is now eligible for feed-in compensations for a period of 15 years. Thus several improvements were adopted in 2004 in order to facilitate the achievement of the old (12.5% by 2010) and new (20% by 2020) RES-E targets.

More recently, the new conservative-social democrat government, which replaced the red-green coalition in November 2005, is expected to "pursue the ambitious goals fur further expansion [of RES-E] in Germany", as stated in the coalition agreement of 11 November 2005. They also committed to maintain the basic structure of the RESA (feed-in tariffs), even if the economic efficiency of the individual rates is expected to be revised by 2007[17].

---

[17] Coalition agreement between CDU, CSU and SPD, 11 November 2005.

Table 4: Feed-in rates under the amended RESA 2004 for new plants commissioned between 1 August 2004 and 31 December 2004 (in €ct)[18]

Source	Installation capacity	Rates**	Capacity range	Degression ***
Hydro	<5MW	9.67	<500kW	-
		6.65	>=500kW – <5MW	
	>=5MW – <150MW	7.67	<500kW	1%
		6.65	>=500kW – <10MW	
		6.10	>=10MW – <20MW	
		4.56	>=20MW – <50MW	
		3.70	>=50MW – <150MW	
Landfill gas, sewage gas, mine gas	unlimited	7.67	<500kW	1.5%
		6.65	>=500kW – 5MW	
		6.65	Mine gas >=5MW	
	unlimited (if innovative technologies used)	9.67	<500kW	1.5%
		8.65	>=500kW – 5MW	
		8.65	Mine gas >=5MW	
Biomass*	<20MW	11.50	<150kW	1.5%
		9.90	>=150kW – <500kW	
		8.90	>=500kW – <5MW	
		8.40	>=5MW – <20MW	
Geothermal	unlimited	15.0	<5MW	1% from 2010
		14.0	>=5MW – <10MW	
		8.95	>=10MW – <20MW	
		7.16	>=20MW	
Wind	onshore	8.7 starting rate 5.5 end rate		2%
	offshore	9.10 starting rate 6.19 end rate		2% from 2008
Solar	on buildings or sound barriers	57.4	<30kW	-
		54.6	>=30kW – <100kW	
		54.0	>=100kW	
	façade systems	62.4	<30kW	
		59.6	>=30kW – <100kW	
		59.0	>=100kW	
	others	45.7		

* Specific higher rates apply for regenerative raw materials, wood combustion, and power from CHP plants.
** All rates are granted for 20 years, except small hydro (30 years) and large hydro (15 years).
*** For newly commissioned installations, the rate is reduced annually (degression) to provide a continual incentive to improve efficiency and reduce costs.
Source: BMU, The main features of the Renewable Energy Sources Act of 21 July 2004.

---

[18] This table presents the main data about the new feed-in tariffs from 2004, for the details, see BMU, The main features of the Renewable Energy Sources Act of 21 July 2004.

V. Interpretation of RES-E Policy Change in Germany

In Germany, RES-E was rapidly considered to be an interesting yet marginal energy option after the oil crisis in the 1970s. Nuclear power and coal remained the German energy policy priorities, until the Chernobyl accident in 1986, and the climate change issue at the beginning of the 1990s changed the status quo. As a consequence, a first series of policy instruments in favour of RES-E were implemented by the German government in 1988/89 (investments programs) and 1990 (feed-in tariff), with the aim to stimulate the development of an RES-E industry and market in Germany. At that time, no ambitious targets were adopted and RES-E was expected to play an increasing but still marginal role in the German energy supply. Nevertheless, the conjunction of the wind power investment program (100/250 MW program) with the feed-in tariff resulted in an unexpected explosion of wind turbine investments during the 1990s (see table 1) and the development of a leading domestic industry.

At the end of the 1990s, the anti-nuclear commitment of the red-green coalition parties (leading to a formal phasing-out agreement in 2001) and the strong political commitment of the German government on climate change, made RES-E appear to be a very promising solution to a German sustainable energy supply. Therefore, the policy objective of the German RES-E policy changed from a marginal energy option to a fundamental alternative in the future German energy market. The turning point of this policy change can be traced back to 2000, with the adoption of the central instrument of the RES-E policy thus far, the RESA. Indeed, the explicit purpose of the RESA was "to achieve a substantial increase in the percentage contribution made by renewable energy sources to power supply in order at least to double the share of renewable energy sources in total energy consumption by the year 2010"[19]. This meant increasing the proportion of RES-E from about 6% of total electricity consumption in 2000 to 12.5% by 2010, according to the European indicative target for Germany in the directive of 2001 (see table 5). The policy instrument used to attain these goals did not change with the adoption of RESA (see table 5), but the settings of the instrument were revised to improve and adapt it to the context of the competitive electricity market: long-term and legally guaranteed feed-in rates, increased rates but scaled down over time and differentiation by location for wind power.

---

[19] Renewable Energy Sources Act, 1st April 2000.

Table 5: RES-E policy change in Germany

	Before 2000	After 2000	Policy change
Policy objective	None	12.5% by 2010 and 20% by 2020	Yes
Policy instrument	Feed-in tariffs: – Target: electricity distributors – Incentive: price (feed-in tariff) – Resource: private budget (electricity bills)	Feed-in tariffs: – Target: electricity distributors – Incentive: price (feed-in tariff) – Resource: private budget (electricity bills)	No

Thereafter, the 2002 energy report from the ministry of the Economy[20] and the 2002 environmental report from the ministry of the Environment[21] investigated the longer-term prospects of RES-E in Germany and reached the conclusion that a target of 20% of RES-E in the total electricity consumption by 2020 would be a suitable scenario for the future sustainable energy supply in the context of the phasing-out of nuclear energy. Some even suggested that, by the middle of the century, RES-E would account for half of the electricity consumption (50% by 2050)[22], but this target remains a political vision (especially shared by the red-green coalition) and not a legally binding policy objective. The amendment of RESA adopted in 2004 then formalised those objectives by stating them explicitly (minimum 12.5% by 2010 and 20% by 2020). The existing feed-in tariff instrument was also improved and most of the compensation rates increased (especially for solar power).

In sum, German RES-E policy has been characterised by an unusual stability in terms of policy instrument (feed-in tariff) during the last decades. However, the policy objectives underlying this instrument changed extensively in the period 2000-2004, which required the settings of the policy instrument to adapt to the new ambitions and constraints of the policy.

---

[20] BMWI, Sustainable Energy Policy to Meet the Needs of the Future. Energy Report, June 2002.

[21] BMU, German Environmental Report 2002, March 2002.

[22] BMU, German Environmental Report 2002, March 2002. BMU, Report from the Federal Republic of Germany pursuant to Article 3 para.2 of EU Directive 2001/77/EC, March 2003.

VI. Explaining RES-E Policy Change in Germany

A. Policy Actors[23]

In Germany, RES-E policy was the competence of the ministry of the Economy until 2002, when it moved to the ministry of the Environment. The ministry of the Economy has traditionally been very close to the big electricity utilities and the coal industry, therefore it has always been very reluctant to significantly support the RES-E sector. The bills that led to the adoption of the 1990 feed-in law or the 2000 RESA originated in RES-E advocated by members of parliament (from different political parties) rather than in the German government[24]. In 2004, the switch of ministry modified the RES-E policy-making process, as the green minister of the Environment took the lead to revise and reinforce the 2000 RESA. So, unlike in most other policy areas, the RES-E policy has been traditionally dominated by the Bundestag, often in opposition with the government (especially the ministry of the Economy) (see table 6). It is only very recently with the ministry of the Environment being charged with the RES-E policy that the German government took over the lead in this policy area, but even then the Bundestag made some revisions to annul concessions to the ministry of the Economy (2004 RESA) (see table 6). Therefore, far from being a "technical" policy issue under the close supervision of the German ministries, RES-E policy has proved to be a much politicised issue in Germany, with lively debates among members of parliament and between parliament and government. The central political supporter of RES-E from the very beginning has been the Green Party, especially in the 2000 and 2004 policy change processes[25]. The Social Democrats joined a little later, but still faced internal divisions on this issue (e.g. the Social Democrat minister of the Economy in 2000 and 2004). However, they proved their strong commitment during the coalition negotiations in 2005[26].

---

[23] See Lauber and Mez, 2004 and Jacobsson and Lauber, 2005, for an in-depth analysis of the RES-E policy actors.
[24] Lauber and Mez, 2004 and Jacobsson and Lauber, 2005.
[25] Survey at 20 German RES-E policy actors with 12 valid answers, June 2005.
[26] Lauber and Mez, 2006.

Table 6: Dominant RES-E policy actor (H1/1)

	H1/1
Dominant policy actor?	Politicians: – Bundestag in 2000 – Government, Minister of Environment in 2004

In addition, first the members of parliament (especially in 2000) and then the ministry of the Environment (in 2004) have been able to rely on a growing coalition of actors from the sector to support their proposals in favour of RES-E[27]. The debate about the policy change was open to debate early in the process, which gave all the actors of the sector the opportunity to influence the policy change at its early stage (see table 7). In 2000, when the RESA was discussed, a large coalition of actors in favour of RES-E got together and successfully challenged the traditional energy policy actors (electricity companies and industrial associations). That coalition included not only the RES-E associations (e.g. wind power and solar associations, BEE) and environmental groups (BUND), but also industrial associations (e.g. VDMA), trade unions (e.g. IG Metall) and even the unexpected utility Preussen Elektra (see table 7)[28]. Among these actors, some (BBE, BWE, BSE, Eurosolar) were more influential than others, but they shared the same preference in favour of the RES-E policy change (see table 7). In 2004, the RES-E coalition grew with the addition of a trade union (ver.di) and a powerful industrial association (BVMW)[29].

So, we observe that the influence of the actors did not depend on the stage of the policy process in which they were consulted, which means that this hypothesis could not be tested (first proposition of H1/2 non-applicable). However, the most influential actors were in favour of the policy change, while the actors who were opposed had less influence, which explains why the policy ultimately changed (second proposition of H1/2 validated). The lack of influence of the electricity companies and the industrial associations was probably due to their internal division (e.g. PreussenElektra and VDMA with different preferences), and the lack of influence of the ministry of the Economy (their traditional politically) on this issue.

---

[27] Lauber and Mez, 2004 and Jacobsson and Lauber, 2005.
[28] Jacobsson and Lauber, 2005, p. 139.
[29] *Ibid.*, p. 142.

Table 7: Influence of the actors of the sector (H1/2)

Policy actors (H1/2)	Early / late	Preferences	Influence
Electricity companies			
PreussenElektra	early	+	-
Association of German utilities (VDEW)	early	-	-
Association of German transmission operators (VDN)	early	-	-
Verband der Verbundunternehmen und Regionalen Energieversorger (VRE)	early	-	0
Industrial associations			
Federation of German Industries (BDI)	early	-	-
Federation of the Engineering Industries (VDMA)	early	+	-
Confederation of small and medium sized enterprises (BVMW)	early	- (+)	0
RES-E associations			
Bundesverband Erneuerbare Energien (BEE)	early	+	+
Bundesverband WindEnergie (BWE)	early	+	+
Fördergesellschaft Windenergie (FGW)	early	+	-
Wirtschaftsverband Windkraftwerke (WVW)	early	+	-
Bundesverband Solarenergie (BSE)	early	+	+
Eurosolar	early	+	+
Deutscher Fachverband Solarenergie (DFS)	early	+	-
Deutsche Gesellschaft für Sonneenergie (DGS)	early	+	-
Solarförderverein (SFV)	early	+	-
Bundesverband BioEnergie (BBE)	early	+	-
Bundesverband Deutscher Wasserkraftwerke (BDW)	early	+	-
Environmental associations			
Bund für Umwelt und Naturschutz (BUND)	early	+	-
Trade Unions			
IG Metall	early	+	-
Ver.di	early	- (+)	0

Legend: Preference (+) favourable, (-) not favourable, (0) indifferent; Influence (+) strong, (-) weak, (0) no significant influence. The (+) for BVMW and Ver.di means that their preference changed to + in 2004 compared to 2000.

In conclusion, in 2000 and 2004, the German RES-E policy was dominated by very committed political actors (especially green and social democrat members of parliament in 2000 and 2004, and the green minister of the Environment in 2004), who fought hard for their proposals (H1/1 validated). In addition, they were supported by a very open and eclectic coalition of actors of the sector who had been involved in the policy change process from the very beginning. In this case, the individual influence of the actors did not result from the stage of the process at which they were consulted, but rather from their political connections with the dominant political parties (die Grüne or the SPD) (first proposition of H1/2 non applicable). Besides, the actors who had the strongest influence on the policy were those who were in favour of the policy change (members of the RES-E coalition) (second proposition of H1/2 validated).

B. Past Policy

Whether past policy has been a success or a failure usually explains why and how policies change. In the German case, RES-E policy was unanimously perceived as a success in 2000 at the time of the revision of the 1990 feed-in law[30]. The German wind turbine manufacturing industry developed intensely during the 1990s, as well as the installed capacity of wind power in Germany. Thus, the system of feed-in tariffs proved to be very effective for wind power, although it was less successful for the other RES-E sectors (solar, biomass, hydro, geothermal), because the compensation rates were not sufficient for those technologies[31]. In addition, the feed-in tariff instrument was generally perceived to be affordable for the final electricity consumers. Therefore, the past policy instrument (feed-in tariff) appeared to be unexpectedly effective during the 1990s (especially for wind power) and relatively affordable; this is why, in 2000, the red-green coalition decided to maintain it (see table 8), while adapting its setting[32].

So, the model of feed-in tariff did not change in 2000, but the settings had to be revised to better fit the new policy objectives and the new configuration of the electricity sector after the liberalisation of the market. For instance, in the light of their past poor performance, the compensation rates had to be increased significantly for solar, biomass, hydro and geothermal power, in order to stimulate their market development and reach the new policy targets. On the contrary, the rates for wind power were to be adapted to avoid overcompensations in very windy sites and to better follow the technological learning curves. In addition, the unequal distribution of financial burden among the utilities was not acceptable anymore in 2000, with the liberalised electricity market, where competition between electricity companies increased and electricity prices fluctuated according to market conditions[33].

---

[30] Survey at 20 German RES-E policy actors with 12 valid answers, June 2005.
[31] Explanatory memorandum of the RESA 2000.
[32] *Idem.*
[33] Explanatory memorandum of the RESA 2000.

Table 8: RES-E past policy success/failure (H2)

	Past policy success/failure (H2)	
Objective	No quantitative target	"Failure"
Instrument	Feed-in tariff: efficient and affordable	Success

Unlike the past policy instrument, the policy objective pursued by the conservative-liberal government was very modest (provided for a few hundred MW of additional generation to calm down dissenters) and no quantitative RES-E targets had been formulated[34]. So, this lack of an ambitious RES-E target was perceived to be "failure"[35] of the past RES-E policy in 2000 (see table 8). Then, the red-green coalition adopted new ambitious targets for the RES-E sector in 2000 and 2004 (12.5% of electricity consumption by 2010 and 20% by 2020). However, the adoption of new targets results less from the "failure" of the past policy than from an increased political priority to RES-E by the new political coalition.

In conclusion, the success and failure of the past RES-E policy explain better the non-change of 2000 (success of the policy instrument in the past) than the change of policy objective of 2000 ("failure" of the past policy objective). However, H2 is validated, as the policy instrument did not change (success) and the policy objective changed ("failure"), which was expected under H2.

C. *Europeanisation*

Germany has been a frontrunner in RES-E policy in Europe. The 1990 feed-in law was one of the first acts in favour of RES-E, and became a reference in Europe. German RES-E policy precedes the formulation of the European policy (see figure 2). Therefore, the interaction between the German and European RES-E policies has been "bottom-up" rather than "top-down" so far, if one considers the "lobbying" of the German government and RES-E associations defending the German feed-in tariff model at the European Court of Justice and the European Commission in 1999-2001.

---

[34] Jacobsson and Lauber, 2005, p. 134.
[35] The fact that there were no quantitative targets does not represent a failure of the policy *stricto sensu* (because there was no point of reference to assess the failure), but a strong loophole that was perceived as a "failure" of the policy.

Figure 2: The European policy and German RES-E policy change

EU policy					
Directive liberalisation + Green paper SER	White paper SER		Working paper E-SER	Draft directive E-SER	Directive E-SER + ECJ judgement
1996	1997	1998	1999	**2000**	2001
				RES-E policy change in Germany	

At the time of adopting the RESA 2000, the European RES-E policy was still in its infancy and very uncertain (see figure 2). The 1997 Commission's white paper on "Energy for the future: renewable sources of energy"[36] laid down an indicative target of doubling the share of RES in the gross European energy consumption by 2010 (12% of RES by 2010), which represented an increase of the RES-E share of the total electricity consumption to 22.1% by 2010. Besides, the white paper provides that every member state was expected to set its own RES-E target, in accordance with the European target. Thus, the European policy did not impose a target to the member states (flexibility for the member states), but still pressed the member states to set targets congruent with the European one. So, in 1997, the German government was encouraged by the European policy to double its share of RES-E by 2010, which was finally adopted (12.5% by 2010). Under these conditions, one cannot talk about an actual misfit of the past RES-E policy with the EU policy (because there was no mandatory EU target), but about a perceived misfit between the ambitious long-term RES-E targets adopted at the European level and the absence of a long-term RES-E target in Germany (misfit, see table 9).

In addition, the white paper of 1997 proposed different categories of instruments for the development of RES-E, but it remained very vague. Then, the 1999 working paper discussed the advantages and disadvantages of the different policy instruments, without pointing to a "best" ideal type, but still showing significant interest in competition-based models (such as the TGC) of feed-in tariff ones. A the same time, there were claims against the feed-in law introduced both at the DG Competition's level by the utilities and at the European Court of Justice in the

---

[36] European Commission, White Paper for a Community Strategy and Action Plan. Energy for the Future: Renewable Sources of Energy, COM(97)599.

PreussenElektra vs. Schleswag case (state aid argument). This reinforced the uncertainty about the position of the European policy on the RES-E policy instruments. Therefore, in this context the German feed-in tariff was perceived to be at odds with the European Commission's preference in terms of RES-E policy instrument, though the European policy was still relatively uncertain at that time (misfit, see table 9). Nevertheless, the Bundestag decided to keep this instrument in 2000, when it adopted the RESA, despite the perceived misfit with the EU policy (H3 invalidated); and in the end, its choice proved to be right, as the perceived misfit turned out to be groundless afterwards.

Besides, the European policy partly influenced the change of the settings of the feed-in tariff system in 2000, as RESA picked up some suggestions of the European DG Competition to avoid overcompensation: differentiated rates for wind, phasing out of rates over time (settings of the instrument).

Table 9: Europeanisation of RES-E policy (H3)

	Fit/Misfit (H3/1)	
Objective	No RES-E target	Misfit
Instrument	Feed-in tariff: uncertainty about state aid, not market-based	Misfit

Finally, the RESA 2000 explanatory memorandum makes it clear that the German government and parliament would defend their policy model against European state aid or competition criticism. During the elaboration of the European Directive on RES-E of 2001, the German government successfully "lobbied" to advocate the effectiveness and efficiency of the feed-in system, while many opponents gave priority to other instruments, such as quotas with tradable green certificates (Lauber, 2002; de Lovinfosse and Varone, 2004). In the end, the judgment of the ECJ in the case PreussenElektra vs. Schleswag (March 2001[37]) as well as the end of DG Competition's questions about the legality of the German feed-in system (May 2002) (Jacobsson and Lauber, 2005, p. 139) proved them right, as those decisions legitimised the German feed-in tariff model and reinforce its position in Germany and Europe. And the 2001 directive finally abandoned any attempt to harmonise RES-E policy instruments in Europe in the near future.

In conclusion, the German policy change of 2000 was determined primarily by domestic factors (e.g. domestic policy actors, past policy, new governing coalition) and the perceived misfit with the EU policy

---

[37] Judgment of the court of 13 March 2001, in Case C-379/98, PreussenElektra AG against Schleswag AG.

did not have a significant influence on the German policy change[38]. So, one cannot talk about a significant Europeanisation of the German RES-E policy (H3 minor explanatory factor). In addition, H3 proves to be invalidated, as the misfit of the policy instrument did not result in the change of the given policy instrument, as would have been expected. On the contrary, the German government confirmed their attachment to the German feed-in tariff model and successfully defended it at the European level, as the increased legitimisation of the German policy instrument at the European level demonstrates.

D. *Political and Institutional Context*

Germany is a consensus democracy usually governed by a two-party coalition. In 1998, the federal election brought to power an entirely new coalition: the red-green coalition (made up of the SPD and die Grüne). Political science theory assumes that major policy changes are more likely when a new governing coalition replaces the existing one. In the case of Germany and even more as far as the RES-E policy is concerned, this assumption has proved to be true (H4/1). In 1998, the new coalition not only included two new governing parties, but also two parties known to be very committed to the development of RES (especially the Green Party). This appears in their coalition agreements of 1998 and 2002, unlike the preceding governing coalition (the large ideological distance with the preceding coalition) (see table 10). In 2002, when it was strengthened by the re-election, the Green Party managed to retrieve the RES-E policy from the ministry of the Economy and transfer it to the ministry of the Environment, led by a Green minister. This shift demonstrates the increased power of the Green Party in the coalition from 2002 and its commitment to facilitate the adoption of RES-E policies in the future (see table 10).

As opposed to majority democracies, policy changes are expected to be more difficult when the political power is divided between different political parties, especially in case of ideologically opposed parties, because a consensus is required to make a decision. However, the red-green coalition (Social Democrats and Greens), in power in Germany from 1998 to 2005, shared a common priority and commitment on the issue of RES-E policy (small ideological distance concerning RES-E between the governing parties) (see table 10). In addition, the governing coalition proved to be very coherent, with the exception of the Minister of the Economy in 2000 and 2004, who acted more as a veto player than a supporter of the RES-E policy change, even if the change was supported by his party (the Social Democratic Party) and the coalition in

---

[38] Survey at 20 German RES-E policy actors with 12 valid answers, June 2005.

which he participated. In addition, on the RES-E policy issue, some conservative and liberal members of parliament (opposition members) have traditionally confronted their party leaders to promote the development of RES-E in Germany (e.g. in 1990, for the feed-in law), and in 2000 and 2004, they supported the RES-E policy change, even if they were in the opposition (Jacobsson and Lauber, 2005; Lauber and Mez, 2004). In addition, the red-green coalition successfully reached an agreement with the opposition parties in order to ensure a long-term security for RESA in 2004[39]. So, the RES-E policy change in 2000 and 2004 was not challenged by many partisan veto players (see table 10).

Table 10: Political and institutional context of the RES-E policy change (H4/1 and H4/2)

Governing coalition (H4/1)	Veto players (H4/2)
In 1998, entirely new governing coalition (Green Party and Social Democrat Party) with large ideological distance with preceding coalition. In 2002, Green Party at Environment and Energy Ministry (more power to Green Party)	– Few though powerful institutional veto players (bicameral parliament and federal state), but small ideological distance over RES-E (Bundestag and Bundesrat traditionally in favour of RES-E), and weak internal coherence (weaker party discipline in Bundestag and Bundesrat about RES-E) – Few partisan veto players (two party government) with small ideological distance between governing parties, strong internal coherence (except Minister of Economic Affairs) and support from opposition MPs

With its fragmented institutional structure, characterised by multi-level governments (federal and Länder) and a bicameral Parliament composed of two powerful Chambers (Bundestag and Bundesrat), the policy-making process in Germany is potentially threatened by significant veto players (see table 10). Indeed, when the Länder are in charge of implementing a federal law, barriers could easily be raised to slow down the process or weaken its content. At the federal level, the Länder are also able to oppose a federal bill in the Bundesrat, which often occurs when the Bundestag and Bundesrat majorities differ. However, such vetoes in the Bundesrat have rarely been raised against RES-E policy propositions (except in 2004, but with limited impact). On the contrary, the Länder have traditionally been very committed to RES-E (there is a small ideological distance between them concerning RES-E). In addition, party discipline among the party groups in the Bundestag is traditionally very strong, while it is less strong in the Bundestag. How-

---

[39] This strategy today proves to be successful because the new coalition (conservatives and social democrats) endorsed the previous coalition's RES-E targets and decided to pursue with the RESA.

ever, on the RES-E issue in the Bundestag, we observed that individual MPs have sometimes adopted positions opposing their party leadership (e.g. the adoption of the EEG in 1990), but this has been to the benefit of the RES-E policy so far, with opposition MPs joining the majority's policy change proposal (see table 10). Thus, veto players in the German RES-E policy are often less aggressive than expected (opposition parties, Bundesrat), but they do not always come from where you expect (e.g. minister of the Economy).

In conclusion, the new red-green coalition has been able to induce a significant shift in the German RES-E policy and reach a large consensus about it, thanks to its internal cohesion (except the ministry of the Economy), the large ideological distance with the preceding governing coalition, the large coalition that supported them on this issue (majority and opposition members of parliament), the division among the opposition and the enhanced position of the ministry of the Environment from 2002 (H4/1 validated). In addition, the absence of veto players to oppose the government policy facilitated the adoption of the policy change (H4/2 validated).

VII. Conclusion

With the arrival of a new red-green coalition in Germany in 1998, the RES-E policy gained more political priority. In 2000, the existing feed-in law was replaced by RESA, which maintained the existing feed-in tariff system (no instrument change), but adapted its settings to the new policy objectives and the requirements of the liberalised electricity market. Besides, in the 2000-2004 period, new RES-E policy objectives were formulated (minimum 12.5% by 2010 and 20% by 2020). In sum, the German RES-E policy changed significantly, with increased ambitions in terms of RES-E market penetration in the medium and long-term, though the policy instrument used to attain these new goals did not change (only the instrument settings were adapted).

Table 11 shows which hypotheses are validated or invalidated in the German case. Firstly, the two hypotheses about the policy actors are validated, though the second one only partially. The actors who dominated the policy making process were the politicians (Bundestag and Ministry of the Environment and Energy), which corresponds to what we expected in case of change of the policy objective (H1/1 validated). In addition, the policy instrument did not change, which also confirms what we assumed about the political actors being dominant in case of change of policy objective and the administrative actors being dominant in case of change of the policy instrument (H1/1 validated). Besides, the influence of the actors of the sector who supported the policy change is demonstrated in the German case (second proposition of H1/2 validated), as a large coalition of actors defended the change of policy objective and the status quo regarding the policy instrument. The link between the earliness of the participation of the actors of the sector in the policy-making process and the degree of influence could not be tested in this case, which means that the first proposition of H1/2 is non applicable. So, H1/2 is only partly validated in the German case. Secondly, the hypothesis about the success or failure of the past policy is validated, as the past policy instrument was successful and did not change and the past policy objective was not ambitious enough ("failure") and thus changed (H2 validated). Thirdly, the hypothesis about the Europeanisation of the domestic policies is invalidated in the German case (H3 invalidated). The change of policy objective confirms the hypothesis about the misfit between the European and domestic policies, resulting in domestic policy changes, but in reality the change of policy objective in Germany was determined by domestic factors rather than European ones. On the contrary, the perceived misfit of the German policy instrument with the European policy in 1999-2000 did not lead the German actors to distance themselves from this instrument, as might have been expected, but, on the contrary, to confirm their commitment to defend it at the European level (H3 invalidated). Finally, the two hypotheses about the influence of the political and institutional system on the policy change are validated, though H4/1 only regarding the change of policy objective. The change of the governing coalition, with two new internally coherent political parties (except the Minister of the Economy) and supporters of an increased political priority to RES-E, resulted in the change of the policy objective, as had been anticipated (H4/1 validated). However, the policy instrument did not change, as would have been expected, since the new coalition supported the existing instrument (H4/1 invalidated). Moreover, there were very few veto players able to oppose the policy change, and they did not do so because of the absence of ideological distance regarding the RES-E policy (large consensus in favour of RES-E in Germany), which confirms our assumption (H4/2 validated).

Finally, table 11 shows the relevance of the hypotheses to explain the RES-E policy change in Germany. We observe that all hypotheses proved to be relevant to explain either the change of policy objective or the non-change of policy instrument, some of them being more significant than others. The major explanations of the change of policy objective are: the policy actors who dominated the policy process (the Greens and Social Democrats in Bundestag in 2000 and the Green Minister of the Environment and Energy in 2004) with the support of a large coalition of actors of the sector (H1/1 and H1/2), as well as the change of governing coalition in 1998 with the Green Party (especially) and the Social Democrats (H4/1). In addition, the success of the feed-in tariff system during the 1990s (H2) and the support of a large coalition of actors in its favour (H1/1 and H1/2) explain why there was no change of policy instrument in 2000. Finally, the misfit with the European policy (H3) and the absence of veto players to oppose the new RES-E objective (H4/2) contribute to explaining the change of policy objective, but to a lesser extent.

Table 11: Validation and relevance of the hypotheses in Germany[40]

	H1		H2	H3	H4	
	H1/1	H1/2			H4/1	H4/2
Validation hypotheses						
Policy objective	V	"V"	V	V	V	V
Policy instrument	*V*	*"V"*	*V*	*I*	*I*	*n.a.*
Explanation policy change/no change						
Policy objective	++	++	0	+	++	+
Policy instrument	*+*	*+*	*++*	*0*	*0*	*0*

Legend: ++ (major explanation of policy change/no change), + (minor explanation), 0 (does not explain policy change/no change). V (hypothesis validated), I (hypothesis invalidated), "V" (hypothesis partly validated), n.a. (hypothesis not applicable).

In conclusion, the hypotheses of our theoretical framework are fully validated in the German case, as far as the change of policy objective is concerned. However, when investigating the validity of the hypotheses by default to explain the non-change of policy instrument, we observe that two hypotheses are not validated (H3 and H4/1) and one is non-applicable (H4/2). In addition, the hypotheses prove to be highly relevant to explaining either the change of policy objective (political actors with support from actors of the sector and the change of governing coalition) or the absence of change of policy instrument (success of the policy instrument in the past).

---

[40] The policy instrument lines are in italics because there was no change of policy instrument in Germany, so the validation and relevance of the hypotheses can only be assessed by default for the policy instrument.

CHAPTER 6

RES-E Policy in the Netherlands

I. Introduction

In the Netherlands, the energy competence is devoted to the ministry of the Economy, through the general directorate of energy. Nevertheless, several specific topics of the energy policy are also managed by other ministries, such as the ministry of the Environment for waste and climate policy; the ministry of Agriculture and Nature Management for biomass and wind energy; the ministry of Finance for ecotax; and the ministry of Science for research. The municipalities and the provinces also have specific competences related to land-use planning and can help to develop local renewable energy projects on their territories. They also play an important role as owners of electricity and gas distribution companies.

The definition of renewable/green energy has been subject to a highly sensitive political debate in the Netherlands. The two main issues are, on the one hand, the definition of the renewable energies that are taken into consideration regarding of the national targets (large definition, "renewable energy") and, on the other hand, the definition of the eligible renewable energy sources/technologies for the policy support mechanisms (more restricted definition, "green energy")[1]. In the Netherlands, the concepts of "sustainable energy", "renewable energy" and "green energy" have been used from the 1980s onwards, both by the Dutch government and by the energy companies. These concepts need to be clarified before tackling further analysis of the Dutch RES-E policy.

The terms "sustainable" and "renewable" energy were predominantly used by the Dutch government at the end of the 1980s and the beginning of the 1990s, while the concept "green energy" appeared only in the middle of the 1990s both in the public and private programs (see figure 1).

---

[1] See Dinica and Arentsen (2001) for further details.

Figure 1: Definitions of sustainable, renewable and green energy in the Netherlands

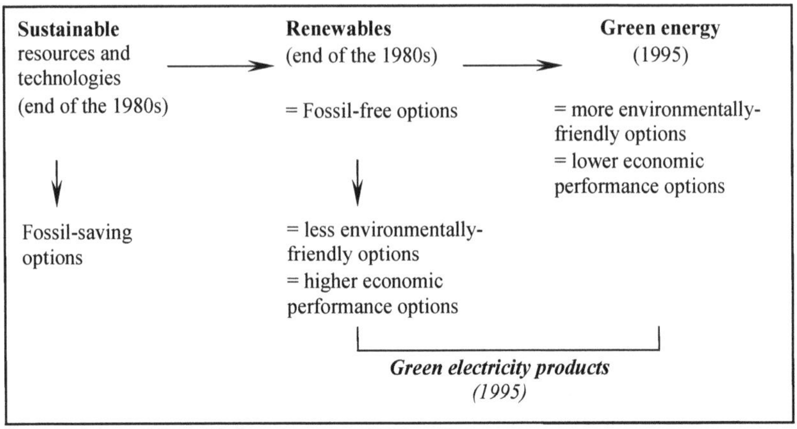

```
Sustainable              Renewables                 Green energy
resources and            (end of the 1980s)         (1995)
technologies      ─────▶                    ─────▶
(end of the 1980s)       = Fossil-free options      = more environmentally-
                                                      friendly options
    │                        │                      = lower economic
    ▼                        ▼                        performance options
Fossil-saving            = less environmentally-
options                    friendly options
                         = higher economic
                           performance options
                         └──────────────────────────────────────┘
                                  Green electricity products
                                          (1995)
```

Source: Adapted from Dinica, Arentsen, 2001, p. 8.

In the government's definitions, the concept of "sustainable" energy is particularly relevant for CO_2 emission reduction targets, since it covers both fossil-fuel saving energy options and fossil-fuel free options (among which the renewable energies)[2]. The term "renewable" energy only refers to the energy produced from renewable energy sources such as wind, hydro, sun, biomass, waste etc.[3]. However, within the renewable energy category, one must distinguish less environmentally-friendly technologies (such as industrial heat pumps, large hydro, waste)[4] from more environmentally-friendly ones (e.g. solar, wind, geothermal energy). The term "green" energy is usually associated with the latter renewable energy sources, clean and environmentally-friendly renewable energy sources or technologies. In addition, the concept "green" energy often refers to technologies that have lower economic performance than conventional ones and therefore need policy support. The concept "green energy" was first used by the Dutch government in 1995 with the so-called "Green funds" policy, which stipulated that the returns from green funds invested by private individuals in Dutch financial institutions are exempt from income tax. The eligible investment projects for these green funds covered all renewable energy projects,

---

[2] Ministry of the Environment, 1989, National Environmental Policy Program.
[3] Ministry of the Economy, 1996, Third White Paper on Energy Policy.
[4] For instance, in the 1997 "Action Program Sustainable Energy in Progress" of the ministry of the Economy, the industrial heat pumps and the synthetic fraction of waste were excluded from the list of renewable energies because of their fossil-fuel background.

excluding waste and any other biomass than wood or energy crops. In 1998[5], only the small hydropower plants with a capacity of less than 15 MW were eligible for the tradable green certificates and the energy tax exemption. Then, in 2002[6], even the small hydro plants were excluded from this financial support. Finally, in 2003, the concept of "sustainable electricity" was reintroduced with the MEP law that encompasses an extended definition of eligible RES-E including hydropower and mixed-biomass[7]. In conclusion, the definition of RES-E changes over time according to political or economic strategies.

The Dutch energy companies often use the same definitions as the government, but not always. For instance, the Green Labels trading system refers to the same definition of the renewable energies as the definition of the 1998 Electricity Act. However, the term "green electricity" has been introduced by the distribution companies from 1995 onwards, as a generic name for special products that differ from conventional electricity in terms of their lower environmental impact (see figure 1). Nevertheless, these "green electricity" products could originate from all kinds of RES-E installations (including large hydro and waste plants) as well as from nuclear power installations. Thus, the energy companies' definition of "green electricity" differs from the government's definition on several aspects.

II. Overview of the Electricity Sector

In the Netherlands, power generation from natural gas is particularly high (see figure 2). This is due to the large domestic reserve of gas that was discovered in Groningen in 1959. The so-called *Groningen field* is considered to be one of the biggest in the world, and is very cheap to exploit. Indeed, gas largely contributes to the Dutch energy independence and security of supply.

---

[5] Electricity Law of 2 July 1998, Staatsblad 1998, No.427.
[6] Decision of the Ministry of the Economy, September 2001.
[7] Law of 5 June 2003 modifying the Electricity Law of 1998 in order to promote the production of environmental-quality electricity (MEP law), Staatsblad 2003, No.235.

Figure 2: Electricity generation by fuel, 1972-2002

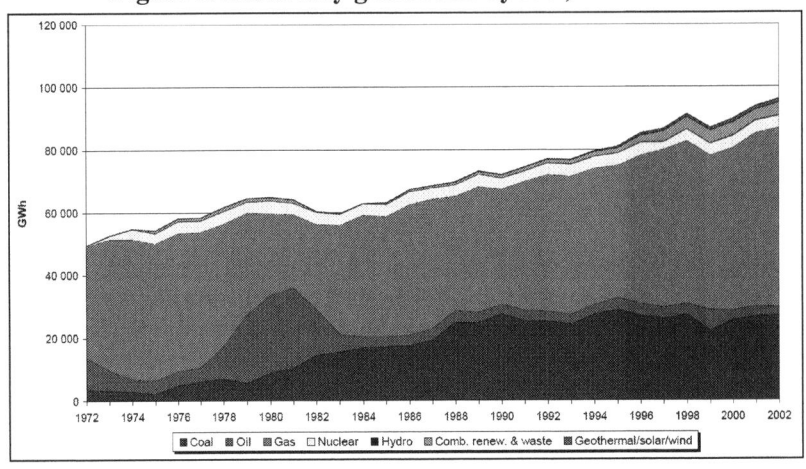

Source: IEA.

The second most important fuel used in electricity production is coal. Even if it has been progressively supplanted by gas, it still remains a very important energy source in the Netherlands. The last Dutch coal pit closed in 1974, but massive imports still provide coal to the Dutch electricity plants. More recently, due to environmental constraints, many coal plants started to co-combust biomass.

Despite domestic reserves, oil has never been used as an important fuel in electricity generation, except around 1980, when there was a sudden in-crease of the use of oil due to the very low prices during the counter oil crisis.

Nuclear power has never been very popular in the Netherlands. The Dutch electricity industry did not receive many incentives to develop a nuclear industry because they had their own reserves of coal and natural gas and no direct access to uranium (as opposed to Belgium, with its former colony of the Congo). Still, two nuclear reactors were erected: one in Dodewaard (a research reactor that was closed in 1994) and one in Borssele (a commercial reactor in operation since 1973). After 1986, the Chernobyl accident reinforced resistance in public opinion and diminished the legitimacy of extending nuclear power. In 1994, the Dutch parliament made a political compromise with the electricity companies to shut down the existing nuclear power plants by the end of 2003. Nevertheless, this decision was subsequently strongly opposed by the nuclear power plant employees who protested and took legal action against the closure of the Borssele plant. In addition, big changes were under way in the Dutch electricity industry at the end of the 1990s, with the liberalisation of the electricity market. Finally, after several legal procedures and demonstrations, the Dutch government decided in November 2002 that it would not pursue the early closure of the Borssele nuclear power plant. The Dutch nuclear phase-out policy was then officially reversed. Nevertheless, it is highly unlikely that there will be any new nuclear power plants in the Netherlands in the future.

Finally, electricity generation from renewable energy sources remains marginal compared to the other fuels. However, it started increasing in the 1990s and is expected to gain more importance by 2015. RES-E originates mainly from biomass, waste incineration and wind. Due to a very low potential, hydro represents a minor source of electricity production in the Netherlands.

Prior to market reform in 1998, the Dutch power market was dominated by four generating companies, namely EPON, EZH, EPZ and UNA, forming the so-called "centralised" market (as opposed to the "decentralised" market, represented by the distribution companies and the industry). They co-operated through an organisation called SEP (Samenwerkende Elektriciteits-Produciebedrijven) which was a joint stock company owned by its members. SEP's most important role was to own and operate the high-voltage transmission grid and it enjoyed a statutory monopoly on imports until 1998. SEP stopped co-ordinating the centralised market after the establishment of the transmission system operator, TenneT, in October 1998 (but SEP still owned TenneT until 2001), and was finally dissolved in 2001, when TenneT was nationalised. Today, three of the four centralised generators have been taken over by foreign energy companies: UNA by Reliant Energy (the US), EZH by E.On (Germany) and EPON by Electrabel (Belgium). EPZ was bought by two Dutch distribution companies (Essent and Delta). In addition, since 1998, the share of the "decentralised" generators, namely the distribution companies (e.g. Nuon, Essent/Delta) and industrial autoproducers, has increased (to more than 50% of total electricity generation); while the share of the "centralised" generators decreased (to less than 50% of the total electricity generation). The increase of the Dutch electricity demand today is essentially covered by the decentralised generators and increased electricity imports.

Before market reform, there were 23 electricity distribution companies in the Netherlands. All of them also distributed natural gas and some of them district heat. Following the introduction of the 1998 Electricity Act, legal unbundling has been implemented at the distribution level as the distribution companies have divided their network and supply activities into different companies. In effect, there are at present 20 distribution companies (in charge of the distribution grid) and three main electricity suppliers (Nuon, Essent and Eneco) in the Netherlands. In reality, most of the electricity suppliers and distribution companies remain vertically integrated (no ownership unbundling) and they are still public-owned (provincial or municipal stakeholders).

In the Netherlands, the share of RES-E generation has strongly increased during the last decade, from a production of about 700 GWh in 1990 to more than 3,500 GWh in 2002 (see figure 3). Biomass and waste incineration take top position in renewable energy options in the Netherlands, followed by wind power. All other options are basically still in the experimentation stage or in initial market introduction.

The co-combustion in coal plants is one of the reasons to expect a massive growth in biomass use. In 2000, the Dutch government and the electricity producers signed a covenant on the reduction of CO_2 emissions from the power plants. One of the actions to reach the reduction targets is the co-combustion of biomass in the coal power plants. This explains why an important increase in biomass incineration from 2000 can be observed. In addition, an interesting side-effect of the co-combustion for the electricity producers is that part of the electricity can now be sold as green electricity.

Waste incineration has always been an important source of electricity production in the Netherlands. Nevertheless, it has lately been supplanted by biomass and wind energy.

Figure 3: RES-E generation by fuel, 1990-2002

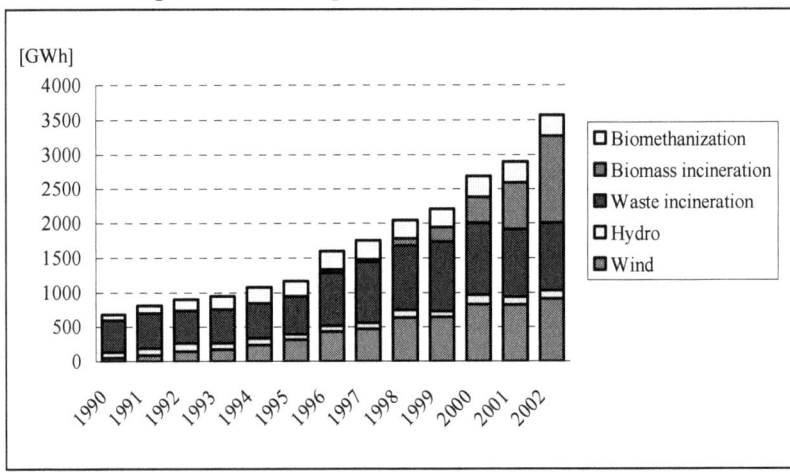

Source: Novem/Ecofys/KEMA.

Wind power has been increasing significantly as from 1990, but still remains rather limited compared to other European countries, like Germany or Spain. However, the important offshore potential of the Netherlands might result in a further increase of wind power in the future. So far, onshore projects have faced considerable opposition from local inhabitants, associations or municipalities (the so-called NIMBY phenomenon). Therefore, the future of onshore wind is very problematic in the Netherlands.

The potential for hydropower is limited in the Netherlands. It is therefore expected to remain marginal in the future. On the contrary, PV power is expected to play a more important role in the medium or long term.

III. Overview of the RES-E Policy History in the Netherlands (1974-1998)

A. 1974-1990

Before 1974, the Netherlands had no specific energy policy. Energy production and supply was regarded as part of industry policy and the main concern was to guarantee a continuous flow of cheap energy for all. The energy sources were mainly coal and oil. From 1960, with the discovery of huge gas reserves, natural gas became the dominant options of the Dutch energy policy. At the same time, nuclear energy also emerged as an important energy option, but with less intensity than in other European countries (e.g. France and Belgium), due to the long-term availability of natural gas. In addition, an energy policy debate started in the Netherlands in the middle of the 1970s, focusing mainly on the role of nuclear energy (Kamp, 2002). Protests against nuclear power postponed the development of the nuclear program that was foreseen by the government and the energy companies. Finally, the discussions on nuclear energy ended in 1986, after the Chernobyl accident. It became politically impossible for the Dutch government to continue with its nuclear program. Therefore, no new nuclear power stations have been built since then and discussions about the phasing out of the only commercial plant in operation (Borssele) are still ongoing.

In 1974, the new Dutch policy goals were presented in the first White Paper on Energy[8]. The emphasis shifted from low-cost energy production to an efficient, social and environmental perspective. The key Dutch policies became energy savings/conservation and diversification. In addition to energy savings, the aim of the Dutch government at that time was to develop alternative energy sources that would have to meet not only the energy demand, but also the new economic, social and environmental criteria, e.g. energy security and independent energy supply. The government encouraged the development of new energy

---

[8] Ministry of the Economy, 1974, First White Paper on Energy.

sources and technologies, but also aimed at ensuring that the Dutch industry and research community would make a large contribution. This marks a shift from a relatively passive energy policy to an active one (Kamp, 2002). Two new councils were created to implement the white paper: LSEO (national commission for energy research) and NEOM (Dutch energy development company). LSEO was responsible for the development of alternative energy technologies in the Netherlands. It selected the research projects for funding, on the basis of their scientific and technical merit. NEOM aimed at bridging the gap between research and the market. It was responsible for transferring the knowledge between research institutes and from these institutes to the commercial sector.

In the first White Paper, renewable energies were regarded to be new solutions for fuel diversification. Several national R&D programs, each concentrating on different energy sources (especially wind turbines, solar technologies, biomass and waste technologies), were set up. The first programs were the NOW program in 1976 (National Research Program on Wind Energy)[9] and the NOZ program in 1978 (National Research Program on Solar Energy). From the beginning, the Dutch government targeted especially the development of wind turbines technology, since wind energy was perceived to have most potential in the Netherlands. Within the Dutch wind turbine programs, two sub-programs were distinguished: the large-scale programs, with an emphasis on large wind turbines with a power capacity of several mega-watts; and the small-scale programs aiming at developing smaller wind turbines with a power capacity of several dozens of kilowatts[10].

In the middle of the 1980s, the development of renewable energy continued to be supported mainly by R&D programs, but those programs were to become more market-oriented. For instance, the new wind energy program, implemented in 1986 – IPW (Integral Programme for Wind Energy, 1986-1990) –, aimed not only at R&D, but also at the market introduction of the wind turbines, especially in the form of demonstrations or pilot projects. The investment subsidies ranged from 35% to 40% of total project costs and all types of projects developers were eligible.

Then, with the growth of the number of wind turbines, the sitting problems began at the end of the 1980s (Kamp, 2002). Several groups of inhabitants organised at the local level against the new projects for wind turbines. Furthermore, the municipalities were reluctant to grant the

---

[9] The NOW programme is divided into NOW 1 (1976-1981) and NOW 2 (1981-1986).
[10] For a detailed analysis of the wind energy programs in the Netherlands, see KAMP, 2002.

building permits that were required to erect wind turbines. Therefore, in 1991, the ministry of the Economy and the ministry of housing, land-use planning and the environment tried to solve the problem by signing an agreement with the Dutch coastal provinces, which are the windiest ones. These provinces committed themselves to the installation of a minimum wind turbine capacity by the year 2000, with a total capacity of 1,000 MW (which was the goal of the previous IPW program). But this agreement did not work out in the end because the institutions that were to grant the building permits and change the local development plans, namely the local councils, did not participate in the negotiations.

B. 1990-1995

At the beginning of the 1990s, the objective of the Dutch government changed. So far it had mainly aimed at developing and installing new technologies (especially wind turbines), but from then on, it developed a more production-oriented RES-E policy. To do so, it used a variety of instruments: feed-in tariffs, investment subsidies and fiscal incentives.

1. Feed-In Tariffs

In 1989, the first Dutch Electricity Act was adopted[11]. This law aimed at reforming the electricity sector in order to prepare for a forthcoming liberalisation of the market, but it also had two important consequences for RES-E generation. Firstly, the law allowed the distribution companies to invest in small-scale production capacity of up to 25 MW (previously only electricity production companies were allowed to produce electricity). This meant that the distribution companies could commission RES-E generation plants on their own. This measure led to a progressive challenge of the monopolistic position of the big centralised generating companies and strengthened the position of the distribution companies acting as decentralised RES-E generators (see details below). Secondly, the law put an obligation on the distribution companies to purchase the RES-E generated by decentralised producers located in their area at a minimum price (feed-in tariffs)[12]. This obligation created a protected market for RES-E with unlimited demand as well as guaranteed minimum prices. The tariffs were fixed in a Standard Arrangements for Re-deliveries (called SAR) and were supposed to be annually reviewed[13]. According to V. Dinica and M. Arentsen (2001), the SAR prices varied between 0.032-0.038€/kWh in the period 1991-

---

[11] 1989 Electricity Act, Staatsblad 1989, No.535.
[12] 1989 Electricity Act, A.42.
[13] For an in-depth analysis of the SAR system, see Dinica, Arentsen, 2001.

1997 (Dinica, Arentsen, 2001, p. 48). The SAR was negotiated between the distribution companies and the decentralised RES-E generators themselves and formalised in an agreement that became directly enforceable without the approval of the ministry of the Economy. Therefore, *de facto* this feed-in tariff system was mostly implemented and managed by the distribution companies with no significant public control on the system, which appears odd, considering that it was based on a law. In addition, the SAR prices were too low to expect a significant development of new decentralised RES-E investments. This is due to the fact that the decentralised RES-E producers had to negotiate their tariffs with their electricity distribution companies, and, since they were not well organised, they did not have powerful arguments.

2. Investment Subsidies and Fiscal Incentives

In addition to the feed-in tariffs, the Dutch government adopted several investment subsidies programs to boost the investments in RES-E technologies during the 1990s (mostly in wind, solar and biomass). In 1991, the multi-annual program "Application of Wind Energy in the Netherlands" (TWIN) replaced the existing IPW program. There was more emphasis on the market introduction of wind turbines[14]. The subsidy covered up to 35% of the investment costs. However, due to sitting problems (the so-called NIMBY phenomenon) and low prices for wind power on the market, these subsidies were not sufficient to reach the government[15]. As far as biomass was concerned, the "Energy Production from Waste and Biomass Program" (EWAB) was created in 1993, but it started to focus on market introduction with consistent investment subsidies only in 1998. Solar PV has long been considered a strategic but longer-term alternative, which is why the first significant solar investment subsidy program only started in 1997, with the so-called NOZ-PV program (1997-2000).

Finally, the Dutch government adopted some fiscal incentives for RES-E investments. From 1995 onwards, returns from investments in "green funds" (namely investments in RES-E technologies) managed by Dutch financial institutions (e.g. Triodos was the first bank to propose such funds) were exempted from income tax. In addition, from 1996, the Dutch government granted a favourable fiscal regime to the private companies investing in RES-E technologies under the "Accelerated Depreciation Scheme on Environmental Investments" (VAMIL). This fiscal incentive allowed the RES-E investors to increase the cash flows

---

[14] See Dinica, 2003, for further details on the development of wind turbines in the Netherlands.

[15] Only 396 MW of wind turbines had been installed by 1998, while the Dutch target by 2000 was 1,000 MW. Kamp, 2002.

of their projects in the first years of investments, by transferring the taxes to the following years of operation. Nevertheless, the non-profit companies (for instance the distribution companies) were not eligible. In addition, a specific program called "Tenders for Industrial Energy Saving Program" (TIEB) was introduced in 1996. It subsidised investment costs for demonstration (up to 30%) or market introduction projects (up to 20%) commissioned by private industrial companies. A similar program, "The Energy Investments Regulation for the Non-profit Sector and Special Sectors" (EINP), was adopted in 1997 for the non-profit companies (for instance the distribution companies) with a subsidy ranging between 14.4% and 20% of project costs.

3. Distribution Companies MAP Program

In parallel with the government's RES-E policy and the related instruments we have just described, the Dutch distribution companies developed their own voluntary RES-E program from 1991, as part of their larger Environmental Action Plan (so-called MAP program[16]) aimed at reducing their CO_2 emissions[17]. In order to finance this program, they negotiated an agreement with the Dutch government that allowed them to increase their tariffs. Therefore, they were able to plan new RES-E investments with public funding, but without actual public control. Just like in the feed-in tariff system, the distribution companies appeared to be the central actors when it came to implementing RES-E policy in the Netherlands during the 1990s. According to the agreement signed with the Dutch government, the distribution companies charged a levy of between 0.5% and 2.5% on the consumers' tariffs (the MAP levy). The consumer electricity tariff had to be annually approved by the minister of the Economy for each distribution company. The MAP levy was then charged by every distribution company on top of this tariff (with an exemption for the consumers buying RES-E). The level of the levy was revised annually the minister of the Economy[18]. The funds collected through that levy were managed locally by the distribution companies, with very little transparency. Most of the RES-E investments that benefited from the MAP investment or production subsidies[19] were managed by the distribution companies themselves, sometimes in joint ventures with industrial developers.

---

[16] It was then revised in 1994 (MAP II) and 1997 (MAP 2000) and ended in 2000.

[17] See Slingerland, 1997, for more details on the MAP program.

[18] In 2000, the MAP levy amounted to between 0.08-0.18€/kWh. Source: Dinica, Arentsen, 2001, p. 21.

[19] Dinica and Arentsen (2001) estimate that the MAP production subsidy varied between 1.36-3.63€ct/kWh in the period 1991-1997.

C. 1996-1998

In 1995, the Dutch government established a long-term target for renewable energies: 10% of final energy consumption was expected to come from renewable energies by 2020 (and 5% by 2010)[20]. This target was presented as a real challenge at the time when it was adopted. In 1997, a new action program for sustainable energy was adopted to reach the target of the third white paper[21]. The strategy of the Dutch government moved from a production-driven strategy, mostly managed by the distribution companies (feed-in tariffs, production subsidies), to a more demand-driven strategy based on a new fiscal incentive (the REB tax system).

1. REB Tax and Production Subsidies

In 1996, as part of the "greening" of the tax system in the Netherlands, the Dutch government decided to levy an energy tax on final energy consumption: the so-called Regulatory Energy Tax (REB)[22]. This energy tax was charged on all kinds of energy consumption (electricity, natural gas, heat oil) at the distribution level. Until 1998 it had to be paid, irrespective of the types of energy used for the generation of electricity. But, as from 1 January 1998, RES-E was exempted from the REB tax.

The aim of this energy tax was twofold. First, like all ecotaxes, it aimed at stimulating energy saving attitudes by the Dutch consumers. Second, it aimed at increasing the production of RES-E because of its environmental benefits. Therefore, the money collected from the REB tax by the electricity distributors (the REB fund) was used to subsidise the generation of RES-E in the form of production subsidies. These subsidies were paid directly by the distribution companies to the renewable electricity generators located in their area. The level of the production subsidy was 0,134€/kWh until 1998 and then increased annually from 1999 to 0.02€/kWh in 2002 (see table 1). In 1996, the renewable energy sources eligible for the REB production subsidy were: wind energy, solar energy, small-scale hydropower (<15MW), woody-biomass and biogas. Both private generators and distribution companies were eligible, including imported RES-E. Indeed, since it was a fiscal instrument, it could not be restricted to domestic electricity production according to European competition law. Therefore, both domestic and

---

[20] Ministry of the Economy, Third Energy White Paper, December 1995.
[21] Ministry of the Economy, Action Program for Sustainable Energy, March 1997.
[22] Law of 13 December 1995 amending the law on environmental taxes concerning the implementation of a regulatory energy tax, Staatsblad 1995, No.662.

imported RES-E were eligible for the production subsidy (and from 1998 onwards, for the REB tax exemption).

Table 1: Regulatory energy tax for electricity per user category and production subsidy (€ct/kWh)

Electricity consumption (kWh)	1996	1997	1998	1999	2000	2001	2002
Ecotax							
0-800	0	0	0	0	0	5.83	6.01
800-10,000	1.34	1.34	1.34	2.25	3.72	5.83	6.01
10,000-50,000	1.34	1.34	1.34	1.47	1.61	1.94	2.00
50,000-10 million	0	0	0	0.10	0.22	0.59	0.61
> 10 million	0	0	0	0	0	0	0
Production subsidy	1.34	1.34	1.34	1.47	1.61	1.94	2.00

Source: VAN SAMBEEK, 2002, p. 6.

From the 1st of January 1998, the RES-E generators twice benefited from this fiscal system. Firstly, they received the production subsidy from the REB fund, just as before. But in addition, they were exempted from paying the REB tax. Therefore, their profit amounted 0.0268€/kWh in 1998 (=1.34+1.34) and it increased to 0.0801€/kWh in 2002 (=2+6.01) for the smaller generating companies (see table 1).

2. Distribution Companies Green Electricity Products

In the same period, the distribution companies also intended to stimulate the demand for RES-E. To do so, they introduced new electricity products on the market with the "green electricity" label. The consumers who were willing to pay a surcharge to receive green electricity were offered green products by their distribution company[23]. The first voluntary green electricity scheme was introduced in 1995 by the distribution company PNEM. Then, several other distribution companies developed similar schemes under different trade names. However, there was no correlation between the trading names and the RES-E products. In some cases, the same trading name was used for different RES-E products and vice versa. In addition, not all "green products" were actually that "green", since they sometimes included nuclear power[24]. Consumers' subscription to the green electricity products remained limited until the end of 1999, when the WWF and the ministry of the Economy organised a national awareness campaign about green energy.

---

[23] For an in-depth analysis of this scheme, see Dinica, Arentsen, 2001.
[24] The main green electricity products offered by the Dutch distributors in 2000 included: wind and solar; wind-only; biomass-only; wind and biomass; wind and hydropower; wind and nuclear; wind, biomass and hydropower. Source Dinica, Arentsen, 2001.

In 2001, about 250,000 green consumers (mainly domestic consumers) were registered for a total traded volume of around 850 GWh (Dinica, Arentsen, 2001, p. 28). Nevertheless, V. Dinica and M. Arentsen observe that this system could not be considered a real "drive" for new investments, but rather a "survival" option for the existing installations (Dinica and Arentsen, 2001, p. 61).

The extra-prices paid by the consumers were negotiated with the ministry of the Economy (because the market was not liberalised yet) and then collected into a fund managed by the distribution companies[25]. The funds were used either as investment/production subsidies for the distribution companies themselves, or as production subsidies for private RES-E generators who did not receive MAP subsidies. From 1998, with the exemption of the REB tax for RES-E, it became possible for some distribution companies to offer green electricity products at the same price as conventional electricity, which also explains the huge increase in the consumption of green electricity products by 2001.

IV. Liberalisation of the Electricity Market in the Netherlands (1998-2002)

A. *The Liberalisation of the Electricity Market*

The policy-making process in the Netherlands is characterised by a consensus-building tradition, called the politics of "accommodation" by Arend Lijphart (Lijphart, 1968). The electricity sector is no exception to this tradition and electricity reforms have always been carried out in mutual agreement between the sector and the government. In addition, electricity policy implementation has traditionally been carried out by the electricity companies themselves with very limited control from the government (e.g. feed-in tariffs, REB tax).

Until the late eighties, the Dutch electricity industry consisted of many relatively small vertically integrated utilities (production, transport, distribution) with regional monopolies. These utilities were owned by municipal and provincial authorities (Arentsen, Kunneke, Moll, 1997, p. 176). Between 1985 and 1989, a first reform of the electricity sector was initiated by the government to achieve a more efficient supply of electricity and low electricity tariffs. In 1985, the electricity companies created their own advisory board (the Brandsma Commission) to organise the reform of the sector. The Dutch government agreed to let the sector manage the reform on a self-regulatory model, but still

---

[25] According to V. Dinica and M. Arentsen, the extra-prices amounted between 2.3 and 4.5€ct/kWh.

prepared a law to be used as a 'stick' in case the self-regulating forces of the sector turned out to be unreliable or ineffective (*Ibid.*, p. 177). Several agreements were made between the sector and the government to reduce the monopoly power of the integrated electricity companies. Two major changes resulted from the reform: the separation between the production and the distribution activities, and the progressive concentration of production from 15 to four generating companies (Koster, 1998).

In 1989, the first Electricity Act[26] was adopted to improve the efficiency and reducing the monopoly of the power industry. First, it imposed a legal separation between production and distribution activities (unbundling), to confront the generating companies with countervailing powers, namely the 29 regional distribution companies. Second, the Act strengthened the existing co-operation among the centralised electricity generators through SEP (the Dutch Electricity Generation Board) and made this co-operation a legal obligation to reduce costs through economies of scale. It also established a strict criterion for new entry into the central generation market: a minimum of 2,500 MW capacity. In addition, every new centralised power station above 25 MW required SEP, as well as ministerial and parliamentary approvals. In contrast, the 1989 Electricity Act allowed the distribution companies to invest in decentralised electricity generation facilities (<=25 MW) without permission from SEP. The decentralised electricity production market had the advantage of a lighter regulatory burden. Therefore, between 1989 and 1996, the production monopoly of SEP was progressively challenged by the decentralised electricity production activities managed by the distribution companies, partly as joint ventures with private industrial firms (mostly in CHP plants). Indeed, the decentralised electricity production grew very fast (25% of electricity production in 1996 (Koster, 1998, p. 664)) and created an overcapacity in central production. Therefore, "the distributors became the dominant actors in the Dutch electricity sector, a dominance which nobody had forecast, but which was embedded in the Electricity Act." (Arentsen, Kunneke, Moll, 1997, p. 182) In fact, distribution companies gained an influence that was not reflected in the institutional structure, while the producers were struggling to maintain their market shares and influence thanks to the monopoly that they were granted by the 1989 Electricity Act. Finally, the Electricity Act opened the first segment of the supply market to competition. The large electricity consumers (>=20 GWh/year) and the distribution companies became eligible to get their supplies from the company of their choice. In practice, very little use was made of this provision because the distributors' prices did not differ much (due to national basic tariffs set by SEP) and the transmission prices (managed by SEP) were not transparent.

---

[26] 1989 Electricity Act, Staatsblad 1989, No.535.

In 1995, the ministry of the Economy published a new White Paper on energy policy (the third one[27]), which revealed the contours of further reform of the electricity sector to be adopted in the Netherlands to anticipate the European directive on the liberalisation of the electricity market (EU directive 96/92/EC[28]). This paper led to the adoption of a new Electricity Act in 1998[29], which transposed the European directive into Dutch law. Four main changes followed.

First, electricity production was no longer considered to be a natural monopoly and became entirely liberalised. Therefore, SEP lost its monopoly on domestic centralised electricity generation and imports.

Second, in the 1998 Electricity Act, a further liberalisation of the market was decided with the immediate eligibility of the large consumers (capacity per connexion > 2MW). Afterwards, the market opening was decided to be implemented in two steps. First, all consumers with a transmission value of more than 3*80 amperes and a capacity per connexion of less than 2 MW were expected to be eligible from January 2002. Then, small consumers (transmission value of less than 3*80 amperes) were expected to remain dependent on their distribution companies until January 2007. However, in 2000, the ministry of the Economy decided to accelerate the process and January 2007 was replaced by January 2004[30]. Finally, in June 2003, the ministry of the Economy realised that the electricity market would not be ready by January 2004 for full opening and that too much uncertainty remained concerning the protection of consumers. Therefore, the deadline was postponed until July 2004 (see table 2).

Table 2: Deadlines of electricity market opening in the Netherlands

Date	Eligibility	Market shares*
1989 Electricity Act	> 20 GWh/year for at least 4,000 hours a year	
1998 Electricity Act	Consumers with a capacity per connection of at least 2 MW	± 35%
01/07/01	Green electricity for all consumers	
01/01/02	> 3*80 amperes transmission value and < 2 MW capacity per connection	± 55%
01/07/04	All consumers with a maximal transmission value of less than 3*80 amperes	100%

* Koster, 1998, p. 666.

---

[27] Ministry of the Economy, Third Energy White Paper, December 1995.
[28] Directive 96/92/EC of the European Parliament and of the Council of 19 December 1996 concerning common rules for the internal market in electricity.
[29] Electricity Law of 2 July 1998, Staatsblad 1998, No.427.
[30] Law of 22 June 2000 on the rules concerning the transport and supply of gas, Staatsblad 2000, No.305.

In addition, according to the proposal of the 1999 Energy Report[31], the ministry of the Economy decided to accelerate the liberalisation of the green electricity market in 2001 with the aim of giving an extra impulse to the development of this market in the Netherlands[32]. In contrast with conventional electricity, the supply of green electricity became eligible from the 1st of July 2001. In fact, the process occurred in two steps. Between the 1st of July 2001 and the 1st of January 2002, the Dutch electricity consumers could already choose a new green electricity supplier, but the prevailing distribution company remained responsible for the physical electricity supply to consumers. On the 1st of January 2002, the Dutch green electricity market became completely liberalised. The green electricity suppliers became responsible for both the physical supply of electricity to the consumers and the purchase of the necessary green certificates. In absolute numbers, at the end of 2000, about 200,000 households were using green power, but after the full opening of the green power market in July 2001, it increased to about 1,000,000 green electricity consumers in 2002 (see figure 4).

Figure 4: Number of renewable electricity customers (1996-2002)

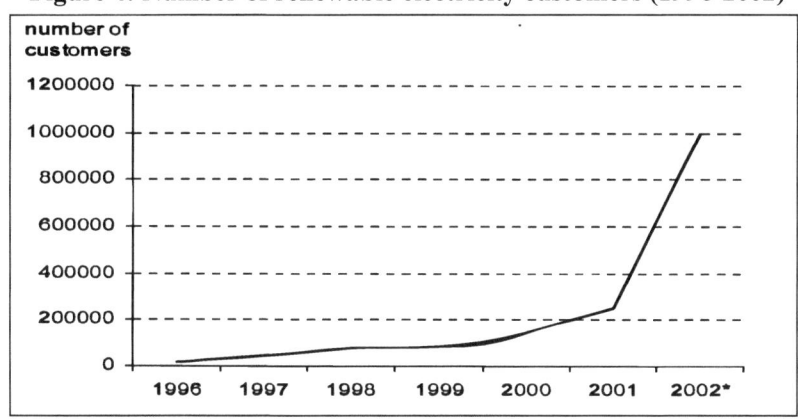

* Mid 2002.

Source: Van Sambeek, 2002, p. 9.

Third, the distribution and transport activities remained monopolistic, since they were considered to be a natural monopoly. Therefore, a strict legal and accounting separation was enforced between the grid activities (national transmission grid and regional distribution grids) and

---

[31] Ministry of the Economy, 1999 Energy Report, 15 November 1999, p. 27.
[32] Letter from the Ministry of the Economy to the Parliament concerning the liberalisation of the green market, 8 March 2001.

the competitive activities (electricity production and supply). It was still possible for production and supply companies to own transmission or distribution grids, but a strict unbundling was to be organised between the different activities in order to prevent cross-subsidisation and to ensure non-discriminatory third party access to the grids. For instance, TenneT, the new transmission system operator, was established as a private company owned by the centralised generating companies through SEP. The new distribution grid operators also remained in the hands of the former distribution companies.

Fourth, a new department was created within the existing Netherlands Competition Authority (NMa), in order to regulate the electricity market and ensure third-party access to the electricity networks: DTe (Directie Toezicht Energie).

B. RES-E policy in the Liberalised Market (1998-2002)

In 1998, the Dutch RES-E policy was revised in the light of new contextual adaptation pressures: the liberalisation of the electricity sector, the Kyoto Protocol (Dutch target of 6% greenhouse gas emissions reduction) and the European white paper on renewable energies[33]. No consensus was reached at that time to significantly change the existing instruments in favour of RES-E, but they were modified in order to comply with the new requirements of the liberalised market and increase financial support to RES-E. The strategy of the Dutch government remained oriented toward the stimulation of the demand for RES-E, assuming that an increasing demand would generate the development of new production capacities. Unfortunately, we will see that this strategy resulted in a massive import of cheap RES-E from other countries, to the detriment of the domestic production.

1. Revised Feed-In Tariffs

In the 1998 Electricity Act, the Dutch government presented three different approaches to promote RES-E in the liberalised market. The first recommended pursuing the feed-in tariff system initiated in the 1989 Electricity Act, but with substantial adaptations to the way it was implemented and the level of the tariffs. The second approach proposed charging the electricity generators and suppliers (>10 GWh/year) to promote the development of RES-E, with a report to be submitted to the ministry of the Economy every two years. It would be financed by an additional levy to be charged by the transmission grid operator (TenneT). The third idea was to impose a quota of RES-E on the final

---

[33] COM(97)599 final of 26/11/1997, Energy for the Future: Renewable Sources of Energy, White Paper for a Community Strategy and Action Plan.

electricity consumers. It would be implemented by means of tradable green certificates to be redeemed by the electricity suppliers on behalf of their customers, and the suppliers who would fail to comply with the quota would have to pay a penalty. The penalties would feed a fund to be redirected to the RES-E generators with unsold green certificates. Although this was a new powerful scheme, there was no political sympathy at that time for mandatory instruments (e.g. quota). Therefore, only the first approach was eventually implemented, while the others remained in the act as security provisions for the future in case the existing instruments were insufficient to reach the European indicative target (9% of Dutch electricity consumption from RES-E by 2010[34]).

In 1998, the feed-in tariff system adopted in 1989 and managed by the distribution companies was transformed into a governmentally guaranteed feed-in tariff in favour of Dutch small RES-E self-generators (not distribution companies). The purpose was to keep the feed-in tariff system. The distribution companies were obliged to purchase RES-E from RES-E generators in their area, but to restrain the scope of beneficiaries and to regulate the tariffs better. Only small- and medium-scale self-generators were eligible: hydro, biomass and CHP self-generators with a capacity of no more than 2 MW (but only until December 2001), as well as wind and solar self-generators with a capacity of no more than 8 MW (600 kW from 2002). The tariffs were to be approved by the ministry of the Economy, but they were not significantly higher than previously (around 0.036€/kWh (Dinica, Arentsen, 2001, p. 49)). The outcome of this system was very limited, especially because the tariffs were too low and conditioned in the medium term.

2. Exemption of REB Tax for RES-E

In the whole range of instruments adopted by the Dutch government to promote RES-E at the end of the 1990s, the most effective one was the REB tax system: production subsidy to RES-E from REB fund (starting in 1996) and exemption from the REB tax for RES-E (starting in 1998). It became especially powerful from 1998, when RES-E started to be exempted from the REB tax[35]. This exemption provided a competitive advantage to RES-E producers on the electricity market, which allowed RES-E to be offered at the same price as conventional electricity by the distribution companies. In addition, the level of the REB tax

---

[34] Directive 2001/77/EC of the European Parliament and of the Council of 27 September 2001 on the promotion of electricity from renewable energy sources in the internal electricity market.

[35] Law of 18 December 1997 on the modification of several tax laws, Staatsblad 1997, No.732. Art.VI f, h. Decision of 26 May 1998 on the deadlines for the enforcement of several articles of the law on environmental taxes, Staatsblad 1998, No.323.

and production subsidy increased from 1999 to 2002 (see table 1), which boosted the support to the RES-E on both sides: demand and production. In 2002, the financial advantage cumulated to 0.08€/kWh for the smaller electricity consumers (0.06€/kWh of REB tax exemption + 0.02€/kWh of production subsidy). However, this instrument also entailed severe adverse effects because of the fact that imports of RES-E were also eligible for ecotax exemption and production subsidy. Massive imports of cheap RES-E hampered the development of domestic production capacity and caused important tax loss for the Dutch government through the subsidies of foreign RES-E producers. Therefore, despite its success in terms of increased demand for RES-E, the REB tax system has been progressively phased out from 2002 and replaced by a more effective support scheme in terms of domestic RES-E production development. For instance, from 2002, hydropower has no longer been eligible for the ecotax exemption because of massive imports of cheap hydropower from Nordic countries. Thereafter, with the adoption of the MEP law in 2003, the REB production subsidy was replaced by the MEP production subsidies (limited to domestic producers), while the REB ecotax exemption decreased progressively until its abolition in January 2005: 0.029€/kWh in 2003, 0.03€/kWh from January to June 2004 and 0.015€/kWh from July to December 2004.

3. Voluntary Green Certificates Trading with REB Tax Exemption

By July 2001, the green electricity market was fully liberalised, which means that the electricity consumers who decided to switch to RES-E could choose their electricity suppliers freely. At the same time, the ministry of the Economy introduced a green certificate trading system to regulate the RES-E market and make it more transparent[36]. The green certificates were issued, tracked and redeemed by TenneT, the national transmission grid operator, which managed the electronic database where all green certificates were registered. The size of the green certificates varied, but in any case, it was a multiple of 1 MWh. The green certificates had a term of validity of maximum one year. The distribution companies were responsible for the technical monitoring of the system: certification of the RES-E installations, installation and follow-up of the metering device, transmission of the data to TenneT.

As from July 2001, the REB tax exemption has been coupled with the green certificates in the way that the green certificates became the proof for RES-E to be delivered in exchange for the exemption. Indeed,

---

[36] Decision of the Ministry of the Economy of 7 May 2001, Regulation on green certificates, Staatsblad 2001, No.87.

the holders of the green certificates were exempted from paying the REB tax by surrendering the green certificates to the tax authority. Therefore, it did not modify the REB tax system but introduced more transparency in the way the system was implemented. The REB tax exemption was the demand-driver for green certificates by electricity consumers or suppliers. Unlike in other countries (e.g. Belgium, UK), the electricity consumers or suppliers were free to participate in the green certificate trading system (voluntary system), but the interaction with the REB tax scheme encouraged them to do so.

In order to receive green certificates, the Dutch RES-E generators had to resign from the other existing support schemes organised by the distribution companies (MAP subsidies or green label trading system) in order to avoid cross-subsidies[37]. In addition, the self-generators received green certificates only for the surplus electricity they fed into the distribution grid, and not for the total amount of RES-E they produced. The eligible RES-E for green certificates were: hydro (<15MW capacity)[38], wind power, solar power, and biomass installations without addition of fossil fuel products[39]. The green certificates were first limited to domestic RES-E. As a consequence, from July 2001 until December 2001, only the Dutch RES-E producers could receive green certificates, and were therefore able to claim the REB tax exemption. However, it was clear from the beginning that such a discriminatory arrangement could not be upheld in view of the European legislation (internal electricity market). In addition, the Dutch government realised that without imports, the European indicative target of 9% of RES-E electricity consumption by 2010 would never be met. Therefore, as of January 2002, imported RES-E has been entitled to Dutch green certificates, but only under the following conditions: reciprocity, no double subsidisation, metering data and plant verification, import capacity and compliance with the Dutch definition of eligible renewable energy sources[40]. Between July 2001 and October 2003, the amount of green certificates issued for domestic RES-E production represented around 5,136 GWh, compared to a production of more than 15,518 GWh coming from imports[41]. This means that less than one third of the Dutch green certificates were granted to domestic RES-E production.

---

[37] This process of resignation was called "unMAPing" in Dinica, Arentsen, 2001, pp. 29-30.
[38] Hydropower did not receive green certificates anymore from January 2002.
[39] It corresponded to the definition of RES-E in the 1998 Electricity Act.
[40] Decision of the Ministry of the Economy of 24 October 2001, Amendment of the regulation on green certificates, Staatsblad 2001, No.208.
[41] www.energie.nl, March 2004.

In January 2004, the Dutch green certificates were replaced by Dutch guarantees of origin (GO), to comply with the European directive of 2001[42]. The purpose of the GO is to harmonise the RES-E labels in Europe and allow a transparent international RES-E trade[43]. In the Netherlands, like in other European countries (Belgium, the UK, Sweden, Italy, etc.), where a system of green certificates was already in operation, the implementation of the GO was easier, but required to adapt the prevailing regulations and clarify the relations between the green certificates and the GO[44]. For instance, CertiQ (a subsidiary company of TenneT) became responsible for the issuance and tracking of both the green certificates and the GO (so-called guarantee management body)[45]. In the GO system, only the Dutch RES-E production could receive Dutch GO. However, the prevailing system of green certificates remained in operation for the imported RES-E. In addition, GO coming from other European countries could apply for the Dutch REB tax exemption under certain conditions[46].

4. The Distribution Companies Green Label Trading System and RECS

In 1997, the distribution companies concluded a new voluntary agreement with the government as part of their Environmental Action Plan (MAP). According to this agreement, the distribution companies adopted a voluntary target of 1,700 GWh of RES-E to be supplied to consumers by 2000 (only domestic RES-E eligible). For each distribution company, a specific target was then calculated on the basis of its volumes sales. But, the RES-E potential differs substantially between the regions; some of them were experiencing difficulties in reaching their targets while others were experiencing overproduction of RES-E. So a flexible instrument was introduced: the green label trading system. The trading system represented a solution to certify the eligible RES-E

---

[42] Law of 20 November 2003 amending the 1998 electricity law concerning the promotion of the sustainable electricity (implementation of the European directive 2001/77/EC), Staatsblad 2003, No.493.

[43] Directive 2001/77/EC of the European Parliament and of the Council of 27 September 2001 on the promotion of electricity from renewable energy sources in the internal electricity market.

[44] Decision of the Ministry of the Economy of 8 December 2003 on the guarantees of origin for sustainable electricity, Staatsblad 2003, No.242.

[45] Decision of the Ministry of the Economy of 8 December 2003 on the guarantee management body, Staatsblad 2003, No.242.

[46] Similar conditions to the rules for imported green electricity in the green certificate system: reciprocity, no double subsidisation, metering data and plant verification, import capacity and compliance with the Dutch definition of eligible renewable energy sources.

production across the country and to organise flexible compliance with the regional targets, the green labels being traded between the distribution companies. This system was a pioneer at that time and it subsequently inspired several different green certificate trading systems (e.g. the Dutch green certificate system in 2001, RECs in 1998).

This system of green labels was initiated and designed by EnergieNed, the federation of the Dutch energy companies. The KEMA Group was the central register authority in charge of maintaining a centralised electronic database with all the movements of green labels (issuance, trade, redemption). The size of a green label was set at 10,000 kWh. Therefore, a demand for 170,000 green labels by 2000 emerged in the Netherlands, according to the overall target adopted by the distribution companies. The green label trading system became operational at the beginning of 1998, but most trades occurred at the end of the MAP period only, in 2000. Any domestic RES-E generator that qualified for receiving an exemption for the ecotax and the production subsidy from the REB funds was also entitled to receive a green label for each 10,000 kWh produced: wind energy, solar PV, small-scale hydropower (<15 MW), biogas, wood-biomass. Thus only domestic RES-E was eligible in the green label trading system, unlike the REB tax system that accepted RES-E imports. RES-E generators had to choose between either receiving the MAP subsidy or switch to the green label trading system. The distribution companies had to finance the purchase of green labels with the MAP levy funds only. It was not allowed to use the revenues from the green electricity product schemes (to which consumers subscribed voluntarily) to pay for these labels. So, cross-subsidies among the distribution companies schemes were prevented (MAP subsidies, green electricity products and green labels), but cross-subsidies between the distribution companies schemes and the public RES-E instruments were not (REB tax exemption and production subsidy, feed-in tariffs).

In practice, this program has been somewhat unbalanced in favour of the distribution companies (versus private or industrial RES-E investors) and it has been more supportive for co-generation and wind technologies than for the other technologies. Nevertheless, it has played a crucial role in the rise and development of RES-E capacity in the Dutch energy system. The total amount of green electricity that the distribution companies have finally recorded at the end of the program (end of 2000) was of 1,500 GWh, which is only 200 GWh short of the initial target of 1,700 GWh set in 1997 (Dinica, Arentsen, 2003, p. 612). The main reasons invoked by EnergieNed were the delays in the licensing procedures and the social opposition to onshore wind turbines projects (the so-called NIMBY phenomenon).

Parallel to the green label trading system, the Dutch distribution companies were the instigators of a European green certificate trading system (RECS), which was set up in 1998, along with other European electricity companies and government agencies. The RECS system is based on a set of common rules (basic commitments), but every country has a specific Issuing Body entitled to grant RECS certificates according to domestic eligibility criteria. RECS works with an encompassing definition of the renewable energy sources, considering all energy sources, with the exception of fossil and nuclear fuels. Therefore, there are often conflicts between the government green certificates (more restrictive) and the RECS certificates (more liberal). The size of the RECS certificates is standard, 1 MWh, and remains valid for five years. RECS certificates are issued only to RES-E delivered to the grid, and the generators who have already benefited from another certificate mechanism are not eligible so as to avoid cross-subsidies. The RECS market works on a purely voluntary basis driven by commercial strategies (environmental image of the company, green electricity products), since no financial compensation applies in so far as the RECS certificates are not recognised by the governments.

V. RES-E Policy Change in the Netherlands (2002-2003)

At the beginning of 2002, the large amount of RES-E imports alerted the Dutch government. For instance, in 2002, imports represent 75% of the RES-E consumption, about 10 TWh (Van Damme, 2005, p. 18). Indeed, the green electricity market has been a success since 2001, with the early eligibility of the green electricity consumers on the Dutch electricity market. The demand for green electricity increased and the domestic production was not sufficient to cover it. In addition, the system of REB tax exemption and production subsidy, for which the imports were eligible, represented a very interesting incentive for foreign RES-E producers to export to the Netherlands. So RES-E imports increased without any significant additional investments in the Netherlands or abroad. In turn, increased imports led to the congestion of the Dutch transmission system and thus to congestion rents that ultimately increased the electricity price to the electricity consumers. In fact, in the absence of an integrated European policy, the Dutch policy was the most attractive for the foreign RES-E producers. As a result, the Dutch government was badly hit by its liberal and favourable RES-E policy: tax loss, transmission congestion, and subsidisation of the foreign RES-E generators on the one hand, ineffectiveness for the development of domestic RES-E production on the other. This situation required a rapid policy change to limit adverse impact on the Dutch economy. A report from the Dutch Audit Office denounces that, in the period 1999-2004, the cost of the REB tax system for the Dutch tax payer amounted 1,340 million €, with 556 million € only for 2002[47]. In addition, the paradox of the REB system was that it was so favourable for imports that it was perceived as politically and financially unsustainable in the long run by the Dutch investors, who anticipated a policy change and therefore postponed their investments.

The Energy report of February 2002[48] presented the adverse effects of the existing RES-E policy. Both the public and private Dutch electricity actors acknowledged those failures and the need to revise the 1998 Electricity Act and the REB fiscal policy. The process of policy change started in the summer of 2002, under the initiative and supervision of the ministry of economic affairs. It was at first kept secret, to avoid interference with the market and rumours that would have a detrimental effect on the investments (no consultation of the actors of the sector)[49]. The ministry of the Economy commissioned ECN to make a comparison of the existing RES-E policy instruments in other European countries. It resulted from this report that the German system was the most appropriate to the Dutch situation, since it specifically targets the production side of the market, which was the weakness of the prevailing Dutch policy. Therefore, the German policy was taken as a model for the design of the new Dutch policy instrument. A first proposal was submitted by the ministry of the Economy to Parliament in November 2002. The final version of the MEP law was adopted in June 2003 and it entered into force in July 2003[50]. The ministry wanted the discussions in parliament to be swift, in order to adopt the law as fast as possible, mainly due to the emergence of debilitating tax loss. But in the end, the discussions in parliament and with the actors of the sector in the autumn 2002 proved to be more controversial than expected by the ministry. A lot of contestation emerged from the actors of the sector and the members of parliament about the policy-making procedure adopted by the ministry (too rapid and not enough consultation with the stakeholders). Indeed, the actors of the sector were consulted only at the same time as the bill was submitted to parliament and therefore most of the details of the system had already been fixed. However, most of the critics focused not so much on the policy instrument itself, but on the method of calculation to

---

[47] Report from the Dutch Audit Office, Groene stroom, June 2004.

[48] Ministry of the Economy, 2002, Energy Report 2002: Investing in Energy, Choices for the Future, 26 February 2002.

[49] Interview with Willem van der Heul (Ministry of the Economy), Den Haag, 17 November 2003.

[50] Law of 5 June 2003 modifying the Electricity Law of 1998 in order to promote the production of environmental-quality electricity (MEP law), Staatsblad 2003, No.235.

set the level of the subsidies (not adapted to the reality of the market) and the proposed levels of subsidies (too low).

Unlike the REB tax system, the MEP system is based on a production-driven strategy. Under the MEP law, Dutch RES-E generators can apply for production subsidies guaranteed for 10 years (MEP subsidy)[51]. The level of the MEP subsidy depends on the production output (kWh generated) and the kind of RES-E technology and sources (see table 3). The MEP subsidies are reviewed annually, taking into account the decline in costs resulting from learning curves. The MEP subsidies are adopted by the ministry of the Economy on the basis of reports from ECN and KEMA. In the annual reviews, tariffs are fixed for the next two to three years. It is a challenging task to set an appropriate level of subsidy for the forthcoming years, owing to the uncertainty of future costs, as these costs will be driven largely by global, not national markets. The government prefers annual reviews over a pre-set reduction scheme to be able to use the latest market parameters to define the appropriate tariff level. The current system of guaranteeing a fixed feed-in tariff for investors for ten years should increase their confidence as they need predictability. In addition, the increase of MEP tariffs are conditioned by the progressive lowering (0.029€/kWh in 2003 and 0.015€/kWh from July 2004) and suppression (as from January 2005) of the REB exemption. This means that, until January 2005, the MEP tariffs were growing according to the decrease of the REB exemption in order to compensate the revenue losses for the Dutch RES-E producers.

---

[51] Note that MEP subsidies are also granted to CHP installations, but this specific case will not be discussed here.

Table 3: The MEP tariffs (2003-2005)

MEP subsidies (€ct/kWh)	2003[52]	January 2004[53]	July 2004[54]	2005[55]
Landfill gas	0	0	0.6	2.1
Sewage gas	0	0	0	0
Mixed biomass streams[56]	2.9	2.9	2.9	2.9
Pure biomass	4.8	4.0	5.5	7
Pure biomass-animal flour	4.8	0	0.6	2.1
Stand-alone bio-energy (<50 MW)	6.8	6.7	8.2	9.7
Wind-on-shore	4.9	4.8	6.3	7.8
Wind-off-shore	6.8	6.7	8.2	9.7
Hydropower, wave and tidal	6.8	6.7	8.2	9.7
Solar PV	6.8	6.7	8.2	9.7

Source: www.enerq.nl/mep/ and Burgers *et al.*, 2004.

An extensive definition of RES-E was chosen (sustainable electricity), which required a modification of the green certificates regulation[57]. For instance, mixed biomass installations and hydropower plants are eligible for the MEP subsidies (see table 3), while they have not been eligible for the green certificates so far. Furthermore, the producer must be connected to the Dutch electricity grid and the installation must be maintained and exploited for at least ten years. Finally, only installations put into use after 1 January 1996 are eligible.

TenneT, the Dutch transmission system operator, which has been appointed to implement the MEP system, created a subsidiary company (EnerQ) to take care of it. EnerQ allocated the MEP subsidies to the Dutch RES-E producers on the basis of the green certificates (until the end of 2003) or guarantees of origin (from January 2004) that they received from CertiQ. So, unlike the German feed-in tariffs, the MEP

---

[52] Decision of the Ministry of the Economy of 6 June 2003 on the MEP subsidies for 2003, Staatsblad 2003, No.121.

[53] Decision of the Ministry of the Economy of 13 December 2003 on the MEP subsidies for 2004, Staatsblad 2003, No.249.

[54] Interim increase of MEP subsidies due to a simultaneous decrease of the REB exemption.

[55] Decision of the Ministry of the Economy of 13 December 2003 on the MEP subsidies for 2005, Staatsblad 2003, No.249.

[56] The MEP subsidy is only guaranteed in proportion to the degree of biologically degradable material, to the installations with a minimum total energy efficiency of 26%. The level of the subsidy does not increase since this category of biomass was not eligible for ecotax exemption.

[57] Decision of the Ministry of the Economy of 6 June 2003 on the amendment of the regulation on green certificates, Staatsblad 2003, No.116.

system does not imply the purchase of the electricity by EnerQ, but only the purchase of the RES-E certificates. The MEP subsidies are financed through an annual MEP levy on the connections to the electricity grid in the Netherlands (34€ per connection in 2003, 39€ in 2004[58], and 40€ by 2006[59]). The MEP levy is collected directly by the distribution companies and then passed on to EnerQ. The financial burden of the MEP levy on the final electricity consumers is compensated by an equivalent reduction in REB tax, so there should not be any impact on the electricity prices. The Dutch energy regulator (DTe) supervises compliance with the provisions of the MEP law.

But the MEP law has already shown some limits. In fact, the minister of the Economy (Laurens Jan Brinkhorst) informed Parliament in November 2004[60] and then again in March 2005[61] that the MEP subsidies had become so popular that the budget was not sufficient to cover all the requests for subsidies and that a serious budget deficit was to be expected. For instance, the budget deficit for 2003-2005 amounted to about 174 million €[62]. At the same time, the share of RES-E in the electricity consumption increased very quickly from 3.3% at the end of 2003 to 4.5% in 2004[63], which means that the target of 9% for 2010 seems reachable. Therefore, this policy proves to be very effective in terms of market development, but represents a serious threat for the public budget. Therefore, in May 2005[64] the minister of the Economy informed Parliament that the MEP law needed to be amended in order to ensure an improved budget control over the MEP scheme and avoid further budget deficits in the future. However, the MEP will remain in force to facilitate the target of 9% RES-E in electricity consumption by 2010 (indicative target of the 2001 directive). Two solutions were proposed by the Ministry[65]: in the short term, the Minister suggests to refuse the MEP-subsidy to any new large biomass or wind offshore projects that would harm the budget, and in the long term, an annual MEP-subsidy ceiling will be adopted to control the budget. Important adjustments to the MEP law are thus expected during the next year, but it will not happen without strong opposition from the electricity compa-

---

[58] Decision of the Ministry of the Economy of 13 December 2003 on the connection tariffs for 2004, Staatsblad 2003, No.249.
[59] IEA, 2004.
[60] Letter from the Ministry of the Economy to Parliament, 19 November 2004.
[61] Letter from the Ministry of the Economy to Parliament, 11 March 2005.
[62] Letter from the Ministry of the Economy to Parliament, 10 May 2005.
[63] Letter from the Ministry of the Economy to Parliament, 11 March 2005.
[64] Letter from the Ministry of the Economy to Parliament, 10 May 2005.
[65] *Idem.*

nies and the RES-E investors who have been relying on the MEP law to ensure long-term security.

VI. Interpretation of RES-E Policy Change in the Netherlands

The Netherlands has been a pioneer in RES-E policy from the 1990s onwards and it proved to be very successful in creating a large demand for RES-E among the electricity consumers by the beginning of the 2000s. This resulted from the joint efforts of the distribution companies (green products, green label trading system) and the Dutch government (REB tax system) in favour of RES-E during the 1990s. However, in 2002 (turning point), the Dutch government realised that severe perverse effects resulted from the REB tax exemption scheme: massive imports of cheap RES-E and tax loss for the Dutch government; which called for the revision of the RES-E policy instrument in the Netherlands. In 2003 (turning point), the Dutch government decided to phase-out the REB tax system and replace it with a production-oriented instrument based on production subsidies (MEP subsidies) in favour of the domestic RES-E producers. Under the new instrument (MEP subsidies), the incentive to RES-E is still a favourable price (€/kWh), though the target group of the policy is no longer the electricity consumers (demand), but the domestic RES-E generators (supply) (see table 4). In addition, financial support to RES-E is no longer funded by the state budget (public budget), but by a levy on the connection to the grid (consumer bills) (see table 4). However, if the total amount of money collected through the levy is smaller than the total amount of MEP subsidies awarded to the RES-E generators by EnerQ, the budget of EnerQ shows a deficit, and since EnerQ is a public company, the financial burden will ultimately fall on the public budget. This is what happened during the first years of operation of the MEP system (2003-2005).

Table 4: RES-E policy change in the Netherlands

	Before 2003	After 2003	Policy change
Policy objective	9% by 2010	9% by 2010	No
Policy instrument	– Target group: electricity consumers (demand) – Incentive: price (REB tax exemption) – Resource: public budget	– Target group: domestic RES-E generators (supply) Incentive: price (MEP subsidy) – Resource: private beget (grid connection levy)	Yes

The policy objective of the Dutch government remained unchanged with the adoption of the MEP law in 2003, only the instruments used to attain them were modified (see table 4). Since the 1995 energy white paper[66], the policy objective of the Dutch governments has been to reach 10% of renewable energies in the final energy consumption by 2020 (which represented a target of 17% RES-E production by 2020). All the subsequent energy papers (1999[67], 2002[68] and 2005[69] energy reports) or action programs (1997[70], 1999[71], 2004[72] action programs) reaffirmed this target and discussed the best strategies to reach it. Then, from 2001[73], a specific target for RES-E of 9% of the electricity consumption by 2010 has been pursued, with an intermediate target of 6% by 2005[74]. Therefore, the overall objective of the Dutch RES-E policy has not been discussed during the 2002-2003 policy change process, but only the means to attain them have been modified in the light of the past policy outcomes.

VII. Explaining RES-E Policy Change in the Netherlands

A. Policy Actors

During the 1990s, the Dutch RES-E policy was characterised by a consensual partnership between the government and the electricity companies. Especially EnergieNed, the federation of the Dutch energy companies, was a powerful actor in the electricity sector and the central actor in the RES-E policy implementation. On the one hand, the distribution companies were in charge of implementing most of the RES-E policy schemes (feed-in tariffs, collect of the REB tax and payment of production subsidies). On the other hand, they adopted their own programs in favour of RES-E (MAP production subsidies, green electricity products, green labels) based on agreements signed with the government (1991, 1994, 1997). However, with the liberalisation of the electricity market and increasing competition between the distribution companies,

---

[66] Ministry of the Economy, Third Energy White Paper, December 1995.
[67] Ministry of the Economy, Energy Report 1999, November 1999.
[68] Ministry of the Economy, Energy Report 2002: Invest in Energy Choices for the Future, February 2002.
[69] Ministry of the Economy, Energy Report 2005: Now for later, July 2005.
[70] Ministry of the Economy, Action Program for Sustainable Energy, March 1997.
[71] Ministry of the Economic, Sustainable Energy in Progress, July 1999. Ministry of the Environment, Climate Policy Implementation Plan, June 1999.
[72] Ministry of the Economy, Biomass Action Plan 2004: working together on bio-energy, 2004.
[73] Directive 2001/77/EC of the European Parliament and of the Council of 27 September 2001 on the promotion of electricity from renewable energy sources in the internal electricity market.
[74] Report from the Dutch Audit Office, Groene stroom, June 2004.

it became more and more difficult to achieve consensual positions within EnergieNed[75] and the MAP program ended in 2000.

The liberalisation of the electricity market urged the Dutch government to exert closer control over the sector. So, new public actors were institutionalised (e.g. DTe, CertiQ, EnerQ) and existing actors were nationalised (e.g. Novem, TenneT) in order to endorse new public responsibilities, especially in the implementation of the RES-E policy. The relations between the electricity companies and the Dutch government thus changed with the latter gaining more authority over the former as a result of the 1998 electricity act.

In the summer 2002, the RES-E policy change process was initiated and dominated by the ministry of the Economy and kept secret in its early stage. Only ECN and KEMA, private consultants on energy issues, participated in the early process, since they elaborated the reports that served as a basis for the formulation of the new policy instrument. Because of their central position in the electricity market and their expertise in the sector, TenneT (the transmission grid operator) and EnergieNed were consulted by the ministry at the end of the summer of 2002 about the policy project, but only informally[76]. The main electricity companies (Nuon, Essent, Eneco) were also informed on the new project, but without having the opportunity to actively participate in the formulation of the policy at this early stage. So the formulation of the new policy instrument was dominated by the ministry of the Economy and private consultants and kept secret from the actors of the sector until quite a late stage[77]. This secrecy was justified by the need to avoid interference with the market and rumors that would harm the investments. This is a typical case of policy change process dominated by civil servants and confined within a closed community of policy experts (see table 5).

The proposal of the ministry was submitted to parliament in November 2002. The discussions in parliament proved to be more controversial than expected by the ministry who at first expected the law to be adopted by the end of the year. A lot of contestation emerged from the members of parliament about the policy-making procedure adopted by the ministry (too rapid and lack of consultation with the actors of the sector), especially because they were surprised to hear the criticism of

---

[75] Interview with Peter Niermejier (RECS), Utrecht, November 2003.
[76] Interview with Willem van der Heul (Ministry of Economic Affairs), Den Haag, 17 November 2003. Interview with Theo de Lange (ECN), Amsterdam, 21 November 2003.
[77] Interview with Willem van der Heul (Ministry of Economic Affairs), Den Haag, 17 November 2003.

the actors of the sector on a proposal that was already so finalised. Apart from that, most of the discussions in parliament focused not so much on the policy instrument itself (the change to a production-driven instrument was widely accepted), but on the settings of the instrument (level of subsidies, long-term security, and eligible RES-E generators), which reflected the concerns of the actors of the sector. Eventually, the new law was adopted in June 2003, later than expected, but still not too late compared to other legislative processes, especially if one considers that the government had changed in the meantime (in May 2003). The two Chambers of the parliament did not fundamentally oppose the proposal of the ministry of the Economy and eventually, the adoption in parliament was consensual and swift. In sum, parliament did not play a major role in the policy change process (H1/1).

Table 5: Dominant policy actor in RES-E policy change (H1/1)

	H1/1
Dominant policy actor?	Civil servants: Ministry of economic affairs with private experts

The consultation of the actors of the sector (energy companies, RES-E associations, environmental associations) occurred late in the policy-making process, in November 2002, when the proposal for the new policy instrument was already finalised. Tennet (the transmission grid operator) and EnergieNed (the association of Dutch energy companies) had already been consulted by the ministry in the summer of 2002, but only informally. The main energy companies (Nuon, Essent, Eneco) had been informed about the new proposal at the end of the summer, but they did not participate in its preparation at this early stage. It seems that the ministry wanted to avoid the participation of the actors of the sector during the formulation of the policy (hence the secrecy), and consult them only at the end because it was necessary to do so, but without allowing them to influence the policy. Table 6 shows that, except for TenneT and EnergieNed, who were consulted a little earlier than the other actors by the ministry of the Economy, the actors of the sector had no significant influence on the policy change (first proposition of H1/2 validated)[78].

In addition, table 6 shows the preferences of the actors of the sector regarding the proposed new policy instrument. The electricity companies seemed to be opposed to the new policy for different reasons: the end of the tax exemption and thus the import of RES-E (unfavourable

---

[78] Interview with Willem van der Heul (Ministry of the Economy), Den Haag, 17 November 2003.

RES-E Policy in the Netherlands

for the green suppliers, Nuon and Eneco, who imported most of their RES-E), the proposed MEP tariffs were too low to cover the costs of the RES-E generation in the Netherlands (especially for Essent[79] and EnergieNed[80]), they were concerned about the continuity and security of the MEP subsidies in the long term as the tariffs were fixed only two years in advance, and finally, Nuon defended the use of another instrument (quota and TGC). However, the electricity companies were not influential enough to modify the proposal of the ministry at this early stage. The RES-E associations and the environmental associations were in favour of the new policy instrument (at the condition that the MEP tariffs were increased), except the representatives of the solar and waste sectors, mainly because they were expected to benefit less from the system than the other sectors. But they did not have a significant influence on the policy formulation, except on the revision of the tariffs later on (implementation of the policy). Finally, it appears that TenneT and EnergieNed, who appear to be the most influential actors, did not actually have any specific preference about the new policy. This is probably because they have less direct interest in the RES-E market: TenneT was consulted especially for its expertise in the market and its role in the implementation of the MEP system (via CertiQ and EnerQ), and EnergieNed was also consulted mainly for its expertise and its knowledge of the sector. So, the second proposal of H1/2 cannot be tested in this case, since the influential actors showed no preference about the policy change (second proposal of H1/2 is non applicable).

**Table 6: Private policy actors
in RES-E policy change process (H1/2)**

Policy actors (H1/2)	Early / Late	Preference	Influence
Electricity companies			
TenneT	early	0	+
EnergieNed	early	0	+
Nuon	late	-	0
Essent	late	-	0
Eneco	late	-	0
Green electricity companies	late	-	0

---

[79] Essent is the biggest RES-E generator in the Netherlands.
[80] Following the liberalisation of the electricity sector, EnergieNed lost most of its power since there are no longer so many issues on which the electricity companies act collectively, except in the field of environmental policies, the only remaining working group of EnergieNed today. Interview with Peter Niermeijer (RECS and Ecofys), Utrecht, 18 November 2003.

RES-E associations			
Duurzame Energie Koepel	late	+	0
Nederlandse Windenergie Vereniging	late	+	0
Platform Bio-Energie	late	+	0
Holland Solar	late	-	0
Vereniging Afval Bedrijven (VVAV)	late	-	0
Environmental associations			
Stichting Natuur en Milieu	late	+	0

Legend: Preference (+) favourable, (-) not favourable, (0) indifferent; Influence (+) strong, (-) weak, (0) no significant influence.

In conclusion, the change of RES-E policy instrument in the Netherlands in 2003 was dominated by the ministry of the Economy, with the support of private consultants. The political parties in parliament raised a lot of criticism on the method used by the ministry to formulate the new policy as well as the settings of the MEP system (level of subsidies, long-term security, eligible RES-E generators), but they did not fundamentally oppose the change of policy instrument and the adoption in parliament was rather easy and swift (H1/1 validated). In addition, we observe that most of the actors of the sector were consulted late in the policy-making process and that they did not have a strong influence on the policy, which validates our assumption about the link between the earliness of the participation and the influence (first proposition of H1/2). In addition, the actors who had the main influence on the policy making process (TenneT and EnergieNed) were those who did not express any specific preferences about the policy change, which means that we were not able to test the hypothesis about the link between the preferences of the most influential actors and the policy change (second proposition of H1/2). In the end, this means that H1/2 is only partly validated in the Dutch case.

B. Past Policy

In 2002[81], the Dutch government acknowledged that the outcome of the past RES-E policy had been mitigated. On the one hand, the demand for RES-E reached unexpected levels in 2002, due to the conjunction of the REB tax exemption scheme with the early eligibility of the RES-E consumers on the liberalised electricity market (in 2002 about 13% of electricity consumption from RES-E[82]). The Netherlands was then seen

---

[81] Ministry of the Economy, 2002, Energy Report 2002: Investing in Energy, Choices for the Future, 26 February 2002.

[82] Report from the Dutch Audit Office, Groene stroom, June 2004.

as an example by most other European countries. But on the other hand, the increase of the demand for RES-E occurred to the detriment of both the domestic RES-E producers (because most of the demand was covered by imports, 10% electricity consumption from imported RES-E[83]) as well as the Dutch tax payers (because of the increased number of REB tax exemptions and subsidy evasion to foreign countries due to imports). In the period 1999-2004, the cost of the REB tax system for the Dutch tax-payer amounted 1,340 million €, with 556 million € only for 2002[84], and a large majority of this budget evaporated abroad to foreign RES-E generators. In sum, the past policy succeeded in creating a demand for RES-E, but failed to develop the domestic RES-E sector (not effective) and caused dramatic budget loss (too expensive) (see table 7).

Table 7: Past policy success/failure (H2)

	Past policy success/failure (H2)	
Objective	Past RES-E target within reach	"Success"
Instrument	– not effective to develop domestic RES-E generation – too expensive for public budget, especially because tax evasion in foreign countries	Failure

In addition, the European Commission has not made clear so far how the indicative target of 9% RES-E in the final electricity consumption should be metered, namely should it take into account imported RES-E or only domestic production, and, considering the level of imports, this is a crucial question for the Dutch policy. In order to avoid double counting at the European level (RES-E counted for the exported and imported country targets) and adopt a safer position in the long term, the Dutch ministry of the Economy decided to consider that the target should take into account only domestic RES-E production. Since the past RES-E policy instrument (REB tax exemption system) contributed mostly to increasing the imports of RES-E (only about 3% of domestic RES-E in total electricity consumption in 2002), the Dutch government was urged to change the policy instrument if it was willing to reach its RES-E policy objective (9% of domestic RES-E in total electricity consumption by 2010. So, in 2002, the policy objective was not really "successful" in the light of the past outcomes (3% in 2002 versus 9% expected in 2010). But, the Dutch government did not consider that the problem of the RES-E policy was the policy objective, since there was a political commitment to maintain it and there was the existing RES-E

---

[83] Idem.
[84] Idem.

potential to reach it ("success")[85]; the problem of the Dutch government was rather the policy instrument used to attain the objective (failure) (see table 7).

In conclusion, the past policy objective was not seen as a problem for the Dutch government in 2003 (high political priority and existing RES-E potential), but the past policy instrument was (H2 validated).

C. *Europeanisation*

At the time of the Dutch policy change (2002-2003), the European RES-E policy had been laid down in a directive of 2001 (see figure 5). However, the European RES-E policy consisted more of a series of propositions (e.g. indicative targets by 2010, increase of the financial and regulatory support to RES-E but without prescribing any policy instrument) than mandatory rules to be implemented in the member states (except the guarantees of origin). So, except the guarantees of origin, the Dutch government was not constrained by European harmonisation rules in 2002, when it started reviewing its RES-E policy, but it was just influenced by European RES-E policy propositions and ideas.

Figure 5: The European policy and Dutch RES-E policy change

EU policy								
Directive liberalisation + Green paper SER	White paper SER		Working paper E-SER	Draft directive E-SER	Directive E-SER + ECJ judgement		Directive liberalisation	
1996	1997	1998	1999	2000	2001	2002	**2003**	
							RES-E policy change in the Netherlands	

In fact, the Dutch RES-E policy change originates primarily in domestic constraints (failures of the past policy) and the European policy had hardly anything to do with it. The Dutch RES-E policy objective and instrument fitted perfectly well with the European policy in 2002 (see table 8), so no policy change was justified from a European perspective (H3). This means that the fit/misfit hypothesis is not validated in the Dutch RES-E policy change case.

---

[85] This is the reason why the "success" is in quotation marks in table 7.

However, we do observe in the Dutch case that the ministry of Energy often presents the change of policy instrument as the solution to make sure that the European indicative target of 9% of RES-E in the total electricity consumption by 2010 will be reached. Indeed, the failure to develop the domestic RES-E sector in the past was perceived as highly critical in 2002 partly because of the indicative target of the directive of 2001 and the first evaluation of this target that was to be reported in 2005. The government's commitment to reach the European target was thus used to legitimise the policy change, but it does not represent an explanation of the policy change itself[86].

Table 8: Europeanisation of RES-E policy (H3)

	Fit/Misfit (H3)	
Objective	9% by 2010	Fit
Instrument	Tax exemption open to imported RES-E	Fit

In addition, some other elements of the European policy help us to understand some indirect reasons behind the Dutch RES-E policy change. For instance, massive imports of RES-E (which caused the past policy failure) were only possible because of the liberalisation of the electricity market, which originates primarily in the European policy. The choice of a feed-in tariff system based on the German model to replace the existing tax exemption was influenced by the effectiveness of the German policy in the past and the decision of the European Court of Justice in 2001 in the PreussenElektra case to legitimise the German feed-in tariff system.

In conclusion, the fit/misfit hypothesis is not validated in the Dutch case, since no misfit (or perceived misfit) between the European and Dutch RES-E policy can explain the change of policy instrument (H3). However, we observe the influence of the European policy as a legitimating reference in favour of the new instrument.

D. Political and Institutional Context

Two changes of government occurred during the policy change process towards the adoption of the MEP law in June 2003. Firstly, in July 2002, a new centre-right coalition emerged from the general elections with the Christian Democrats (CDA), the Conservative-Liberals (VVD) and the Populist Party (LPF). It resulted from the dramatic defeat of the

---

[86] Interview with Willem van der Heul (Ministry of Economic Affairs), Den Haag, 17 November 2003. Interview with Peter Niermeijer (RECs), Utrecht, 18 Novembre 2003.

Labour Party (PvdA), who had been leading the previous governing coalition (PvdA, D66, VVD), and the incredible success of the populist leader Pim Fortuyn (LPF). Thus, the new governing coalition included two new political parties (CDA and LPF) with very different ideological priorities from the past governing party PvdA. However, this ideological difference was not significant as far as the RES-E policy was concerned, since the position of CDA and PvdA do not differ much on this issue (small ideological distance with the preceding government), and it was not a political priority for the LPF, which focused on immigration and security issues.

Secondly, the Populist Party (LPF) proved to be unable to govern and the government already resigned in October 2002. Then, the following elections allowed the Christian Democrats (CDA) and the Conservative-Liberals (VVD) to remain in government and they were joined by a smaller party, the Progressive Liberals (D66), to retain a comfortable majority in Parliament. So the new governing coalition (starting in May 2003) did not change significantly from the past one, with only one new party (D66), which was very close to the other in terms of ideological preferences and did not had much power in the coalition.

In sum, these government changes are not significant to understand the RES-E policy change, because they did not result in a context of a large ideological rift between the new and the old governing coalitions on the RES-E policy (H4/1 invalidated). Indeed, the need for RES-E policy change was widely acknowledged and a large consensus among majority and minority political parties supported it. The issue had already been raised by the PvdA, D66 and VVD government in the 2002 Energy Report. Therefore, it was not a new policy issue adopted specifically by the new governing parties in 2002 or 2003 (see table 9). In addition, in the Dutch policy-making system, the bills do not die with the dissolution of the Parliament, which means that government changes have had less impact on the legislative process than in other countries.

Table 9: Governing coalition (H4/1) and veto players (H4/2) in the Netherlands

Governing coalition (H4/1)	Veto players (H4/2)
– Change of governing coalition in July 2002 and May 2003 – But no ideological distance with past coalition about RES-E policy – And large consensus between political parties in favour of RES-E policy change, including opposition parties	– Few institutional veto players: bicameral parliament but one chamber more powerful than the other (2nd Chamber) and Queen no veto power; with small ideological distance over RES-E and party discipline – Many partisan veto players (multi-party coalition), but small ideological distance over RES-E (large consensus in favour of RES-E policy change) and strong party discipline

The Netherlands is a consensus democracy, in which coalition governments traditionally include at least two or three political parties (many partisan veto players). However, in the Netherlands (like in Belgium, for instance), two characteristics of the political system help to facilitate the policy-making process and guarantee the stability of the governing coalition: the coalition agreement and party discipline. Indeed, the coalition agreement signed at the beginning of the government's term presents the policy agenda on which the governing parties have agreed to focus. This is thus a very important document that gives an insight into the policy changes to be expected during the term and that binds the coalition partners. In addition, Dutch political parties traditionally impose strong party discipline on their members, which means that the governing parties and their members of parliament are closely linked, which tends to facilitate the policy-making process. As far as the RES-E policy was concerned, a consensus in favour of policy change was easily reached in 2002-2003, in the light of the tremendous budget loss that threatened the government if the fiscal incentives were to be continued (partisan veto players with no significant ideological distance over RES-E policy). So, the presence of many potential partisan veto players in the Netherlands did not prevent the RES-E policy change from being adopted thanks to the small ideological distance between them on the RES-E policy and the strong party discipline (see table 9).

As far as the institutional veto players are concerned, the Dutch parliament is composed of two Chambers (bicameral system): the first Chamber (Senate) and the second Chamber (House of Representatives). The second Chamber is elected directly through general elections and shares the same majority as the government (with strong party discipline), while the first Chamber is composed of representatives of the Provinces (potentially other majority). Those two Chambers represent two distinct institutional veto players. However, in reality, the second Chamber is by far the more important one and the veto power of the first Chamber is thus limited. In addition, the two Chambers' positions regarding the RES-E policy usually does not significantly differ (small ideological distance). The Queen also represents an institutional veto player in the Netherlands, but her power is so limited in practice (like in Belgium) that it does not really count as an effective veto player. Finally, the Netherlands is a decentralised unitary state characterised by a significant degree of decentralisation, with the provinces and the municipalities enjoying strong independence and policy autonomy on some issues (e.g. spatial planning, electricity distribution and production), but the conduct of the energy policy remains in the hands of the central government with no veto power to the decentralised body. The municipalities and provinces may represent strong opposition to the government, especially on energy policy issues, but they do not have the power to veto the policy in the end (no veto players).

In conclusion, the successive changes in the Dutch government in 2002 and 2003 are not significant to understanding the RES-E policy change, because there was no significant ideological distance between the governing coalitions regarding the RES-E policy. The need for RES-E policy change was widely acknowledged and a large consensus among majority and minority political parties supported it (H1/1 not validated). Besides, the Dutch political system is characterised by many partisan veto players, but since the ideological distance between them is not significant, the RES-E policy change is easier to achieve (H4/2 validated).

VIII. Conclusion

The Dutch RES-E policy experienced serious drawbacks in 2002, which led the ministry of the Economy to design a new policy instrument, the MEP subsidies, to solve the failures of the past and make sure that the Dutch RES-E policy objective (9% of total electricity supplies from RES-E) could be reached by 2010. Therefore, the policy change in the Netherlands concerned only the policy instrument, while the existing policy objective was not modified.

Table 10: Validation and relevance of the hypotheses in the Netherlands[87]

	H1		H2	H3	H4	
	H1/1	H1/2			H4/1	H4/2
Validation hypotheses						
Policy objective	*V*	*n.a.*	*V*	*V*	*V*	*n.a.*
Policy instrument	V	"V"	V	I	I	V
Explanation policy change/no change						
Policy objective	*+*	*0*	*+*	*+*	*0*	*0*
Policy instrument	++	0	++	0	0	+

Legend: ++ (major explanation of policy change/no change), + (minor explanation), 0 (does not explain policy change/no change). V (hypothesis validated), I (hypothesis invalidated), "V" (hypothesis partly validated), n.a. (hypothesis not applicable).

Table 10 summarises the findings of this research on which hypotheses of our theoretical framework are validated or invalidated in the Dutch RES-E policy change case. First, we observe that the relation between the dominant policy actor (the civil servants) and the policy

---

[87] The policy objective lines are in italics because there was no change of policy objective in the Netherlands, so the relevance and validation of the hypotheses can only be assessed by default for the policy objective.

change (policy instrument) confirms our hypothesis (H1/1 validated). Moreover, the absence of change of policy objective also confirms our hypothesis, as the dominant policy actors are not political actors but civil servants in the Dutch case (H1/1 validated). Second, the first proposition of H1/2 assumes that if the actors of the sector are consulted early on in the policy change process, they have more influence than if they are consulted late, and this is validated regarding the change of policy instrument in the Netherlands (first proposition of H1/2 validated). However, the second proposition of H1/2, about the link between the preference of the most influent actors and the policy change, cannot be tested in the Dutch case (second proposition of H1/2 non-applicable). So H1/2 can only be partly validated in this case. Third, the relation between the evaluation of the success/failure rate of the past policy and the policy change is validated in the Dutch case, as the policy objective (success in the past) did not change while the policy instrument (failure in the past) did (H2 validated). Fourth, H3 expects that if the European and domestic policy fit, no policy change should happen. However, in the Dutch case, the past Dutch RES-E policy objective and instrument fitted with the European policy, but the Dutch RES-E policy instrument changed anyway because of domestic constraints (H3 invalidated). Fifth, H4/1 assumes that when the government coalition changes and there is a significant ideological distance between the past government and the new one, public policies are expected to change as well. In the Dutch case, we observed two changes of government over the period 2002 and 2003, but there was no ideological distance between those governments on the RES-E policy issue, because the RES-E policy change was not a controversial issue among the political parties (large consensus). Therefore, the status quo in terms of policy objective validates our assumption, while the change of policy instrument does not (H4/1 invalidated). Finally, the absence of a significant ideological distance between the partisan veto players on the RES-E issue and the large political coalition in favour of the change of policy instrument confirms our hypothesis (H4/2 validated).

Finally, table 10 looks at whether the hypotheses are relevant or not to explaining the policy change or the absence of policy change in the Dutch case. Some of them have proved to be very relevant (H1/1 and H2), while others have proved to be less significant (H3 and H4/2) or irrelevant (H1/2 and H4/1) in the Dutch case. First, the fact that the policy change process was dominated by the ministry of the Economy and not by politicians, with very limited consultation of the actors of the sector, explains why only the policy instrument changed and not the objectives of the policy (H1/1). In fact, the ministry did not want to question the overall objective of the RES-E policy but just to revise its instrument and it wanted to do this swiftly. In addition, the actors of the sector did not have a significant influence on the ministry; therefore they do not explain the policy change (H1/2). Second, what explains the change of policy instrument in the Netherlands best is the failure of the past policy instrument (H2). Indeed, we observe that the main reasons why the policy instrument had to be revised lie in the serious perverse effects of the past instrument (budget loss, massive RES-E imports, no domestic RES-E generation increase). In addition, the status quo regarding the policy objective is also linked to the fact that it was not perceived to be a problem and that it appeared within reach. Third, the relation with the European policies partially explains why the policy objective did not change (fit with the EU target), but does not contribute to explaining the change of policy instrument (H3). Finally, the change of the governing coalition does not explain the RES-E policy change in the Netherlands (H4/1), but the fact that the RES-E policy change was supported by a large coalition of political parties (potential partisan veto players) contributes, but to a small extent, to explaining the change of policy instrument (H4/2).

In conclusion, our theoretical framework is less robust in the Dutch case than in the others, as two hypotheses are not validated and not significantly relevant to explaining the change of RES-E policy instrument in the Netherlands (the relation with the European policies and the influence of the change of governing parties). However, the two central hypotheses of our theoretical framework regarding the change of policy instruments (the policy actors and the past policy) are validated in this case and have proved to be significant explanatory factors to account for the change of RES-E policy instrument in the Netherlands. Therefore, our theoretical framework is useful to explain the policy change in the Netherlands, but it is not entirely validated in this case.

CHAPTER 7

RES-E Policy in the United Kingdom

I. Introduction

The United Kingdom of Great Britain and Northern Ireland is made up of three separate but overlapping jurisdictions: England and Wales (grouped as one jurisdiction for electricity policy), Scotland and Northern Ireland. Devolution has created greater political independence for both Scotland and Northern Ireland, but energy policy has remained a national competence determined by the Department of Trade and Industry (DTI) since the closure of the Department of Energy in 1992. As far as RES-E is concerned, the Department of the Environment (Department for Environment, Food and Rural Affairs – DEFRA – since 2001) also shares some competence with the DTI (e.g. planning system, biomass and waste policy, climate change policy). The implementation of the RES-E policy for Scotland and Northern Ireland is the responsibility of the devolved bodies. Consequently, specific Orders have been adopted in Scotland and Northern Ireland for the implementation of the Non-Fossil Fuel Obligation (NFFO) in the 1990s and the Renewables Obligation (RO) since 2002. However, they did not differ significantly from the England and Wales Orders. The following chapter focuses on England and Wales, and this for three reasons: England and Wales represent by far the largest jurisdiction in the UK, it provided a model for electricity reform in the other jurisdictions in the 1990s, and the RES-E policies in Scotland and Northern Ireland have been very similar to the England and Wales one since the 1990s with comparable NFFO and RO schemes. Finally, the local authorities have a significant influence on the implementation of the UK RES-E policy due to their responsibility to grant the planning permits.

The British energy policy was dominated by the priority to fossil fuels (domestic coal, oil and gas) and nuclear power until the end of the 1990s. In addition, the privatisation process that started at the beginning of the 1980s, progressively covered all energy sectors by the end of the 1990s, with the privatisation of the nuclear industry as the last energy privatisation in 1996. In this context, the promotion of RES-E had not been a priority of the British government until the beginning of the

2000s, when the climate change issue and the reduction of the CO_2 emissions became an increasing political concern under the new Labour government. In terms of RES-E policy instruments, the NFFO contracts introduced the first financial support to RES-E on the electricity market from 1990, but it did not prove very successful in the end due to numerous obstacles in the system itself (priority to nuclear power, low premium prices, bidding procedure) and in the overall context (e.g. electricity pool market, planning permits). With the arrival of the Labour government in 1997 and the increased concerns about the climate change issue, the development of RES-E in the UK gained more political priority and new policy instruments were adopted to support the RES-E generators (RO, CCL, investment grants). However, even if the British government seems to have initiated a substantial change in its RES-E policy since the end of the 1990s, its ambitions remain very limited compared to the huge opportunities of the UK in terms of RES-E potential.

During the 1990s, RES-E technologies eligible for the NFFO Orders were determined based on technology bands revised in each Order. Only the most competitive RES-E technologies were considered and competition between different projects in a given technology band applied. The main technology bands over the 1990s were: hydropower, waste combustion (municipal and industrial), sewage gas, landfill gas, onshore wind power. Other technologies like wave power, tidal or solar power were not considered under the NFFO system (Ross, 2000). In 2000, the Utilities Act gave an encompassing definition of RES, including all "sources of energy other than fossil fuels or nuclear power" (Helm, 2003, p. 363). However, the definition of the RES-E technologies eligible to the RO system or the Climate Change Levy (CCL) exemption from 2001 was further restricted (Helm, 2004; Dinica, 2004). For instance, large hydropower plants are not eligible for the CCL exemption (> 10 MW) as well as for the RO quota (> 20 MW). In addition, only the non-fossil part of waste is eligible for the RO quota, but electricity generated from waste is eligible for the CLL exemption under certain conditions[1].

II. Overview of the Electricity Sector

The energy fuel mix for electricity generation in the UK was dominated by coal until the beginning of the 1990s (over 60% of the total electricity generation) (see figure 1). This is due to the existence of important domestic coal reserves and the powerful coal industry lobby

---

[1] DTI, "New and Renewable Energy: Prospects for the 21st Century: The Renewable Obligation: Preliminary Consultation", London: TSO, October 2000.

to ensure long-term contracts for the extracted coal with the electricity companies. With the coal crisis in the 1980s and 1990s (social, political and economic issues) and the privatisation of the electricity market in 1989, the domestic production of coal decreased steadily and so the use of coal in electricity generation. However, coal still remains one of the main fuels for electricity generation in the UK nowadays (about 30%), but it is now imported, since most of the British coal mines have been closed and imported coal is cheaper.

Figure 1: Electricity generation in the UK (GWh), 1972-2002

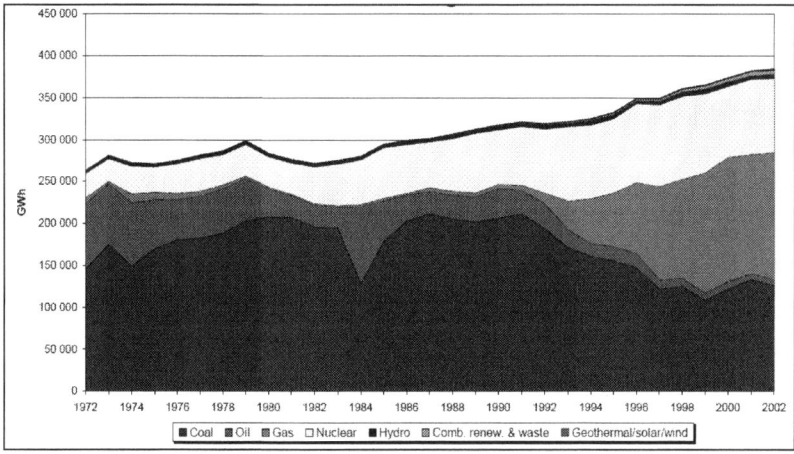

Source: IEA.

Electricity generation from oil represented about 25% of the market in the 1970s, before the oil crisis. It was slowly reduced during the 1980s, with a short burst in 1984-1985, during the counter-oil choc (see figure 1). It was replaced by coal and nuclear power. But, due to the exploitation of the domestic oil reserves in the North Sea, the electricity companies were not as constrained as in the other European countries to reduce their oil consumption at once. It is only during the 1990s that we observe the most significant reduction in oil consumption for electricity generation, mostly due to the rapid increase of gas.

Nuclear power developed in the UK in the 1960s and it has been growing since then to account for about 20% of the electricity generation today (see figure 1). Unlike in many other European countries, the nuclear program has not led to massive opposition in the UK and no phasing-out has been on the political agenda so far, the nuclear option has been kept open in the last energy white papers of the DTI.

Since the beginning of the 1990s, electricity generation from gas has been increasing substantially in the UK (about 40% of total electricity generation in 2000) (see figure 1). This period has come to be known as the "dash for gas". With the increase of gas, the importance of oil and coal decreased significantly during the 1990s. This fundamental change in the UK electricity supplies also corresponds to the development of independent power generators owned by the regional electricity distribution companies, as opposed to the incumbent large electricity generators. However, by the end of the 1990s, the "dash for gas" ceased, as the electricity market was fully open to competition and the investments in new generation capacity could not be passed directly on the electricity bills (Helm, 2003).

So far, RES-E has been very marginal in the UK (under 4% of total electricity generation in 2004, see table 1), despite huge potential in terms of wind, hydro, solar and biomass energy sources (Brennand, 2004). The main reasons are the existence of domestic reserves of other less expensive energy sources (coal, oil and gas), and the lack of political commitment in favour of RES-E by the Conservative government between 1979 and 1997. Since the Labour party came to power, the political conditions for RES-E have improved significantly, but financial and regulatory support to RES-E remains very unambitious compared to the existing potential, mainly for economic reasons linked with the privatisation and liberalisation of the electricity market.

At the beginning of the 1990s, hydropower was the only significant RES-E sector in the UK (over 90% of RES-E generation in 1990), mainly because of the competitiveness of the large plants (see table 1). The other RES-E developed only during the 1990s, when the first financial support to the RES-E market started (the NFFO scheme). The production of hydropower then stagnated, because no more significant investments were made in this sector apart from demonstration projects in wave or tidal plants. The other RES-E sectors where most investments have been increasing during the last decade are wind power, biomass and waste (see table 1). Solar PV is still insignificant because it remains by far uncompetitive with conventional energy sources in the UK.

Table 1: RES-E generation in UK[2] (GWh) and percentage of total electricity generation (%), 1990-2004

	1990	1992	1994	1996	1998	2000	2002	2004
Wind and wave	9	33	344	488	877	946	1,256	1,935
Solar PV	0	0	0	0	0	1	3	4
Hydro	5,207	5,431	5,094	3,392	5,117	5,085	4,788	4,930
Waste	2,80	554	966	1,197	2,034	3,028	3,586	4,975
Biomass	3,16	380	553	608	620	854	1,494	2,328
Total	5,812	6,398	6,956	5,685	8,648	9,914	11,127	14,171
%	1.92	2.11	2.26	1.74	2.59	2.90	3.14	3.95

Source: Dukes 7.1.1; Dukes 5.1.3.

The fastest growing RES-E during the 1990s was waste[3] (especially landfill gas), which accounts for over 30% of all RES-E generation today, the same level as for hydropower (see table 1). Electricity generated from biomass has also increased significantly during the last decade (see table 1), especially since the co-combustion of biomass with fossil fuels has been recognised as RES-E. Recent efforts of the British government focus on the development of energy crops[4], but the effects are not yet observable in the statistics.

The other growing RES-E lately has been wind power, but to a smaller extent (see table 1). In the light of the abundant resources both onshore and offshore, the current development of wind power in the UK is only marginal. Onshore wind power projects have faced several issues that prevented their development in the UK so far: commercial competitiveness, grid connection and planning permission. Offshore wind farms represent a major potential in the UK given its abundant coastal areas. New government capital grants for offshore wind power started in 2002, but as long as it remains uncompetitive without financial support, its development in the UK remains uncertain.

III. Overview of the RES-E Policy History in the UK (1979-1997)

Before 1979, the British energy policy was largely conducted through the public energy companies which dominated the energy sectors (coal, electricity and gas) following the nationalisation in the late 1940s (Helm *et al.*, 1989; McGowan, 1996). Until the privatisation of the electricity sector at the end of the 1980s, the generation and transport

---

[2] Includes only the biodegradable part of municipal solid waste.
[3] Only the biodegradable fraction of wastes is considered as RES-E.
[4] Note that energy crops are also used for other energy outputs than electricity, namely fuel for transport and heating.

of electricity in England and Wales were centralised in the Central Electricity Generating Board (CEGB)[5]. The CEGB was actually the most powerful actor in the electricity market. The distribution and supply of electricity was under the responsibility of fourteen Area Electricity Boards (AEB, twelve in England and Wales and two for Scotland). The AEB were able to influence the action of the CEGB only slightly. The AEB were not allowed to generate electricity and therefore depended heavily on the CEGB. Following privatisation, the organisation of the sector changed drastically, with the de-concentration of the generation sector (many different electricity generators), the concentration of the supply market (about seven large suppliers in 2002), the separation of the network activities (transmission and distribution) from the generation and supply companies (but still ownership links), and the authorisation for the new distribution and supply companies (Regional Electricity Companies – RECs) to own generating capacity (origin of the dash-for-gas).

Before 1979, energy policy was thus a very important element of the British economic development and emphasised centralised planning of the allocation of resources (hence the CEGB and the AEB). The main priorities of the energy policy were the security of energy supplies, to be ensured through the development of the North Sea resources (gas and oil), the expansion of nuclear power and the exploitation of domestic coal (especially in the electricity sector). So the main actor of the electricity policy was the CEGB (and its counterparts in Scotland and Northern Ireland), whose priority was the security of electricity supplies. By 1979, the British energy policy proved to be very effective, notably in the development of the North Sea gas and oil resources, and the UK became self-sufficient in energy supplies. However, over that period, the interests of the consumers and local authorities were largely ignored, as well as the decentralised energy options such as RES-E and CHP (McGowan, 1996).

The British RD&D programs for RES-E technologies started in the mid-1970s, following the first oil crisis. A major review was undertaken by the Energy Technology Support Unit (ETSU) in 1982[6], after the second oil crisis (Mitchell, 1995; Ross, 2000; Connor, 2004). ETSU, a public agency linked to the Department of Energy, was in charge of

---

[5] In Scotland, there were two central organisations: the North of Scotland Hydro-Electricity Board (NSHEB, only for hydro-electricity generation) and the South of Scotland Electricity Board (SSEB, in charge of the electricity generation, transmission, distribution and supply within its operating area). And in Northern Ireland, the Northern Ireland Electricity (NIE) also took care of the entire electricity sector (generation, transmission, distribution and supply).

[6] ETSU, Strategic Review of Renewable Energy Technologies, London, HMSO, 1982.

assessing the new energy technology and "pick the winners" eligible for the government RD&D programmes. In its 1982 report, ETSU suggested that the potential of RES-E in the UK was limited (maximum 5.8% of electricity supplies by 2025), despite the huge resources available (Connor, 2004). The main reason for such a pessimistic statement is that the British government was willing to support only the technologies that could exploit the UK RES-E resources the most economically. Non-competitive technologies were thus considered having no potential in the future UK electricity market. The priority between the technologies changed overtime and many programs were discontinued before they could show their effects because the assessments of ETSU were regularly revised and often soundly based (e.g. wave, tidal and geothermal programs) (Elliott, 1994; Mitchell, 1995; Connor, 2004). During the 1980s, priority was given to large wind power plants and some biomass technologies (e.g. landfill gas). The orientation of the RES-E R&D program during the 1970s and 1980s in favour of large and competitive plants was not surprising, given the central position of the CEGB (and its counterparts) in the UK electricity sector at that time.

In 1979, the Conservatives came back to power with a new priority: the privatisation and liberalisation of the British public utilities (e.g. telecom, rail, energy). They started to promote market forces over state intervention in the energy sector and so the energy policies of the past were progressively discontinued and left to the market (e.g. security of energy supplies, RD&D, grid expansion), except for the nuclear power sector, which remained heavily subsidised until its privatisation in 1996, and the coal industry, which was granted long-term contracts with the electricity companies (Helm *et al.*, 1989; McGowan, 1996; Helm, 1993). The speech given by Nigel Lawson, the new Secretary of State for Energy, in June 1982, clearly outlined the liberal philosophy of the new government to be implemented in the energy sector during the following decades "I do not see the government's task as being to try to plan the future shape of energy production and consumption. (…) Our task is rather to set a framework which will ensure that the market operates in the energy sector with a minimum of distortion and that the energy is produced and consumed efficiently"[7]. This new orientation of the British energy policy was clearly damaging to the RES-E that were not competitive compared to the other energy sources and that could not expect to develop without financial and regulatory support from the government.

---

[7] The text of the speech can be found in Helm *et al.*, 1989, pp. 23-29.

The liberalisation of the British electricity market was a very long process (Helm, 2003). In 1983[8], the British government started its first attempt to liberalise the electricity sector by abolishing the public electricity companies' monopoly in the production and supply of electricity (no longer a legal monopoly). However, no unbundling had then been enforced between the grid and commercial activities of the electricity companies. As a consequence, the 1983 proved to be very ineffective in terms of increased competition due to the dominant position of the vertically integrated public electricity companies in the British market (CEGB and AEBs) (Helm *et al.*, 1989; McGowan, 1996). In addition, under the 1983 Energy Act, the AEBs were obliged to purchase RES-E produced from decentralised electricity generators. But, on average, they were paid about 30% less than the electricity produced by the centralised generators (CEGB), so it did not contribute effectively to developing the RES-E sector (Mitchell, 1995; Ross, 2000).

Then, in the second half of the 1980s, the British government started to privatise other monopoly or near-monopoly industries (e.g. telecom, gas) without restructuring the industry beforehand, namely without liberalising the market (e.g. unbundling). This turned out to be very unsuccessful in terms of market competition, since the privatised monopoly prevented as much as they could any newcomers from entering the market. So privatisation and liberalisation appeared at odds with, which proved to be inconsistent with the rhetoric of the British government in favour of the market forces (McGowan, 1996).

In order to avoid such problems in the electricity sector, the British government decided to undertake the privatisation and the liberalisation policies at the same time, against the interests of the public electricity companies, to retain the existing monopolistic structure. In 1988, the White Paper on Electricity Privatisation[9] proposed several modifications to the structure of the electricity industry to increase competition in the electricity market, especially at the electricity generation level (e.g. the unbundling between network and sales activities, the break-up of the CEGB). In addition, the British government and the AEBs reached an agreement on the progressive liberalisation of the electricity supply market in three stages: the 1 MW market was opened up immediately, the 100 kW market would be open in 1994, and finally the whole market would be opened up in 1998. Those proposals were then endorsed by the British parliament in the 1989 Electricity Act. This act organises the privatisation of the public electricity companies and sets the conditions

---

[8] 1983 Energy Act.
[9] Department of Energy, Privatising Electricity: The Government's Proposals for the Privatisation of the Electricity Supply Industry in England and Wales, London: HMSO, February 1988.

for increased competition in the electricity market (both on the generation and supply levels).

Firstly, the CEGB was broken down into three different companies: two electricity generation companies (National Power and PowerGen) and one independent transmission grid operator (National Grid Company (NGC). The two generation companies were privatised while the NGC's ownership was transferred to the regional distribution companies (Regional Electricity Companies (RECs) – formerly AEBs). The generation market in England and Wales remained very concentrated at first, but following further reform and adaptation then developed into a market with many diverse generating companies. In England and Wales, in 2002, 38 companies were regarded as major power producers. The largest were: British Energy (19% market share), PowerGen (13%), AES (10%), Innogy (10%), Electricité de France (9%), TXU Europe (6%), Edison Mission Energy (6%), and BNFL (3%) (IEA, 2002)[10]. Thus, following the reform of the electricity sector in 1989 and notably the breaking up of the CEGB, the electricity generation companies lost their dominant position in the electricity sector.

Eventually, the nuclear plants were removed from the privatisation process in 1989, due to its unattractiveness to the private sector. The nuclear generation remained in public hands and was transferred to two new public companies (Nuclear Electric and Scottish Nuclear). The privatisation of those companies took another seven years to be completed in 1996. During that period, they benefited from additional financial support through the Non-Fossil Fuel Obligation (NFFO) scheme, which provided that nuclear power was to be purchased at the cost-price (higher than the pool price) by the distribution companies (McGowan, 1996).

Following the 1989 Electricity Act, the 14 AEBs were privatised into Regional Electricity Companies (RECs), and the new RECs had to unbundle (only accounting unbundling) their distribution network operations (remaining a natural monopoly) from their supply sides (open to competition). However, the RECs remained vertically integrated until 2000 (Utilities Act), which made it very difficult for new suppliers to enter the market. The liberalisation of the supply market was decided to be progressive with a transition period going on until 1998, which slowed down the opening of the market and reinforced the monopolistic position of the RECs. In addition, the RECs were allowed to own generating capacities from 1989, which increased their independence from the large electricity generators and increased their influence in the sector. In 1991 there were 14 RECs in Great Britain (England, Wales and Scot-

---

[10] *Energy Policies of IEA Countries: the United Kingdom*, Paris: IEA.

land). In the next decade, electricity supply went through consolidation the previous RECs formed seven major supply groups following takeovers and mergers. Among the seven large suppliers, four major retailers covered two-thirds of the British market: Innogy, British Gas Trading, TXU Europe, and Scottish and Southern Energy (IEA, 2002)[11]. Moreover, consolidation also took place in the electricity distribution (today, nine distribution companies).

Finally, recent years have also seen a trend towards the vertical integration of generation and supply. Generators have been prompted to diversify into supply business, due to increased competition in the generation market.

Secondly, the Electricity Act created the role of Director General of Electricity Supply in charge of an independent regulator institution, the Office of Electricity Regulation (OFFER), whose duties are to ensure competition in the generation and supply of electricity (first duties), and promote consumer protection, promotion of RD&D, economic efficiency in supply and transmission and promotion of heath and environment (second^d duties) (Helm, 2003). The role of competition regulator was clearly recognised as the priority of OFFER and the other duties were largely disregarded during the 1990s. One of the main tasks of OFFER, under its consumer protection duty, was to set a price cap to the electricity sold by the RECs to the electricity consumers as long as they were not eligible to choose their supplier (until 1998). In reality, this price cap was subject to an agreement between OFFER and the RECs, since it was set at a level that allowed the RECs to transfer the costs of their investments in new gas power plants (dash for gas) to the non-eligible customers. The balance between increasing competition (new generation capacities) and protecting consumers (price cap) appeared biased in favour of the former during the 1990s (Helm, 2004).

Thirdly, a pool system based on prices rather than costs was created to organise the electricity market. The principle of the pool is that the electricity generators bid the price at which they would be willing to supply electricity and their availability a day ahead. Participation to the pool was compulsory for the electricity generators and any supplier could buy electricity at the Pool price. The price paid by the electricity suppliers to purchase electricity from the Pool (Pool Selling Price – PSP) is open and transparent. It is composed on the one hand of the Pool Purchase Price – PPP (equal to the system marginal price in a competitive market), and, on the other hand, it includes an uplift element representing the additional costs of reserve, availability and services for the capacity that is not used, only in times of higher demand. So the PSP =

---

[11] IEA, 2002, *Energy Policies of IEA Countries: the United Kingdom*, Paris: IEA.

PPP + uplift (up to 10% of the PSP) (Helm, 2003). However, the SMP was sapped by problems of market power and the availability mechanism was subject to significant gaming from the electricity generators. Therefore, the abuse of their dominant position by the electricity generators resulted in high PSP prices that did not reflect the actual costs. However, the PSP was still too low for the RES-E generators to be competitive. In addition, the charging structure for connection to the UK electricity distribution network penalised the decentralised RES-E generators, because it was designed for the traditional large generators. As a consequence, an RES-E generator who wished to connect to the distribution grid must pay the costs in full, including necessary upstream grid reinforcement, which can be very expensive in some cases. The regulation of the Pool by OFFER was very light, as competition on the generation market was expected to be brought by the market forces, so the perverse effects of the Pool were not forcefully addressed by the regulator.

Finally, the 1989 Electricity Act endorsed the Energy Minister with the power to oblige the electricity suppliers to purchase a certain amount of their electricity from non-fossil sources, namely nuclear power or RES-E. This is known as the Non Fossil Fuel Obligation (NFFO). The main aim of the NFFO was clearly to protect the nuclear industry, which had been excluded from the privatisation process and needed a higher price than the PSP to remain competitive on the electricity market (Mitchell, 1995; Ross, 2000). So the NFFO guaranteed nuclear generators that a minimum quantity of nuclear power (in fact equal to the entire expected capacity of the nuclear generators) would be purchased by the RECs at a premium price, and this until 1998[12] (Mitchell, 1995; Ross, 2000). The RECs established the Non-Fossil Purchasing Agency (NFPA) to take care of their obligations in the NFFO: purchase of the nuclear and RES-E electricity and management of the FFL. The renewables NFFO did not result from an intentional policy of the British government in favour of RES-E, but rather as a by-product of the primary need to support nuclear power and legitimise it in front of the European Commission. Therefore, the nuclear obligation was called the "non-fossil fuel" obligation and included also the RES-E generation (Mitchell, 1995, 2000).

The renewable NFFO was expected primarily to contribute to the development of commercially attractive RES-E technologies. It was not designed to help the non-competitive RES-E technologies to enter the market; this was the role of the RD&D programs. The renewables

---

[12] The European Commission's approval of the NFFO system requested that the nuclear obligation should end in 1998. However, the renewables obligation could continue after 1998.

obligation under NFFO operated through a bidding system with successive calls for tenders/Orders enacted by the British government (from NFFO-1 in 1990 to NFFO-5 in 1998) for a determined declared net capacity of RES-E to be installed by technology bands (hydro, waste, sewage gas, landfill gas, wind bands)[13]. Some RES-E technologies were non-eligible for the NFFO, such as wave, geothermal, tidal and offshore wind power. A premium price for the RES-E delivered to the grid was guaranteed to the winning electricity generators for up to 15 years, according to the contracts signed between the NFPA and the RES-E generators. The initial objective of the Conservative government with regard to RES-E in 1990 was the installation of 600 MW by 2000, which was revised to 1,000 MW in 1992[14] and then finally in 1994[15] to 1,500 MW by the year 2000 (corresponding to about 3% of the total electricity consumption). This was a low and unambitious target compared to the potential for RES-E in the UK (Ross, 2000). In the end, the NFFO scheme contributed to support mostly large and small hydropower plants, landfill gas installations and wind power, but progress has been very slow, since in 1990 RES-E represented less than 2% of the UK electricity market, and in 2000 it represented only about 3% (Smith and Watson, 2002).

The NFFO was financed through a Fossil Fuel Levy (FFL) charged against all electricity consumption from fossil fuel sources and paid by the electricity consumers. The FFL was set at between 10% and 11% of the electricity price in 1990. After the privatisation of British Energy[16] in 1996, the FFL fell drastically to 3.7% and then to 2.2% in 1997, 0.7% in 1998 and 0.3% in 1999 (Mitchell, 2000). The reduction of the FFL after 1996 and even further after 1998 clearly shows the nuclear dimension of the NFFO system versus its renewables dimension. The nuclear obligation was awarded the lion's share of the FFL until 1996, while the renewables obligation only benefited from a marginal part of the budget (between 1% and 8.6% until 1996) (Mitchell, 2000; Ross, 2000). After 1996 and 1998, the RES-E benefited from the biggest slice of the FFL budget, but since the budget itself was heavily reduced, it did not make a significant difference in the end.

---

[13] See Mitchell (2000) and Ross (2000) for a detailed account of the content of the five NFFO Orders and associated contracts.

[14] DTI, Renewable Energy Advisory Group: Report to the President of the Board of Trade. Energy Paper 60, London: HMSO, 1992.

[15] DTI, New and Renewable Energy: Future Prospects in the UK, Energy Paper 62, London: HMSO, 1994.

[16] In 1996, British Energy was a takeover of the two nuclear energy companies of the UK: Nuclear Electric (England and Wales) and Scottish Nuclear (Scotland). It was privatised in June 1996.

The NFFO scheme had mixed success as regards the development of RES-E in the UK. It was partly responsible for reducing the cost of renewables generation thanks to the bidding procedure, but a number of contractors had planning, technical and commercial problems in developing their project. At the end of the NFFO, the generating capacity remained 25% short of the policy target of 1,500 MW[17]. The main problem of the NFFO system is that some RES-E projects bided too low and therefore they were never constructed. In addition, the restriction of the local planning authorities was a major obstacle to the construction of the projects selected for NFFO contracts. A significant share of the NFFO projects were not commissioned, because they were refused planning permission or permission was subject to long planning disputes (the NIMBY phenomenon), especially landfill gas and wind projects (Mitchell, 2000; Smith and Watson, 2002).

Following the privatisation policies and the NFFO, the British government decided in 1988[18] to reduce its support to RD&D in new energy technologies during the 1990s (Helm *et al.*, 1989; McGowan, 1996). This reduction in RD&D funds proved to be dramatic for the RES-E technologies that were not competitive. The RD&D budget was expected to peak in 1993 (around £25 million in 1993) and then be phased out gradually by 2000, as the private electricity companies were expected to take over the development of new RES-E technologies. However, funding from non-governmental sources never grew as originally expected and the RD&D programme was revised at the end of the 1990s under the Labour government (Elliott, 1994; Mitchell, 1995; Connor, 2004). Then, in 1994[19], a further revision of the RD&D program confirmed the reduction of the budget and again changed the priorities between the RES-E technologies. Three categories of RES-E technologies were identified according to their competitiveness and each was eligible for different financial support: the close-to-market technologies received NFFO contracts and RD&D funds (e.g. waste, wind power onshore, small hydropower), assessment technologies receiving only RD&D funds (e.g. solar PV), and watching technologies were only expected to play a role in the UK energy system in the long-term and thus received no support (e.g. wind power offshore, wave, tidal energy) (Elliott, 1994).

---

[17] National Audit Office, DTI Renewable Energy, Report by the comptroller and auditor general, 11 February 2005.

[18] Department of Energy, Renewable Energy in the UK: the Way Forward, Energy Paper 55, London: HMSO, 1988.

[19] DTI, New and Renewable Energy: Future Prospects in the UK, Energy Paper 62, London: HMSO, 1994.

IV. RES-E Policy Change in the UK (1998-2004)

In 1997, the Labour party won the general elections and formed the new government. Compared to its predecessors, it put greater emphasis on environmental and social concerns, but at the same time largely embraced the ideas of the Conservatives in favour of market-mechanisms and reduced public intervention in the energy sector. So, the Labour government tried to marry the potentially conflicting objectives of maintaining a competitive energy market driven by market forces and ensuring that the energy markets enhance environmental and social protection (Helm, 2003). In addition, the Labour government committed itself to protecting the coal industry by extending the coal contracts with the electricity companies. The pursuit of these objectives required a much more active energy policy from the government, with increased intervention in the energy sector. As far as RES-E is concerned, the Labour Manifesto of 1997 mentioned that 10% of electricity supplies were expected to originate from RES-E by 2010. This target appeared to be ambitious compared to the past objective (1,500 MW installed capacity by 2000 corresponding to about 3% of total electricity supplies) and the current level of RES-E (less than 3% of total electricity supplies in 1997) in the UK, but not in the light of the RES-E potential[20]. It required the adoption of new policy instruments to support RES-E, but RES-E policy change took some time to become effective (Climate Change Levy, Renewable Obligation, investment grants) (Helm, 2003; Mitchell and Connor, 2004).

The new Labour government initiated a number of energy reviews, the results of which were set out in the 1998 energy white paper[21] and were to become the key principles for the future UK energy policy (Mitchell, 2000). The 1998 energy white paper presented the reform agenda of the new government, which was expected to deliver social (keeping the electricity prices low for households) and environmental improvements (reaffirming the 10% of RES-E by 2010 goal), to protect the domestic coal industry (new contracts with electricity generators) and at the same time to keep a competitive energy market (reform of the Pool). But this paper tried to reconcile the irreconcilable (protection of the coal industry, increased competition and social and environmental improvements) and it did not provide many concrete actions. Its main

---

[20] The ETSU report R82 "An Assessment of Renewable Energy in the UK" of 1994 suggests that RES-E could be supplying up to 60% of UK electricity by 2025, under specific policy conditions (e.g. carbon tax) that would allow non-commercial RES-E technologies (e.g.) tidal, offshore wind, wave) to play a major part in the UK electricity sector (Elliott, 1994).

[21] DTI, The White Paper – Conclusions of the Review of Energy Sources for Power Generation, London: TSO, 1998.

contribution was to launch a series of energy issues that were to be addressed by the government thereafter, but following a rather eclectic approach (Helm, 2003).

At the same time, the new climate change policy of the Labour government was presented in the 1998 paper, "UK climate change programme" of the Department for the Environment, Transport and the Regions (DETR)[22]. It succeeds the 1990 environmental white paper[23], which acknowledged the climate change issue and set an aspired target of reducing the CO_2 emissions to their 1990 level by 2005, but it did not develop an effective climate change policy to achieve it. However, the 2005 CO_2 target proved to be easier to achieve than expected, even in absence of a dedicated policy, thanks to the privatisation of the coal industry in 1994 and the significant substitution of coal by gas in electricity generation that made the UK CO_2 emissions drop significantly. The 1998 white paper reaffirms the UK commitment to the Kyoto target of reducing the GHG emissions by 12.5% (by 2008) compared to the 1990 level, and it also sets a new domestic target of CO_2 emissions reduction of 20% by 2010 compared to the 1990 level. The paper suggested relying on economic instruments to reach this target (e.g. carbon tax or energy tax). However, the process of designing the new tax was politically sensitive, because that tax could not formally fall on the domestic sector (keeping low energy prices). This meant it had to respect the competitiveness of the British industry, and avoid direct damages to the coal industry (context of coal crisis) (Helm, 2003). The solution the government finally adopted in 2000 was an energy tax (and not a carbon tax, because that would have damaged the coal industry too much) on business energy users (and not on the domestic sector), which was called Climate Change Levy (CCL). In order to protect the energy-intensive users from the damaging effect of this tax, negotiated agreements with the government allowed discounts of the CCL of up to 80%, in return for new energy efficiency reduction measures (Helm, 2003). The CCL started in April 2001, at a rate of 0.43p/kWh for electricity consumption[24]. Most of the revenue from this levy is recycled to the business via other tax reductions (e.g. in national insurance contributions), which means there should not be any gain to the public finances from this levy. Moreover, about 50 million pounds of this levy revenue is used to support energy efficiency, low carbon technologies and RES-E programs (Brennand, 2004).

---

[22] DETR, "UK Climate Change Programme", London: TSO, 1998.
[23] Department of the Environment, "This Common Inheritance: Britain's Environmental Strategy", London: HMSO, 1990.
[24] Other rates applied to the consumption of other energy sources (oil, gas); see Helm, 2003.

The CCL forms an important element of the new British RES-E policy, given that the use of RES-E (except hydro > 10MW and waste incineration under specific conditions) is exempted from the levy. The exempted RES-E generators receive Levy Exemption Certificates (LEC) that the business energy users can handle to the Treasury to claim exemption from the CLL. Unlike the renewable obligation certificates (see below), the LEC cannot be decoupled from the physical delivery of RES-E. So the business electricity consumers who want to claim a CCL exemption have to switch their electricity supply to RES-E. This provides a strong incentive for business to opt for RES-E.

Many of the principles laid down in the 1998 energy white paper were enshrined in the Utilities Act of 2000[25]. With respect to the electricity policy, this act and its following statutory instruments introduced four main changes: the New Electricity Trading Agreement (NETA), the merge of OFGAS and OFFER into OFGEM, the legal unbundling of distribution and supply (dissolution of the RECs), and finally the Renewable Obligation (RO) to replace the NFFO.

Firstly, much criticism was raised during the 1990s about the Pool, especially against the dominant position abuse of the two main electricity generators (National Power and PowerGen). Its revision appeared to be inevitable to the new government. In April 2001, the New Electricity Trading Arrangement (NETA) replaced the existing Pool system. Unlike the Pool, the NETA is based on pay-as-bid auctions from the electricity generators and the electricity suppliers and not on system-marginal prices. In NETA, generators, suppliers and customers are able to contract as and when they choose. This seeks to remove unnecessary restrictions to the market actors. When the amount of electricity involved in the bilateral contracts involves too much or too little power compared to the overall demand, the system operator, the NGC, is responsible for making the balance. In order to cover the provision of electricity at the margin (balancing mechanism), the NGC requires the participants to the NETA to notify their expected physical positions for each half-hour trading period (initial physical notifications or IPNs) a day-ahead and final physical notifications (FPNs) when the balance mechanism for a half-hourly trading period opens]. The rules of the balancing mechanism are set out by the NGC in the Balancing and Settlement Code (BSC) under the review of OFGEM. The BSC imposes heavy penalties on the participants that do not respect their notifications in order to encourage

---

[25] The Utilities Act was supposed to deal with several utilities sector (telecom and water) but ended up dealing only with energy issues (electricity and gas). Therefore it is *de facto* an "Energy Act" rather than a "Utilities Act" (Helm, 2003; Mitchell and Connor, 2004).

them to contract. Under NETA, balancing is expected to play a residual role, with strong incentives to contract.

However, NETA did not prove to resolve the issue of dominant position abuse for two reasons. On the one hand, the large players possess an informational advantage which helps them establish a dominant position in the bidding system and manipulate the balancing mechanism (Macatangay, 2001). On the other hand, the movement of mergers and acquisitions in the electricity sector resulted in an oligopoly, a vertically integrated industry by 2001, which made it more difficult for small players to contract freely (Helm, 2003). In addition, the operation of NETA heavily hampers the development of RES-E in the UK in two ways: first, due to the scarcity of long-term contracts in the electricity market, and second, due to the high penalties to the generators with intermittent supplies (many RES-E generators, especially wind power) in the balancing mechanism (Helm, 2002). Indeed, RES-E generators are unreliable sources of electricity in a market which requires generators to notify their physical supplies in advance. Therefore, the marginal cost of balancing the system with RES-E is greater than other reliable and predictable sources of power. The solution forwarded by OFGEM and DTI is to develop consolidation arrangements between RES-E generators in order to spread the risk of unpredictability, but it did not receive positive feedback from the RES-E sector (Macatangay, 2001; Smith and Watson, 2002; Helm, 2002, 2003). In the 2004 British Energy Act, a single wholesale electricity market for Britain (England, Wales and Scotland), the British Electricity Trading and Transmission Arrangements (BETTA), replaced NETA, but no significant change to the market mechanism was operated.

Secondly, following the Utilities Act, the Office of Gas and Electricity Markets (OFGEM) was installed as the new regulator for the electricity and gas industries, as OFGAS and OFFER merged. The duties of the new regulator remained largely the same as before (namely ensuring competition), but they were slightly altered in favour of more social and environmental guidance to be developed by OFGEM.

Thirdly, the Utilities Act of 2000 enforced legal unbundling between distribution and supply, which resulted in the dissolution of the RECs. This facilitated the reorganisation of the electricity industry towards an increased vertical integration of supply with generation and horizontal mergers of the distribution networks (Helm, 2003). As a consequence of the dissolution of the RECs, the NFFO had no legal basis any longer, since it was organised around the RECs and the NFPA, and the Utilities Act therefore required that a new mechanism was put in place.

Finally, the Utilities Act sets the legal basis for the replacement of the NFFO by a new mechanism to support RES-E. The Utilities Act sets the obligation for the licensed electricity suppliers to supply a proportion of their electricity supplies to their customers from RES-E, but it left to the Energy Minister the task of defining how this obligation would be organised. After much consultation and reflection over the shortcomings and achievements of the former instrument, the Renewables Obligation (RO) started in May 2002, two years after the Utilities Act had been adopted and four years after the last NFFO Order had been launched (NFFO-5 in 1998).

The consultation process about the replacement of the NFFO occurred in three phases, starting in 1999 with the DTI paper "New and Renewable Energy: Prospects for the 21st Century"[26]. This first paper launched a series of questions about what the RES-E technologies are, why they should be supported and how, but it remained very elusive on the answers and short on definitive proposals (Helm, 2003). It acknowledged that the development of RES-E is a necessary part of the achievement of the UK climate change policy (the 1998 Climate Change Program) but it did not propose any concrete measures to address it. At the end of the consultation period and in parallel with the adoption of the Utilities Act in 2000, DTI published its conclusions in response to public consultation[27]. This paper officially confirmed the objective of the UK government to attain 10% of electricity supplies from RES-E by 2010 and laid down the key instruments to be used to reach this target: the RO, the CCL exemption, new RD&D programs and improved regional and local planning support. The main principles of the RO were already presented in this paper, but a second consultation process on the RO mechanism started later in 2000[28]. The content of this new consultation paper did not add much to the conclusions of the previous 2000 paper, but aimed specifically at hearing the comments of the actors of the electricity sector about how the RO should operate. The conclusions of the preliminary RO consultation[29] led in turn to the publication of a last statutory consultation in 2001[30], which presented the conclusions of the government about how the RO was expected to operate. It was then

---

[26] DTI, "New and Renewable Energy: Prospects for the 21st Century", London: TSO, March 1999.

[27] DTI, "New and Renewable Energy: Prospects for the 21st Century: Conclusions in Response to the Public Consultation", London: TSO, January 2000.

[28] DTI, "New and Renewable Energy: Prospects for the 21st Century: The Renewable Obligation: Preliminary Consultation", London: TSO, October 2000.

[29] DTI, "New and Renewable Energy: Prospects for the 21st Century: The Renewable Obligation: Preliminary Consultation: analysis of the responses to the consultation paper", London: TSO, March 2001.

[30] DTI, "New and Renewable Energy: Prospects for the 21st Century: The Renewable Obligation: Statutory Consultation", London: TSO, 2001.

submitted to Parliament to become part of the 2000 Utilities Act as the 2002 RO Order.

The RO entered into force in April 2002. The RO mechanism provides that the eligible RES-E generators (excluding large hydro >20MW) receive Renewable Obligation Certificates (ROCs) for the amount of RES-E they generate (1 ROC = 1kWh). The ROCs serve as proofs for fulfilling their obligation to the electricity suppliers and as additional income to the RES-E generators. Indeed, ROCs are traded on a specific market, separated from the electricity market, and therefore the RES-E generators earn revenue from two markets: the electricity market (at the wholesale market price) and the ROCs market (ROCs value <= 3p/kWh). The suppliers can meet their obligation in tree ways: firstly, by buying RES-E and ROCs from RES-E generators; secondly, by purchasing ROCs on the ROCs market (without RES-E); or thirdly, by paying the buy-out price (3p/kWh). In order to provide incentives for the suppliers to meet the obligation instead of paying the buy-out price, a recycling mechanism is formulated to distribute the receipts of the buy-out prices among the suppliers who complied with their obligation.

The quota of RES-E to be met by the electricity suppliers starts at 3% for the period 2002/2003 and rises gradually to 10.4% by 2010/2011 (RO Order 2002) and 15.4% by 2015 (RO Order 2005). The RO scheme is expected to last until 2027, but the quotas of RES-E to be supplied are set on a shorter-term basis. Unlike the NFFO (technology bands), the RO introduces competition between the different RES-E generators on the ROCs market, which is in line with the UK government's objective to especially encourage cost-effective technologies.

The additional cost of supplying electricity from RES-E may be passed onto the electricity customers. In order to provide a safety net so that the costs for the consumers of the RO do not rise out of control, if there were to be serious delays in the development of the industry, electricity suppliers are allowed to buy out all or part of their obligations as an alternative to demonstrating ROCs to OFGEM. The buy-out price (3p/kWh) therefore theoretically sets a ceiling to the value of the ROCs and to the impact of the RO on the consumer prices (price cap), given that the electricity suppliers would better pay the buy-out price than purchasing ROCs if the ROCs prices rise above the level of the buy-out price. However, this proved to be only theoretical, since at the early stage of the RO operation already the price of ROCs rose above the 3p/kWh limit. The conjunction of different factors can explain this: the shortage of RES-E capacity compared to the demand for ROCs (especially because large hydro is excluded), the benefit from the exemption from the CCL that "consumes" most of the RES-E capacity, and the high demand for ROCs from the electricity suppliers who want to maximise their share in the recycling of the buy-out fund (Smith and Watson, 2002).

In parallel to the introduction of the RO, the Royal Commission on Environmental Pollution (RCEP) suggested, in a new report published in 2000[31], that about 60% reduction in CO_2 emissions would be required by 2050 and that the current British energy policy was at odds with this objective. The Labour government replied with a report from the Cabinet Office's Performance and Innovation Unit (PIU) in 2002[32], which recognised the problems but did not provide any concrete answers. Following the PIU report, the DTI published a new energy white paper in 2003, "Our energy future – a low carbon economy"[33]. This paper presents new ambitious targets for the UK, but remains short in concrete new measures. The white paper endorses the RCEP's CO_2 target of 60% reduction by 2050 and proposes to deliver this through increased RES-E and energy efficiency efforts in the future. It reaffirms the target of 10% of RES-E to be supplied by 2010 and proposes an additional aspiration target of 20% RES-E by 2020, but without laying down new measures to achieve it, except an additional 60 m£ to be provided by 2006 for RES-E capital grants.

The RO and CCL exemption schemes do not distinguish between the RES-E technologies and in this context only the most competitive RES-E technologies are expected to be supported by the RO and CCL exemption (e.g. onshore wind power, landfill gas). However, the British government recognised that moving beyond the 10% target for 2010 requires significant innovation in RES-E technologies and the RO system does not support long-term innovation in new RES-E technologies (e.g. solar PV, wave power, offshore wind farms, energy crops)[34]. Therefore, at the beginning of the 2000s, the government introduced new programs in favour of less competitive (capital grants) and new RES-E technologies (RD&D grants) that complement the RO and CLL exemption[35].

Firstly, the government put in place a substantial RES-E support program worth in total 250 m£ between 2002 to 2006 (plus 60 m£ after

---

[31] RCEP, Energy-The Changing Climate, London: HMSO, 2000.
[32] PIU, The Energy Review, London: PIU, February 2002.
[33] DTI, Our energy future-creating a low carbon economy, Energy White Paper, London: TSO, February 2003.
[34] DTI, Renewables Funding, London: HMSO, May 2002. DTI, Our energy future-creating a low carbon economy, Energy White Paper, London: TSO, February 2003.
[35] For a detailed description of the capital grants and RD&D programmes, see: DTI, Renewables Funding, London: HMSO, May 2002.

2003)[36]. It grants capital subsidies to new RES-E investments in selected technologies. The priority between the prioritised technologies can change over time and currently focuses especially on offshore wind turbines, energy crops and solar PV. A Renewables Advisory Board has been established as an independent, non-departmental public body sponsored by the DTI. It brings together government departments, the renewables industry and the unions. It aims to provide advice to the government on a wide range of RES-E issues, including the priorities of the RES-E support program. The RES-E capital grants are financed and jointly managed by different institutions: DTI, DEFRA (for the energy crops program), and the New Opportunities Fund (NOF) of the National Lottery. In addition, the Carbon Trust[37] also participates in the effort to support new RES-E investments through its Low Carbon Innovation Programme.

Secondly, government support to RD&D in new RES-E technologies has also been extended since 2002[38]. The DTI launched a new research program for pre-competitive RES-E technologies, the New and Renewable Energy Programme (about 18 m£/year), which is managed by Future Energy Solutions (formerly ETSU until 2001[39]). Two calls for proposals are launched every year with technological priorities determined by Technological Route Maps, developed by DTI, in consultation with the industry and academia. This program currently supports biomass, solar PV, wind energy, wave energy, tidal stream and small-scale hydro technologies in priority[40]. In addition, the budget of the Research Councils has been increased (an extra 10 m£), specifically for RES-E research to 2006[41]. Finally, the Carbon Trust also funds RD&D programs under the Low Carbon Innovation Programme.

Finally, the British government improved its planning policy in 2004, with an updated Planning Policy Statement (PPS) on "Renewable Energy". It replaces the former Planning Policy Guidance "Renewable

---

[36] DTI, Our energy future-creating a low carbon economy, Energy White Paper, London: TSO, February 2003.

[37] The Carbon Trust was created in 2001 as an independent company funded by the government following the introduction of the CCL. It aims to help business and the public sector to reduce their CO_2 emissions and provides support programmes for low carbon technologies (capital grants, direct investments or RD&D grants).

[38] DTI, Renewables Funding, London: HMSO, May 2002.

[39] ETSU was created in 1974 by the government as the technological assessment arm of the Energy Department (then the DTI). In 2001, it was privatised as Future Energy Solutions, but it remains involved in the governmental energy RD&D programmes.

[40] DTI, Renewables Funding, London: HMSO, May 2002.

[41] DTI, Our energy future-creating a low carbon economy, Energy White Paper, London: TSO, February 2003.

Energy" of 1993, which set the first rules to improve the consideration of national energy policy objectives in the local planning permission permit procedures. The new PPS improves the planning conditions for RES-E projects (especially wind power and biomass plants, the more conflicting ones) with new principles to be followed by the regional and local planning authorities in their approach to planning for RES-E. It also requires the regional and local bodies to establish regional RES-E targets in line with the government's objectives (10% RES-E by 2010 and 20% by 2020) and derived from assessments of the region's RES-E potential. Improving the planning policy in the UK has been a central component of the new RES-E policy, since it proved critical to the failure of the NFFO scheme during the 1990s.

V. Interpretation of RES-E Policy Change in the UK

The RES-E policy has not been a priority for the British government during the last decades. R&D programs in RES-E technologies were launched after the first and second oil crises, but they were not very consistent in the long term (frequent revisions from ETSU) and they focused only on large-scale and close-to-market technologies. The main priorities of the Conservative government during the 1980s and 1990s was on the security of electricity supplies (until 1989 through the CEGB), the privatisation of the energy industries (gas, electricity, coal, nuclear), and, to a lesser extent, environmental issues, such as climate change. During the 1990s, the renewables NFFO scheme represented the main RES-E policy instrument in favour of RES-E market penetration in the UK, but it resulted less from a dedicated RES-E policy strategy than from a by-product of the nuclear policy. In addition, the outcome of the NFFO in terms of increased RES-E installed capacity was very poor, especially in comparison with the huge RES-E potential of the UK.

With the change in government in 1997, the British RES-E policy progressively took a new turn. A new policy objective of supplying 10% of electricity from RES-E by 2010 was announced in the Labour Manifesto of 1997 and finally adopted as a government target in 2000[42]. An aspiration target of 20% RES-E by 2020 was then formulated in the 2003 energy white paper[43], but without having been formally endorsed by the government so far. The formulation of a new RES-E policy was presented as an important dimension of the British climate change program, which sets ambitious targets for CO_2 emissions reduction (20%

---

[42] DTI, "New and Renewable Energy: Prospects for the 21st Century: Conclusions in Response to the Public Consultation", London: TSO, January 2000.

[43] DTI, Our energy future-creating a low carbon economy, Energy White Paper, London: TSO, February 2003.

reduction by 2010[44] then 60% reduction by 2050[45]). With the Utilities Act of 2000, the foundations for the replacement of the NFFO by the RO were put in place. A long process of public consultation (1999-2002) preceded the introduction of the RO as the central policy instrument of the new RES-E policy in April 2002. Other instruments in favour of RES-E were adopted in parallel with the RO to complement the latter: the CCL exemption for business, the new capital grant programs for near market RES-E technologies, increased RD&D budgets directed to new RES-E technologies, and improved regional and local planning policy for RES-E projects.

Table 2: RES-E policy change in the UK

	Before 2000	After 2000	Policy change
Policy objective	1,500 MW installed capacity by 2000	10% by 2010 and 15% by 2015	Yes
Policy instrument	NFFO: – Target group: electricity suppliers (NFFO purchase obligation) and RES-E generators (NFFO calls for tender) – Incentive: quantity (kW declared net capacity) – Resource: private (consumer bills, FFL)	RO: – Target group: electricity suppliers (quota of RES-E) – Incentive: quantity (kWh supplied) – Resource: private (consumer bills)	Yes

The turning point for the British RES-E policy change can be traced back to the year 2000, when the new policy target of 10% was formally endorsed, the conclusions of the first round of consultations over the RES-E policy options for the future were presented[46], and the Utilities Act was adopted with the foundations for the new RO. Both changes of RES-E policy objectives and instruments occurred at that time. Firstly, a new ambitious policy target (10% of the electricity supplies by 2010) was adopted by the Labour government in 2000, and then extended to 20% by 2020 (but not yet formally adopted). Secondly, the RO was adopted in 2000 as the central instrument of the new RES-E policy. Both the resources used to finance the RO (private budget from electricity consumer bills) and the type of incentive for the RES-E sector (quantity) did not differ significantly from those of the renewable NFFO, though the quantity of RES-E expected from the instrument was no longer expressed in declared net capacity, but in actual RES-E supplies

---

[44] DETR, "UK Climate Change Programme", London: TSO, 1998.
[45] RCEP, Energy-The Changing Climate, London: HMSO, 2000.
[46] DTI, "New and Renewable Energy: Prospects for the 21st Century: Conclusions in Response to the Public Consultation", London: TSO, January 2000.

under the RO (see table 2). Finally, we acknowledge that the actors targeted by the RO differ from those of the NFFO, though this is not so obvious at first sight. Under the NFFO, the electricity suppliers were forced to purchase the electricity of the RES-E generators who won the NFFO calls for tender. Thus there were actually two target groups under this system: the electricity suppliers (demand actors), under the NFFO purchase obligation, and the RES-E generators, under the NFFO calls for tender (supply actors). However, since the core of the RES-E policy were the NFFO calls for tenders rather than the NFFO purchase obligation (also for nuclear power), we do consider that the main targets were the RES-E generators (or more precisely the "potential" RES-E generators) who were competing in the calls for tender (supply actors). Under the RO, the target group is the electricity suppliers (demand actors), who are forced to buy a quota of ROCs every year. Therefore, we consider that there was a change of target group between the NFFO and the RO.

VI. Explaining RES-E Policy Change in the UK

A. Policy Actors

The endorsement of the new RES-E policy objective (10% of RES-E in total electricity supplies by 2010) by the Labour government in 2000 results from a political decision (central actors are politicians, see table 3). The target of 10% was first presented in the 1997 Labour Manifesto, then endorsed by the Labour government in 2000, but without being openly questioned or discussed (no participation of the actors of the sector).

Table 3: Dominant policy actors in the UK (H1/1)

	H1/1
Dominant policy actor?	– Policy objective: politicians (Labour Party) – Policy instrument: civil servants (DTI)

On the contrary, the formulation of the new RES-E policy instrument (namely the RO) results from an administrative decision in which the DTI was that central actor (civil servants, see table 3). Since the closure of the Department of Energy in 1992, energy policy has been taken care of by the larger DTI. At the political level, five different Ministers were in charge of energy from 1997 to 2003, with very broad attributions behind energy. Because of such a turnover at the political level, the major actor of the change of policy instrument was not the Minister for Energy, but the DTI, where all the energy competence resides (see table 1). Two other public actors were closely involved in the formulation of the new policy instrument due to their interest in the sector: the

Department for Environment, Food and Rural Affairs (DEFRA), and the energy regulator (OFFER then OFGEM). The discussions about the RO in Parliament (House of Commons and Lords) were not controversial, neither in 2000, for the adoption of the Utilities Act, nor in 2002, for the approval of the RO Order. On the contrary, the parliamentary process was rather quick and easy.

In addition, the formulation of the RO included the consultation of all the actors of the sector, from its early stages (see table 4). The first consultation (March-July 1999) aimed at assessing the NFFO system, as its pursuit with a sixth call for tenders was considered by DTI, and, proposing an alternative mechanism, the RO. At this time, the RO was only an idea. However, by the end of the consultation process, it became evident to the DTI that the NFFO could not be continued due to its problems (lack of commissioning of the projects) and the new electricity trade arrangement (NETA), which was to be adopted in the Utilities Act[47]. The answers to the consultation[48] and independent reviews of the NFFO (e.g. Elliott, 1994; Mitchell, 1995 and 2000) reinforced this observation. Therefore, between July 1999 and February 2000, over a very short space of time, the idea of the RO stood out as the best policy instrument and the DTI prepared it to be included in the Utilities Act of July 2002. In the end, due to the time constraint, the Utilities Act of 2000 contains only the preliminary ideas about the RO, and the design of the system was prepared afterwards. So, the main principles of the RO were included in the Utilities Act of 2000, but the actual formulation of the system and its details must be found in the RO Order of 2002.

The preparation of the RO Order was very open to the actors of the sector through two additional written consultations (of 2000 and 2001 respectively) and the more active participation of some actors in the process (AEP, individual energy companies, BWEA)[49]. The first RO consultation (in fact the second consultation but first specifically about the RO) discussed the details of the RO system and raised a lot of comments from the interested parties about how to implement the RO[50]. The second RO consultation[51] presented the RO Order submitted to

---

[47] Interview with Peter Stephens (DTI), 5 April 2006. And results of a survey at 18 UK actors with 9 valid answers, June 2005.

[48] DTI, "New and Renewable Energy: Prospects for the 21st Century: Conclusions in Response to the Public Consultation", London: TSO, January 2000.

[49] Interview with Peter Stephens (DTI), 5 April 2006.

[50] DTI, "New and Renewable Energy: Prospects for the 21st Century: The Renewable Obligation: Preliminary Consultation: analysis of the responses to the consultation paper", London: TSO, March 2001.

[51] DTI, "New and Renewable Energy: Prospects for the 21st Century: The Renewable Obligation: Statutory Consultation", London: TSO, 2001.

Parliament. Due to the tight deadlines (the RO Order was introduced in Parliament before the consultation was over), it was less determining than the first one.

Table 4: Influence of the actors of the sector (H1/2)

Policy actors (H1/2)	Early / late	Preferences	Influence
Electricity companies			
Association of Electricity Producers (AEP)	early	+	+
Electricity Association	early	+	-
Individual energy companies (e.g. TXU, Scottish and Southern Energy)	early	+	+
RES-E associations			
British Hydropower Association	early	+	-
British Biogen	early	+	-
British Photovoltaic Association	early	+	-
British Wind Energy Association (BWEA)	early	+	+
The Combined Heat and Power Association	early	+	+
Confederation of Renewable Energy Associations (CREA)	early	0	0
Energy from Waste Association	early	-	-
Association of Coal Mine Methane Operators (ACMMO)	early	-	-
Environmental associations			
Greenpeace	early	+	-
WWF	early	+	-
Country Guardian	early	-	0
Electricity consumers			
Confederation of British Industry (CBI)	early	-	-
National Electricity Consumer Council	early	0	0
Major Electricity Users' Group	early	-	-

Legend: Preference (+) favourable, (-) not favourable, (0) indifferent; Influence (+) strong, (-) weak, (0) no significant influence.

Table 4 summarises the preferences and the influence of the main actors of the sector involved in the formulation of the RO[52]. Since they were all consulted early, we are not able to test the hypothesis on the link between the earliness of their implication and their influence (first proposition of H1/2 non-applicable). In addition, we observe that most actors were in favour of the RO (+), except the waste and the coal associations (because they were not included in the RO), the large electricity consumers (because of fear of price increase), and the local environmental associations (because of the NIMBY syndrome); however, those actors were not very influential in the decision-making

---

[52] Interview with Peter Stephens (DTI), 5 April 2006. And results of a survey at 18 UK actors with 9 valid answers, June 2005.

process. Indeed, the actors who were the most influential supported the RO (second proposition of H1/2 validated).

In conclusion, the change of the RES-E policy objective in the UK was dominated by the political actors (the Labour Party), it was a political decision with not much debate with the actors from the sector that dates back to the Party Manifesto of 1997 (H1/1 validated). On the contrary, the change of policy instrument was dominated by the administration (DTI) and involved a large consultation of the actors from the sector, who proved to be influential in the design of the RO (H1/1 validated, H1/2 partly validated).

B. Past Policy

Under the Conservatives, during almost two decades, the RES-E policy was not a priority compared to the other energy issues (e.g. privatisation, coal crisis, nuclear support). The RES-E policy objective of the Conservative government amounted to no more than the short-term targets of the NFFO Orders (1,500 MW declared installed capacity by 2000[53]). In 1997, when the Labour Party won the elections, this objective appeared to be far too low compared to the RES-E potential in the UK and the increased political priority in favour of RES-E (see table 5). So, the Labour Party committed itself to "a new and strong drive for new and renewable sources of energy" in their 1997 Manifesto and adopted a new target of 10% of the total electricity supplies, covered by RES-E by the end of 2010[54]. So an in-depth RES-E policy change was put in place by the Labour government, starting in the 1998 energy white paper and leading in 2000 to the formal endorsement of the 10% target, the conclusions of the first RES-E policy consultation, and the adoption of the Utilities Act. Further improvements and developments of the RES-E policy followed in the 2003 white paper, but without significant changes to the original RES-E policy from 2000 onwards.

Table 5: RES-E past policy success/failure (H2)

	Past policy success/failure (H2)	
Objective	Past policy objective too low	Failure
Instrument	NFFO: – ineffective to reach target – affordable cost	Failure

---

[53] DTI, New and Renewable Energy: Future Prospects in the UK, Energy Paper 62, London: HMSO, 1994.

[54] DTI, "New and Renewable Energy: Prospects for the 21st Century", London: TSO, March 1999.

Besides the adoption of new policy objectives for the British RES-E policy, the Labour government questioned the performance of the existing instrument, the NFFO, and its compatibility with the new policy objectives. In terms of cost-efficiency, the NFFO has been widely praised for its success in driving down the costs of generating RES-E under the different NFFO contracts (see table 5). Indeed, the prices of the NFFO contracts were halved over the last decade and the more mature RES-E technologies (e.g. landfill gas) were almost competitive by the end of the scheme (DTI, 1999 and 2000; Mitchell, 1995 and 2000; Ross, 2000). Nevertheless, the NFFO proved to be a lot less successful in terms of effective RES-E development effectiveness (see table 5). By the year 2000 (two years after the end of the NFFO), only 331 projects, with an aggregate capacity of 834 MW declared net capacity, had been commissioned under the various NFFO, SRO and NI-NFFO Orders[55]; this is far from the 1,500 MW target. The weakness of the NFFO system was its low commissioning rate. A large share of NFFO contracts were actually never commissioned, and this for three main reasons: too low bidding price, no penalty for failure to commission, and delays to obtain planning permissions (Mitchell, 2000; Ross, 2000). The issue of low commissioning is also directly linked with the formulation of the NFFO itself, as the obligation is formulated in declared net capacity and not in actual RES-E to be supplied. In addition, the funds available for the renewables NFFO were too small for new RES-E technologies to penetrate the market (e.g. offshore wind power, tidal energy, waveenergy, solar PV), which significantly constrained the potential RES-E development under this scheme (Elliott, 1994; Mitchell, 1995 and 2000; Ross, 2000; Helm, 2003; Mitchell and Connor, 2004).

Therefore, by the year 2000, it emerged from the RES-E policy consultation carried out by DTI[56] and from the academic assessment of the NFFO system (e.g. Elliott, 1994; Mitchell, 1995 and 2000; Ross, 2000), that the NFFO was not able to fulfill the new RES-E policy objectives and that a change of the RES-E policy instruments was necessary. The RO was designed to address the main issue of the NFFO, namely the low commissioning rate, as it imposes a quota of RES-E to be delivered by the electricity suppliers (10.4% by 2010), with penalties for non-compliance (buy-out price) and rewards for compliance (recycling of the buy-out fund).

---

[55] DTI, "New and Renewable Energy: Prospects for the 21st Century: The Renewable Obligation: Preliminary Consultation", London: TSO, October 2000.

[56] DTI, "New and Renewable Energy: Prospects for the 21st Century: Conclusions in Response to the Public Consultation", London: TSO, January 2000.

Finally, with the introduction of NETA (end of the pool prices) and the dissolution of the RECs (legal unbundling of supply and distribution), it became impossible for the NFFO system to continue[57]. Indeed, the NFFO contracts were tied to the pool prices and managed by the NFPA on behalf of the RECs. The suppression of those elements meant that the NFFO was not adapted anymore with the new context of the electricity market (misfit) and needed to be entirely revised.

In conclusion, both the past RES-E policy objective and instrument were evaluated as unsuccessful by the Labour Party in the period 1997-2000, though only partially regarding the past policy instrument (NFFO), which explains why the policy changed (H2 validated).

C. *Europeanisation*

Traditionally, the British policies are less willing to cooperate on the European policies than the other European countries and this is especially true for the energy policy (Connor, 2004). The position of the UK at the forefront of the liberalisation of the electricity market in Europe led it to disregard the European liberalisation policy and adopt a passive position, while it could have been more active in promoting its model and serve as an example for the other European countries (McGowan, 1996; Helm, 2003). By the end of the 1990s, the isolation of the UK on the energy policy issues eroded progressively, due to the increased dependence of the UK energy supplies on imports, the new international commitments to protect the environment (e.g. Kyoto Protocol), and the arrival of the Labour Party in government. However, the influence of the European policy on the British energy policy remained marginal (Helm, 2003).

Figure 2: The European policy and RES-E policy change in UK

EU policy					
Directive liberalisation + Green paper SER	White paper SER		Working paper E-SER	Draft directive E-SER	Directive E-SER + ECJ judgement
1996	1997	1998	1999	**2000**	2001
				RES-E policy change in UK	

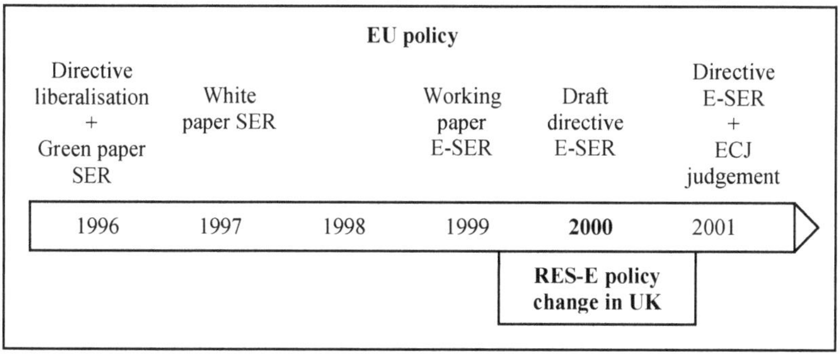

[57] DTI, The White Paper – Conclusions of the Review of Energy Sources for Power Generation, London: TSO, 1998. DTI, "New and Renewable Energy: Prospects for the 21st Century", London: TSO, March 1999.

During the 1980s and 1990s, the UK was a laggard member state on the environmental aspects of the energy policy. At the end of the 1990s, increased concerns and new policies about environmental protection in the energy sector appeared. The RES-E policy change of the beginning of the 2000s contributed to this process. The DTI paper of 2000 about the new RES-E policy for the future[58] acknowledged that full collaboration with the European institutions is desirable and that the UK RES-E policy was expected to contribute to the European White Paper on RES of 1997[59], but other references to the European policy were very rare in the UK. In fact, RES-E policy changes in the UK seem to have been driven by domestic factors rather than by European determinants. Both the UK RES-E policy change process and the development of the European directive on RES-E occurred in the same period (see figure 2), but, from the UK perspective, they appear to have been conducted rather independently, without much consideration for the European policy at the domestic level, and consequently without much intervention of the domestic actors at the European level (see table 6). The European White Paper of 1997, as well as the RES-E directive of 2001 probably represented strong incentives for the UK government to revise and improve its RES-E policy, but no direct influence can be observed. Whether the past policy fitted or not with the European RES-E policy under construction was not regarded by the DTI or the other actors in the UK (see table 6). The adoption of the British RES-E target (10% by 2010) and the change of policy instrument (RO) were national policy choices and the hypotheses about the influence of the European policy are non applicable in this case.

Table 6: Europeanisation of the RES-E policy (H3)

	Fit/Misfit (H3)	
Objective	1,500 MW installed capacity by 2000	n.a.
Instrument	NFFO: – target: electricity suppliers (NFFO purchase obligation) and RES-E generators (NFFO calls for tender) – budget: consumer bills (FFL)	n.a.

Legend: n.a. (non applicable).

In conclusion, we do not observe such a thing as the Europeanisation of the UK RES-E policy. On the contrary, the RES-E policy change that

---

[58] DTI, "New and Renewable Energy: Prospects for the 21st Century: Conclusions in Response to the Public Consultation", London: TSO, January 2000.

[59] European Commission, White Paper for a Community Strategy and Action Plan. Energy for the Future: Renewable Sources of Energy, COM(97)599.

occurred in 2000 results from domestic factors and was elaborated rather independently from the European RES-E policy process. So, in this case, we are not able to validate or invalidate our hypotheses about the Europeanisation of policy changes (H3 non applicable).

D. *Political and Institutional Context*

The UK has a majoritarian political system characterised by single party governments. Two main political parties alternate in government: the Conservative Party and the Labour Party. From 1979 to 1997, the Conservative Party ruled in the UK during four successive terms. In 1997, the Labour Party won the general elections and the government majority changed hands (see table 7). Due to the polarisation of the party system in the UK (a two-party system), government changes often result in significant policy changes, and this was also the case for the RES-E policy from 1997 onwards. This is due to the fact that, unlike in consensus democracies, the party which wins the elections can dominate government and parliament without having to make compromises with other political parties (no partisan veto players, see table 7). In addition, internal party discipline among MPs in the House of Commons and loyalty to the government are strict. Of course, policy legacies often prevent the new government from radically reversing the existing policies (e.g. energy industry privatisation), but policy changes are still easier to implement than in other political systems (e.g. RES-E policy changes). Early in its term, the new Labour government started extensive revisions of the past energy policy, including the RES-E policy, which resulted in significant policy changes at the beginning of the 2000s, notably in the RES-E policy.

Table 7: Change in government (H4/1) and veto players (H4/2) in the UK

Governing party (H4/1)	Veto players (H4/2)
– New party in government: Labour Party – Large ideological distance with Conservatives in the past – No coalition, full power on policy (majoritarian democracy with two party system)	– Few institutional veto players: House of Commons and House of Lord (less powerful), Queen (but no veto power in reality) with strong party discipline in House of Commons and small ideological distance between them – No partisan veto players: no multi-party coalition and party discipline

In addition, the UK is characterised by a rather non-fragmented institutional system, with few institutional veto players, which makes policy changes easier to achieve. The Parliament is bicameral, including the House of Commons (lower house) and the House of Lords (upper house). They both participate in the legislative process, but the House of Commons has the final word, even if the House of Lords disagrees with the bill. In fact, the House of Lords is not very powerful since it can only delay a bill but not withdraw it (see table 7). Unlike the House of Commons, its members are not elected but co-opted and many of the Lords are independent from any political party, which makes it more independent from the government. However, since its actual veto power is limited, the House of Lords is not a significant veto player in the UK. In addition, the royal assent of the Queen is necessary for legislation to enter into force, but this is usually a formality, since the monarchy lost most of its political prerogatives. So the institutional system in the UK is very homogenous, with few institutional veto players, which makes policy changes very likely when decided by the governing party.

In conclusion, the UK political and institutional context facilitates policy change, thanks to the alternation of the two main political parties in government and the presence of very few veto players opposing the government. With the Labour Party's election victory in 1997, the conditions were fulfilled for a revision of the past RES-E policy and no significant veto players were able to oppose it (H4/1 and H4/2 validated).

VII. Conclusion

Following the arrival of the Labour Party in government in 1997, both the RES-E policy objective and instrument changed in the UK. Firstly, the Labour government adopted a new policy objective: 10% of RES-E in the total electricity supplies by 2010, which was then extended to 15% by 2015. Secondly, the past policy instrument, the NFFO, was replaced by the RO, which, like the NFFO, is financed by the electricity consumer bills, but, unlike the NFFO, it targets the electricity suppliers (and no longer the RES-E generators) and provides an incentive in the form of quantities of RES-E to be supplied to the electricity consumers (and not declared capacities).

Table 8: Validation and relevance of the hypotheses in the UK

	H1		H2	H3	H4	
	H1/1	H1/2			H4/1	H4/2
Validation hypotheses						
Policy objective	V	n.a.	V	n.a.	V	V
Policy instrument	V	"V"	V	n.a.	V	V
Explanation policy change						
Policy objective	++	0	++	0	++	+
Policy instrument	++	++	++	0	+	+

Legend: ++ (major explanation of policy change), + (minor explanation), 0 (does not explain policy change). V (hypothesis validated), I (hypothesis invalidated), "V" (hypothesis partly validated), n.a. (hypothesis non applicable).

Table 8 summarises the findings of this case study about the validation of the hypotheses of our theoretical framework. To begin with, our first assumption about the policy actors (H1/1) is fully validated in the UK case, as the dominant actors of the change of policy objective were political actors (the Labour Government) and the dominant actors of the change of instrument were administrative actors (DTI). The second hypothesis, about the influence of the actors of the sector on the policy change (H1/2), has turned out to be non-applicable regarding the change of policy objective and only partly validated regarding the change of policy instrument. Indeed, in the case of the change of policy instrument in the UK, the assumption about the relation between the earliness of the participation in the policy-making process and the influence of the actors of the sector could not be tested (first proposition of H1/2 non-applicable), and the assumption about the preferences of the most influential actors and the policy change was validated (second proposition of H1/2 validated), which means that, in the end, the hypothesis is only partly validated. Secondly, the relation between the success or failure of the past policy and the policy change is validated, for both the policy objective and the policy instrument, as they had both been unsuccessful in the past (H2). Thirdly, the hypothesis about the Europeanisation of the domestic policy (H3) was impossible to validate or invalidate in the UK case, because the process of RES-E policy change was conducted independently from European policy. So, H3 is non-applicable in this case. Finally, the hypotheses about the influence of the political and institutional context on the policy change are both validated. H4/1 is validated, because the RES-E policy change happened under a new government, with full power on the policy change (no coalition with other parties). In addition, the absence of veto players to oppose the government's policy change validates H4/2.

Table 8 also shows the relevance of the hypotheses to explaining the RES-E policy changes in the UK (objective and instrument). On the one hand, the change of policy objective is best explained by three factors: the change of the party in power (H4/1); the fact that the change of policy objective was dominated by political actors (H1/1); and the failure of the past policy objective (H2). Indeed, the change of policy objective resulted from the combination of those factors: the Labour Party came to power in 1997 (H4/1) and introduced a new RES-E target to be reached by 2010 (H1/1), because the past target was not ambitious enough (H2). Besides, because of the political and institutional context of the UK, no veto players opposed this decision, which facilitated the policy change process (H4/2). Finally, we observe that the European policy did not play a role in this policy change, partly because the European policy was still very uncertain at that time, and also because of the traditional independence of the UK government vis-à-vis European policy. Thus European policy is not relevant to explaining the change of policy in the UK (H3). Moreover, the definition of the new policy objecttive did not lead to a large debate with the actors of the sector, which explains why H1/2 is not relevant to explain the change of policy objective.

On the other hand, the change of policy instrument is primarily explained by the failure of the past instrument and its incompatibility with the new electricity trade arrangement in the UK (H2). In addition, the fact that the formulation of the policy instrument was piloted by the DTI (civil servants), with the support of the actors of the sector, also significantly explains the change of policy instrument (H1/1 and H1/2). The political and institutional context (the new Labour government (H4/1) and few veto players (H4/2)) favoured the change of policy instrument, but it represents a minor explanation compared to the other ones. Finally, as for the change of policy objective, the European policy does not seem to have influenced the change of policy instrument in the UK (H3).

In conclusion, except for the Europeanisation hypothesis, which was not applicable in this case (H3), the hypotheses of our theoretical framework are validated in the UK case, though one only partially (H1/2), which confirms the soundness of our theoretical framework. Moreover, the hypotheses prove to be relevant to explaining the RES-E policy changes in UK (except the Europeanisation hypothesis). The only missing element is the fit/misfit with the new context of the NETA, which contributes to explaining the change of policy instrument. However, that is not covered by our framework. We will discuss this later in the comparison.

CHAPTER 8

Comparison

In the last five chapters, we have presented the RES-E policies of Belgium, Denmark, Germany, the Netherlands and the UK. For each case, we ended up with the analysis of the policy change (dependent variable) and the test of the hypotheses that we formulated in the theoretical chapter (independent variables). In this chapter, we will not enter into the details of the case studies anymore, as this has been dealt with in the previous chapters, but we will compare the results of the case studies in order to generalise the findings of this research. Firstly, we look at the different types of policy change that occurred in the different countries (policy objective and/or policy instrument) and we observe if the RES-E policies tend to converge or diverge. Secondly, we compare the results of the test of the hypotheses in each country in order to validate/invalidate the hypotheses, and we propose ideas about how to improve our theoretical framework when necessary. Finally, we compare the relevance of the hypothesis in explaining the policy changes in the five countries in order to point out the most significant explanatory factors (necessary factors).

I. Comparison of the Policy Changes

The patterns of policy change differ in the five countries (see table 1). Some, like Belgium and the UK, experienced changes in the policy objective and instrument. In others, like Denmark and the Netherlands, only the policy instrument changed. And finally, in Germany, the policy objective changed but not the policy instrument. According to Peter Hall (Hall, 1993), Belgium and the UK represent cases of third order change (change of the policy objective and instrument), and Denmark and the Netherlands cases of second order change (only change of the policy instrument), but the case of Germany (change of objective without change of instrument) is atypical in Hall's typology. This demonstrates that there is no hierarchical link between the two types of policy change and that the objective of a policy can change while the instrument does not. However, we observe in the German case that the settings of the instrument can change. So the instrument itself remains the same, but its

How and Why Do Policies Change?

settings are modified in order to be more adapted to the new policy objective.

Table 1: Policy changes in the 5 countries

Country	Turning point	Change policy objective	Change policy instrument
Belgium	1999	Y	Y
Denmark	1999	N	Y
Germany	2000	Y	N
Netherlands	2003	N	Y
UK	2000	Y	Y

Table 1 also shows that the turning points of the five countries are very similar (1999-2000), except for the Netherlands. So it might be interesting to wonder why most of the RES-E policy changes occurred in 1999-2000. What stands out from our case studies is that in all countries, the policy changes were primarily driven by domestic factors (change in government or past policy failure), which does not explain why they occurred at the same time in four out of five countries. The main factor that could explain the convergence of timing is the European policies. In 1999-2000, the European RES-E directive was under discussion, so it certainly gave an impulse to the member states to revise their RES-E policies, but we will see in the next section that it was not a major explanation for the policy changes. In addition, the liberalisation of the electricity market and the corresponding reform of the electricity sectors in the member states in that period also encouraged the reform of the RES-E policies, though again it was not a major explanation for the policy change (except for Belgium).

In three out of the five countries (Belgium, Germany and the UK), the objective of the RES-E policy changed (see table 2). The three of them had no or small- and short-term RES-E targets previously, which explains why they changed. On the contrary, Denmark and the Netherlands already had long-term significant targets prior to policy change. In table 2, Belgium and Germany are both characterised by no previous quantitative target (none). However, the RES-E policy objectives in both countries were very different before the policy change, as in the German case there was an objective to increase the share of RES-E, even if limited and not quantified in specific target; while in Belgium there was no significant objective to increase the share of RES-E, hence no quantitative target. So, the policy change was more significant in Belgium than in Germany, but since we do not consider ordinal policy changes but dichotomous policy changes in this research, this distinction is not emphasised in table 2. In addition, we observe that in all countries, the RES-E targets are congruent with the indicative targets of the 2001

Directive on RES-E (see table 3 in chapter 3). So they tend to converge towards the European target, even if there was no obligation to do so. However, in absolute terms, the policy objectives are very different, due to various reasons that differ according to each country: the existing RES-E production, the political priority in favour of RES-E, the physical potential and the industrial background.

Table 2: Change of policy objective in the 5 countries

Country	Turning point	Target before	Target after	Change of policy objective
Belgium	1999	None	3% by 2004 and 6% by 2010	Yes
Denmark	1999	20% by 2005, 29% by 2010 and 68% by 2030	20% by 2003, 29% by 2010 and 68% by 2030	No
Germany	2000	None	12.5% by 2010 and 20% by 2020	Yes
Netherlands	2003	9% by 2010	9% by 2010	No
UK	2000	1,500 MW installed capacity (± 3%) by 2000	10% by 2010 and 15% by 2015	Yes

The pattern of change regarding the policy instruments is more complex (see table 3). Four out of the five cases experienced a change in their policy instrument (Belgium, Denmark, the Netherlands and the UK). However, the attributes of the policy instrument that changed differ among these countries. New target groups were appointed in Denmark, the Netherlands and the UK. The incentive used to promote RES-E was modified in Belgium, Denmark and the Netherlands. And, finally, the resources used to finance the instrument were privatised in Denmark and the Netherlands. So, the three attributes prove to be relevant to characterise the change of policy instrument. Moreover, whether only one (Belgium) or the three attributes (Denmark) change is not critical in this research, because we do not differentiate according to the extent of the policy instrument change.

Table 3: Change of policy instrument in the 5 countries

	Before policy change			After policy change		
	Target group	Incentive	Resources	Target group	Incentive	Resources
Belgium	demand	Price	private	demand	quantity	private
Denmark	supply	Price	public	demand	quantity	private
Germany	demand	Price	private	demand	price	private
Netherlands	demand	Price	public	supply	price	private
UK	supply (demand)	quantity (price)	private	demand	quantity	private

Regarding the convergence/divergence of the policy instruments in the five countries, we observe some interesting trends. Firstly, in the past and still today in most countries the target group of the RES-E instruments have been the demand actors, namely the distribution or supply companies or the electricity consumers (except in Denmark and the UK in the past and today in the Netherlands). Moreover, today only the Netherlands focuses on the supply actors as target group (namely the Dutch RES-E generators), while Denmark and the UK, which focused primarily on the supply in the past, now target the demand actors. So, we observe a convergence among these countries in favour of the demand side of the market being chosen as the target group of the RES-E policy instruments. Secondly, in the past, most of the instruments used the price as an incentive to support RES-E (except for the UK), but today some of them have changed to quantity incentives (Belgium and Denmark). However, we do not observe a significant convergence among these countries regarding the incentives, as the diversity of incentives remains predominant (see table 3). In addition, in Belgium the quantity incentive is the dominant RES-E instrument (quota), but it has been complemented by subsequent price incentives (minimum prices for the TGC). In Denmark, the new instrument based on quantity incentive has never been implemented and so far the price incentives of the past remain operational. So the picture is a lot more divergent about the incentives than about the target groups. Finally, we observe a privatisation of the resources used to finance the RES-E in all countries. The two cases in which the RES-E instrument was financed by the public budget in the past moved to a private funding through the electricity bills (Denmark and the Netherlands). So, a strong convergence towards the funding of the RES-E instruments by private funds (electricity bills) is demonstrated in our cases. In conclusion, there is still a diversity of RES-E policy instruments in those countries (and in Europe in general), but they tend to converge on some aspects (e.g. resources, target group).

We used specific indicators to characterise the change of policy objective (target of RES-E) and the change of policy instrument (target group, incentive and resources). These indicators proved to be very relevant in our cases, as they capture the RES-E policy changes in those countries fairly well. However, other indicators could have been selected, as the research would have focused on other aspects of the RES-E policy change. For instance, RES-E policy objective could have been characterised by: the relative priority in favour of the different RES-E technologies; the relative priority to RES-E R&D support, RES-E industry support, and RES-E market development; the importance of the RES-E objective in the hierarchy of the goals of the broader energy policy (e.g. energy efficiency, nuclear energy, fossil fuels). Regarding the policy instrument as well, other indicators could have been used: the market-based feature of the instrument; the constraining versus voluntary aspect of the instrument; the administrative and political costs associated with the instrument. In addition, the settings of the instruments could have been considered, such as: the level of the incentive awarded to RES-E, the category of RES-E technology eligible, the technical or geographic conditions to be eligible, the length of time of the support. However, in the end, we still believe that we have selected relevant, interesting and consistent indicators, especially from a comparative perspective.

II. Test of the Hypotheses

In the theoretical chapter of this paper, we identified four main explanatory factors of the policy change: the policy actors (H1), the past policy (H2), the European policy (H3) and the political and institutional context (H4). For each factor, we specified one (H2, H3) or two (H1, H4) specific hypotheses about how we assume these factors explain the policy changes. Then, in the last sections of each country chapter, we tested the hypotheses and we discussed the relevance and the significance of each factor.

In this section, we compare the results of the five case studies in order to see if the hypotheses of our theoretical framework can be validated from a comparative and thus more general perspective (see table 4). According to Popper's idea about the role of the falsifiability in the evaluation of theories, when a hypothesis is invalidated in one case, it is considered to be invalidated in the end. However, this criterion proved to be too strict and very simplistic in reality, and therefore we consider that when a hypothesis is confirmed in a majority of cases, then it means that the hypothesis is validated, though in this context we also consider the cases in which the hypothesis was not confirmed empirically in order to improve the hypothesis. Besides, we do not only carry out an accurate and systematic test of the hypothesis (see table 4), but we also consider the relevance and the significance of the hypothesis to explain the policy changes in the five cases (see table 13). So, at the end, we end up with a fairly accurate picture of the hypotheses that are empirically confirmed and e relevant to explaining the policy changes (see table 14).

Table 4 summarises the test of the hypotheses in the five countries and presents the results of the comparison regarding the validity/invalidity of the hypotheses. We observe that our theoretical framework is largely validated, though some hypotheses seem to be more confirmed than others. In addition, it appears that some hypotheses are validated in cases of change of policy objective and not in cases of change of policy instrument, which confirms the need to distinguish those two types of policy changes. Now, let us summarise the details of the test of each hypothesis and propose a certain number of amendments when necessary.

Table 4: Test of the hypotheses[1]

	Policy change	H1		H2	H3	H4	
		H1/1	H1/2			H4/1	H4/2
Belgium	Objective	V	n.a.	V	V	V	V
	Instrument	I	"V"	V	V	V	V
Denmark	*Objective*	*V*	*n.a.*	*V*	*V*	*V*	*n.a.*
	Instrument	V	V	V	V	I	V
Germany	Objective	V	"V"	V	V	V	V
	Instrument	*V*	*"V"*	*V*	*I*	*I*	*n.a.*
Netherlands	*Objective*	*V*	*n.a.*	*V*	*V*	*V*	*n.a.*
	Instrument	V	"V"	V	I	I	V
UK	Objective	V	n.a.	V	n.a.	V	V
	Instrument	V	"V"	V	n.a.	V	V
Comparison	**Objective**	**V**	**n.a.**	**V**	**V**	**V**	**V**
	Instrument	**V**	**V**	**V**	**I**	**I**	**V**

Legend: V (hypothesis validated), I (hypothesis invalidated), "V" (hypothesis partly validated)[2], n.a. (hypothesis not applicable)[3].

The first hypothesis (*H1/1*) assumes that the type of policy actor that dominates the policy-making process determines what type of policy change is expected to happen. We observe that this hypothesis is empirically validated in our five cases, with the exception of the change of policy instrument in Belgium. As was to be expected, the dominant policy actors of the change of RES-E policy objective in Belgium, Germany and the UK were political actors (governing political party(ies), energy minister(s), or members of parliament)) (see table 5). In addition, we observe that the hypothesis is also confirmed by default in the cases of non-change of policy objective (Denmark, the Netherlands), since the dominant actors of the policy making process are civil servants and not political actors (see table 5). Regarding the change of policy instrument, we assumed that the dominant actors were civil

---

[1] The lines in italics correspond to cases of no policy change in which the test of the hypothesis could only be done by default.

[2] Regarding H1/2, the hypothesis can be "partly" validated because it is composed of two distinct propositions. This means that when one proposition is validated while the other one is invalidated or non applicable, the hypothesis is "partly validated".

[3] In some cases, the hypothesis cannot be tested since the relation that we intend to characterise is not relevant (e.g. H3 in UK). So in those cases the hypothesis is neither validated nor invalidated, but just not applicable (n.a.).

servants. This was validated in Denmark, the Netherlands and the UK, but not in Belgium (see table 5). Indeed, in Belgium, the change of policy instrument was managed by the ministers of energy and their cabinets. The administration for energy was associated in the process, but they did not lead it, as in the other countries. This is a characteristic of the policy making process in Belgium, where the Ministers and their cabinets usually formulate the policies, then submit them to parliament, while the administration is often confined to implementation tasks. Therefore, it is not a surprise to observe that our hypothesis H1/1, which assumed that changes of policy instruments are managed by civil servants, is not fully validated in the Belgian case (see table 5).

Table 5: Details about the test of H1/1 (dominant policy actors)[4]

		Policy change/ no change	Dominant policy actors	H1/1
Belgium	Policy objective	Yes	Political actors	V
	Policy instrument	Yes	Political actors	I
Denmark	*Policy objective*	*No*	*Civil servants*	*V*
	Policy instrument	Yes	Civil servants	V
Germany	Policy objective	Yes	Political actors	V
	Policy instrument	*No*	*Political actors*	*V*
Netherlands	*Policy objective*	*No*	*Civil servants*	*V*
	Policy instrument	Yes	Civil servants	V
UK	Policy objective	Yes	Political actors	V
	Policy instrument	Yes	Civil servants	V

The second hypothesis (*H1/2*) looks at the influence of the actors of the sector in the process of policy change. It refers to two different propositions: one about the influence of the earliness of the participation of the actors of the sector in the policy change (if earlier, more influence), and another one about the influence of the preferences of the most influential actors of the sector on the policy change (if in favour of policy change, policy change more likely). We observe in table 4 that this hypothesis is less validated empirically than the other ones, as it turns out to be non-applicable in a number of cases and only partly validated in most other cases. The reason why this hypothesis is only partly validated in most cases is that it is composed of two different propositions, which means that if one is validated while the other is not, the hypothesis as a whole is only partly validated. To solve this issue, we should split this hypothesis into two distinct hypotheses: H1/2 and

---

[4] The lines in italics correspond to cases of no policy change in which the test of the hypothesis could only be done by default.

H1/3 (see figure 1 in red, which amends figure 3 of the theoretical chapter). By doing so, we will be able to consider the test of the two propositions separately, in order to gain a better understanding about what is validated and what is not in H1/2 (see table 6).

Figure 1: Amended hypotheses H1/2 (actors of the sector)

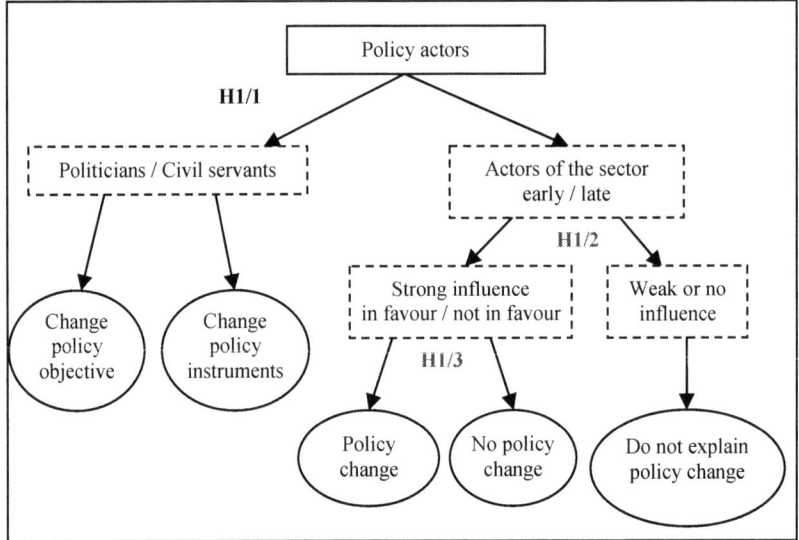

The first proposition of H1/2 (H1/2 in figure 1) assumes that if the actors of the sector are involved in the policy change process in its early stages, they are expected to have more influence on the policy change than if they are involved later on in the process, when the formulation of the policy is already in an advanced stage. Table 6 shows that this hypothesis turned out to be empirically non-applicable in most of the cases, either because the actors of the sector were all consulted early on in the process, or because they were not consulted at all. In the two cases where the hypothesis was applicable, it proved to be validated (Denmark and the Netherlands, see table 6). Therefore, in the end, the hypothesis is validated only in two cases, which leads us to question its relevance. Indeed, in the light of the cases where it was not applicable, we believe that the earliness of the participation in the formulation of the policy is only one among other explanations of the variation in the influence of the different actors of the sector on the policy change. For instance, we mentioned in the German case that the political connections of the actors with the governing political parties who dominate the policy change is a better explanatory factor of the influence of the actors than the earliness of their implication in the formulation of the policy.

Table 6: Details about the test of H1/2 (actors of the sector)[5]

	Policy change /no change	H1/2		H1/2
		Earliness (1st prop)	Preference (2nd prop)	
Belgium	Policy objective	n.a.	n.a.	n.a.
	Policy instrument	n.a.	V	"V"
Denmark	*Policy objective*	*n.a.*	*n.a.*	*n.a.*
	Policy instrument	V	V	V
Germany	Policy objective	n.a.	V	"V"
	Policy instrument	*n.a.*	*V*	*"V"*
Netherlands	*Policy objective*	*n.a.*	*n.a.*	*n.a.*
	Policy instrument	V	n.a.	"V"
UK	Policy objective	n.a.	n.a.	n.a.
	Policy instrument	n.a.	V	"V"
Comparison	**Policy objective**	**n.a.**	**n.a.**	**n.a.**
	Policy instrument	**n.a.**	**V**	**"V"**

The second proposition of H1/2 (H1/3 in figure 1) assumes that if the preferences of the most influential actors are congruent with the ideas of the dominant policy actors regarding the policy change (hence in favour of the policy change), then the policy change is expected to be achieved; whereas when the preferences are not congruent (hence not in favour of the policy change), the policy change is expected to fail to be carried out or to be adapted. Table 6 shows that this assumption is non-applicable in most cases of change of policy objective, because the actors of the sector were not involved in the formulation of the policy objective. The only exception is the change of policy objective in Germany, where the hypothesis is validated, since the actors of the sector who had the biggest influence were in favour of the change of policy objective. Therefore, we question the relevance of this hypothesis regarding the change of policy objective, as it proves to be non-applicable in most cases. Regarding the change of policy instrument, this hypothesis appears to be more relevant as it is applicable and validated in most cases (except for the Netherlands), as the actors of the sector who had the strongest influence on the formulation of the policy were in favour of the policy change (see table 6). In the Dutch case, the actors who had a strong influence on the policy-making process had no preference regarding the policy change, which means that the hypothesis could not be tested. So, in the light of these cases, this hypothesis turns out to be rather irrelevant regarding the changes of policy objective, but it proves to be

---

[5] The lines in italics correspond to cases of no policy change in which the test of the hypothesis could only be done by default.

validated when changes of policy instrument are concerned (see table 6). However, in order to validate it further, it would be interesting to confront it to cases of change of policy instrument where the most influential policy actors' preferences are not in favour of the policy change. We would see if the dominant policy actors adopt the policy change anyway (the hypothesis would be invalidated) or if they adapt or withdraw it (the hypothesis would be validated).

In conclusion, we observe that hypothesis H1/2, about the influence of the actors of the sector on the policy change, could not be tested empirically in most of the cases, especially in cases of change of policy objective, which leads us to question its relevance (see table 6). It proves to be more relevant and partially validated in the cases of change of policy instrument, but still to a small extent.

Finally, when he considers the influence of the policy actors on the policy change, Peter Hall (Hall, 1993) assumes that when the dominant policy actors are political actors (third order policy changes), the policy change process involves a broad debate with the actors of the sector (and even more actors); whereas when the dominant policy actors are civil servants (second order policy changes) the policy change process remains confined within a closed group of policy experts and is not open to a large debate with the actors of the sector. As we mentioned in the theoretical chapter, we do not believe that these factors are strongly linked, which is the reason why we formulated two distinct hypotheses under H1. Now, in the light of the comparison of our five case studies, we observe that Peter Hall's assumption is validated in three cases out of five, which means that it is not as bad as we expected. Indeed, two cases show results that do not conform to what Peter Hall's assumption had predicted (Belgium and the UK). In Belgium, the change of policy objective was dominated by the political actors, but without a large debate taking place on this issue, and the change of policy instrument was also dominated by the political actors (unlike Peter Hall assumed), but this time with a large consultation of the actors of the sector (see table 7). In the UK, the change of policy objective was also dominated by political actors, without debate with the sector, while the change of policy instrument was dominated by the civil servants, but with a large consultation of the actors of the sector at different stages of the policy process (see table 7). So, in these two cases Peter Hall's assumptions are not validated, though they are validated in Denmark, Germany and the Netherlands (see table 7).

Table 7: Test of Peter Hall's hypothesis about the policy actors

	Dominant policy actor	Large debate with actors of the sector	Validation of Hall
Belgium	Political actors (objective and instrument)	No (objective); Yes (instrument)	I
Denmark	Civil servants	No	V
Germany	Political actors	Yes	V
Netherlands	Civil servants	No	V
UK	Political actors (objective) Civil servants (instrument)	No (objective); Yes (instrument)	I

The third hypothesis of our theoretical framework (*H2*) assumes that if a policy was evaluated as a success (versus failure) in the past, then it is not likely to change (likely to change). Table 4 shows that this hypothesis is validated in all five cases, which confirms Peter Hall's assumption that the failure of the past policy is a necessary condition for policy changes. Table 8 shows the details of the test of the hypothesis in the five cases. On the one hand, in every case of past policy success, there was no policy change, and on the other hand, when policy failures were demonstrated, the policy changed (see table 8). However, we observe that sometimes the failure or the success is not straightforward (policy objective in Belgium, Germany and the Netherlands). In Belgium and Germany, there were no quantitative targets of RES-E in the past, which demonstrates the lack of long-term ambition of the government for the RES-E policy. Therefore, one cannot talk *stricto sensu* about the failure of the past policy target. However, the absence of target was clearly perceived to be a failure of the past policy by the policy actors, and something that needed to be changed. This is the reason why we can still talk about the "failure" of the past policy objective and validate the hypothesis that assumes that such a failure leads to a policy change (see table 8). Moreover, in the Netherlands, the same logic applies to the success of the past policy objective. In the light of the development of the past RES-E domestic generation, the Dutch RES-E target did not appear very successful at the time of the policy change, but the policy actors did not consider it as such, as they maintained that the past policy objective was within reach and that it should not be revised ("success"). So, the perception of the policy actors about the past policy objective was positive and therefore did not change but, on the contrary, was given more political priority, which confirms what we expected according to H2. About the failure of the policy instruments, we identified two main indicators of failure: the ineffectiveness of the instrument to reach the RES-E target and the unaffordability of the instrument for the public or private budget used to finance it. Both of them proved to be useful to characterise the failure of the policy instruments in those cases: the ineffectiveness in Belgium, the Netherlands and the UK; and the unaffordability in Denmark and the Netherlands.

Table 8: Details about the test of H2 (past policy)

Case studies		Past policy	Policy change	H2
Belgium	Policy objective	"Failure"	Yes	V
	Policy instrument	Failure	Yes	
Denmark	Policy objective	Success	No	V
	Policy instrument	Failure	Yes	
Germany	Policy objective	"Failure"	Yes	V
	Policy instrument	Success	No	
Netherlands	Policy instrument	"Success"	No	V
	Policy instrument	Failure	Yes	
UK	Policy objective	Failure	Yes	V
	Policy instrument	Failure	Yes	

Apart from the idea of success/failure of the past policy, which is validated in those cases and proves to be very significant to explain the change of policy instrument, we think that an additional hypothesis should be formulated from the past policy perspective. Looking at those five case studies, we identify a factor that is not included in our theoretical framework (or at least not directly) but proves to be useful to explain the change of policy instrument in some cases: the fit/misfit of the past policy with the reform of the electricity sector (H2/2, see figure 2). In all five cases, the change of policy instrument occurred at the same time as (Belgium, Denmark and the UK) or short after (Germany, the Netherlands) a reform of the electricity sector that significantly modified the context in which the RES-E policy was going to be implemented (e.g. liberalisation of the electricity market, unbundling of the vertically integrated companies, new market trading arrangement).

Figure 2: Amended hypothesis H2 (past policy)

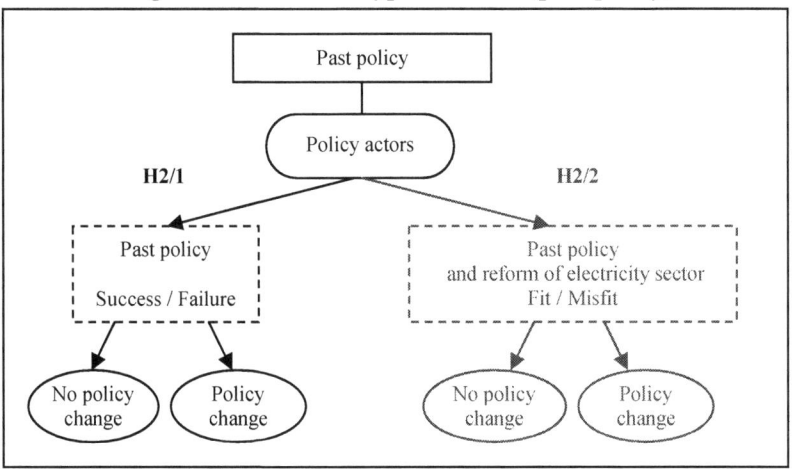

H2/2 assumes that if the past policy instrument fits in the context of the reform of the electricity sector, then no change of the instrument is to be expected (Denmark, Germany and the Netherlands); whereas if the past policy instrument does not fit in the context of the reform of the electricity sector, it is expected to be modified to be more adapted to the new context (Belgium and the UK) (see table 9). In Belgium, the dissolution of the CCEG in charge of the past RES-E policy instrument resulted inevitably in the demise of the past instrument and this called for a policy change. In the UK, the replacement of the pool system by NETA as well as the unbundling of the RECs and the dissolution of the NFPA made it impossible for the past policy instrument to continue, so the change of instrument was inevitable. However, in the Danish and Dutch cases, this hypothesis is not validated, since the past instruments fitted into the new context. So, the change of instrument was due to other factors than the reform of the sector in those cases (hence the failure of the past policy)[6]. Finally, in Germany, the reform of the electricity sector did not call for the change of the past RES-E policy instrument, only the adaptation of some of its settings. In conclusion, this additional hypothesis is useful to explaining the change of RES-E policy instrument in Belgium and the UK, but only three cases out of five confirm its assumption (see table 9), which means that it is only just validated.

---

[6] Nevertheless, in the Netherlands the liberalisation of the electricity market indirectly explains the change of instrument, as it represents one cause of the past policy failure (increase of RES-E imports and subsidisation of foreign RES-E generators by Dutch public budget).

Table 9: Test of the additional hypothesis about past policy (H2/2)

	Past policy instrument fit/misfit with reform of electricity sector	Policy instrument change	H2/2
Belgium	Misfit	Yes	V
Denmark	Fit	Yes	I
Germany	Fit	No	V
Netherlands	Fit	Yes	I
UK	Misfit	Yes	V

The fourth hypothesis of our theoretical framework (*H3*) concerns the influence of the European policy on the domestic policy by looking at the congruence between the European policy and the domestic policy. It assumes that if the domestic policy fits with the European policy, then no policy change is expected; whereas when they do not fit, the domestic policy is expected to change. Table 4 shows that this hypothesis is fully validated regarding the changes of policy objective, but it is just invalidated regarding the changes of policy instrument as two out of the four cases where the hypothesis is applicable did not confirm what we would have expected (Germany and the Netherlands). Besides, we observe that this hypothesis was non-applicable in the UK case, as the domestic policy change happened rather independently from the European policy.

Because the European RES-E policy was still under construction and thus very uncertain at the time of the RES-E policy change in those countries (except for the Netherlands), the domestic policy actors were not confronted with an actual fit or misfit of the domestic policy with the European policy, but they could still perceive whether the domestic policy and the European policy fitted or not in the light of the propositions or ideas developed at the European level. Therefore, the fit or misfit hypothesis relates to the perceived fit or misfit by the domestic policy actors and not the actual fit or misfit in this research. But the perception of the actors is what matters to explain the policy changes, so this is coherent with our theoretical framework. In addition, the directive of 2001 specifies that most of the European RES-E policy neither prescribes any binding policy objectives (only indicative targets), nor requests the harmonisation of the policy instruments according to a specific model, which means that there is no pressure to adapt to the European RES-E policy from the members states (except the guarantees of origin). The European RES-E policy has thus been very soft so far. It gives an undoubted incentive to the improvement of the RES-E policies in Europe, but leaves the member states free to implement it in their own way.

Comparison

However, the fit/misfit hypothesis is validated for the change of policy objectives and not for the change of policy instrument, which means that the domestic policy actors feels the European influence differently, depending on whether the policy objective or the policy instrument is concerned. Indeed, the RES-E targets adopted in the directive of 2001 were set up at realistic levels with the support of the member states in order to ensure their compliance. They are only indicative (no sanction), but the Commission evaluates the compliance of the member states on a regular basis, which encourages the member states to comply and creates positive emulation among the member states (no one wants to be the laggard). On the contrary, as far as the policy instrument is concerned, the European policy has been much more uncertain so far (no prescribed or indicative instrument) and the divergence between the member states' instruments is even somewhat encouraged by the Commission in its attempt to evaluate the successes and failures of each type of instrument in the perspective of a future harmonisation. Therefore, the European RES-E policy is more certain (fixed targets) and the perceived pressure to adapt is more salient (if not compulsory) as far as the policy objective is concerned than regarding the policy instrument; which probably explains why the fit/misfit hypothesis is more strongly validated for the change of policy objective than for the change of policy instrument.

Table 10: Details about the test of H3 (European policy)

Case studies		Fit/Misfit	Policy change	H3
Belgium	Policy objective	Misfit	Yes	V
	Policy instrument	Misfit	Yes	V
Denmark	Policy objective	Fit	No	V
	Policy instrument	Misfit	Yes	V
Germany	Policy objective	Misfit	Yes	V
	Policy instrument	Misfit	No	I
Netherlands	Policy objective	Fit	No	V
	Policy instrument	Fit	Yes	I
UK	Policy objective	n.a.	Yes	n.a.
	Policy instrument	n.a.	Yes	n.a.

It is evident from this comparison that in most policy change cases, domestic factors prevail over European factors, which put into perspective the explanatory value of the Europeanisation literature (see table 13). In the Netherlands, the European policy does not explain the change of policy instrument, because the past instrument and the European policy fitted perfectly well (see table 10). The origin of the policy change must be found in domestic factors (failure of past instrument). In Germany, the policy instrument did not change, though it was perceived to be incongruent with the European policy at that time (see table 10), which proves that domestic factors in favour of the status quo (success of past instrument) were more powerful than the influence of the European policy. Moreover, the perceived misfit between the German policy instrument and the European policy was not coupled with any pressure to adapt from the EU, so the German government was not obliged to modify its policy instrument accordingly. Besides, even if the fit with European policy seems to explain the change of policy instrument and the stability of the policy objective in Denmark, the change of policy objective in Germany, and the stability of the policy objective in the Netherlands (see table 10), it represents only a minor explanatory factor compared to other domestic factors (see table 13). Belgium is the only case where we observe that the European influence on the policy change is validated (see table 10) and significant (see table 13). Finally, the UK case shows that in some cases the domestic policy changes are undertaken rather independently from the European policy.

Therefore, what we suggest to improve H3 in the light of the German and Dutch cases (see figure 3) is, on the one hand, to emphasise the perceived nature of the misfit and add the dimension of perceived adaptation pressure (the German case), and, on the other hand, to modify the second proposition of H3 about the link between the fit and no policy change, as the misfit with the European policy is not a necessary condition for policy change (the Dutch case). So, the new H3 would be: "If the domestic policy objective or instrument is perceived by the domestic policy actors to be a misfit with the European policy and if pressure to adapt to the European policy is also perceived, then domestic policy change is expected" (see figure 3). Therefore, domestic policy change is expected only in case of a perceived misfit AND perceived adaptation pressure. If no pressure to adapt to the European policy is perceived from the domestic policy actors, then they are not expected to modify the domestic policy according to the European policy (e.g. Germany). In addition, the misfit with the European policy is not a necessary condition for domestic policy changes, which means that policy changes can occur even if the domestic and European policy fitted perfectly well. This explains why in case of policy fit, we cannot predict whether the domestic policy will change or not.

Figure 3: Amended hypothesis H3 (European policy)

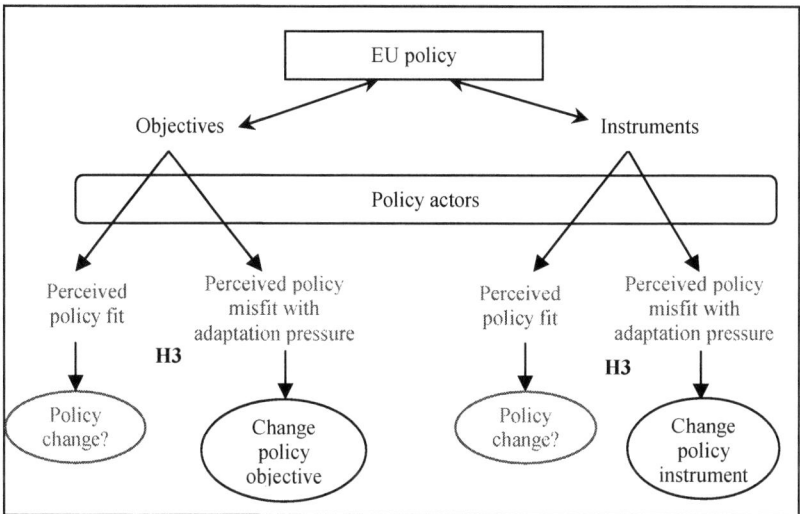

Apart from this hypothesis about the top-down influence of the European policy on the domestic policy change (down-loading Europeanisation process), in all cases we observe (except for the UK) bottom-up relations between the domestic policy actor and the European policy (up-loading Europeanisation process), which confirms what we assumed in the theoretical chapter (see the bi-directional arrows in figure 3). For instance, in Germany, instead of changing the policy instrument according to the perceived misfit with the European policy, the German policy actors "lobbied"[7] at the European level to modify the European policy and reduce the misfit with their own preferences (bottom-up process). Besides, in Belgium, Denmark and the Netherlands the policy actors used the European policy to legitimise the policy change (legitimisation process). In Belgium and Denmark, the change of policy was presented as congruent with the preferences of the European Commission at the time, which reinforced the legitimacy of the new policy. In the Netherlands, the change of policy was presented as more effective to ensure the compliance of the Dutch government with its European commitment over the RES-E indicative target to be reached by 2010. In conclusion, even if the policy change is not explained by top-down Europeanisation processes, bottom-up or instrumentalisation processes of Europeanisa-

---

[7] The term lobby is in quotation marks here because it is usually used to talk about the influence of private actors on governments and not from governments to other public actors.

tion can happen, which does not explain the domestic policy change directly, but contributes to legitimise it.

Finally, *H4* looks at the influence of the political and institutional context on the policy change and it is composed of two hypotheses: one draws from the partisan theory (H4/1) and the second from the veto players theory (H4/2). The first assumes that the governing parties matter to public policies and thus when the governing coalition changes, the policy is expected to change. However, this is conditioned by the fact that the new coalition is ideologically different from the previous one (H4/1). The second considers the influence of the veto players (institutional and partisan veto players) in opposing the adoption of legislation that departs from the status quo (policy change), and assumes that the more numerous the veto players, the less the policy change is likely to happen, on condition that the veto players are not ideologically close and/or weakly internally cohesive (H4/2).

Table 4 shows that the hypothesis about the influence of the change in governing parties on the policy change (*H4/1*) is confirmed in all instances of change/non change of the policy objective, but in only half of the cases of change/non change of policy instrument. So, it appears that this hypothesis is validated regarding the changes of policy objective, but not regarding the changes of policy instrument (see table 4). This finding is not so surprising, as we already observed that the political actors (hence the governing political parties) are more determining to explaining the change of policy objective than the change of policy instrument (H1/1).

Table 11 shows the details about the test of the hypothesis in the five cases of RES-E policy change/non change. In Belgium, Germany and the UK, the elections brought to power new political parties short before the RES-E policy change. In addition, the new governing coalitions (Belgium and Germany) or the new governing party (the UK) and the past ones differed significantly over the RES-E policy issue (large ideological distance). The new parties in government (the Green parties in Belgium, the Social Democrats and the Green party in Germany, and Labour in UK) benefited from significant power (or absolute power in Germany and the UK) over the RES-E policy in the government (see table 11). So those three cases fully validate the assumption of H4/1 as far as the policy changes are concerned (objective and/or instrument). However, in the German case, the absence of change of policy instrument does not confirm what we would have expected, because the new governing coalition supported the existing instrument and decided to keep it instead of changing it along with the new policy objective. In addition, we observe that the explanation of the change of RES-E policy in those countries is not only the presence of Green Parties in government, but primarily the presence of political parties with a significant commitment towards RES-E (the Social Democrats in Germany and Labour in the UK) as opposed to the past governing parties (the Christian Democrats in Germany and the Conservatives in the UK). Besides, in the Danish and Dutch cases, the assumption of H4/1 is not validated, since the policy instrument changed without the governing coalition being modified (see table 11).

Table 11: Details about the test of H4/1 (political and institutional context)

	Change governing parties	Policy change		H4/1
Belgium	Yes: Green parties with large ideological distance and power	Policy objective	Yes	V
		Policy instrument	Yes	V
Denmark	No change in governing parties + non-majoritarian democracy	Policy objective	No	V
		Policy instrument	Yes	I
Germany	Yes: Red-Green coalition with large ideological distance and power	Policy objective	Yes	V
		Policy instrument	No	I
Netherlands	Yes but no ideological difference with past coalition about RES-E policy	Policy objective	No	V
		Policy instrument	Yes	I
UK	Yes: Labour party with large ideological distance and power	Policy objective	Yes	V
		Policy instrument	Yes	V

In Denmark, there was no change in the governing parties short before the RES-E policy change, which means that changes in the governing parties is not a necessary condition for policy change (see table 11). Moreover, in the Danish political context, even if there is a new governing coalition, with a significant ideological difference from the past one, it is very unlikely that a policy change would be explained by this factor, because of the fact that most Danish governments are minority governments and thus policy changes require a large consensus between the governing and the opposition parties (e.g. the change of the government coalition in 2001). So, in the case of non-majoritarian democracies like Denmark, the partisan theory is invalidated. Finally, in the Netherlands, the new governing coalition did not differ much from the previous one regarding the RES-E policy preferences (small ideological distance), so the policy change is not linked with the change in the governing parties (see table 11). This case confirms what Schmidt defended in his theory (Schmidt, 1996) and other scholars had already empirically observed (Varone, 1998). The most relevant factor of policy change is not the change of governing parties, but the ideological distance between the new and the past governing parties, because if the new and past governments share the same preferences in terms of RES-

E policy, the change of government does not explain the policy change. In the Dutch case, the change of RES-E policy was driven by factors independent from the change of governing parties, and, besides, the policy change was supported by a large coalition of political parties (majority and opposition), so the parties did not matter to policy change in this case.

In conclusion, H4/1 is validated in this research for the change of policy objective, as policy objectives changed only in cases where the governing parties changed with a large ideological distance from the past. This means that, according to these findings, the change of governing parties is expected to be a necessary condition for changes of policy objectives. However, the hypothesis is not fully validated regarding the change of policy instrument, since only two out of the five cases of change of policy instrument confirm our assumption (see table 11). Indeed, we observe that the policy instrument did not change when the governing parties changed (Germany) and that the policy instrument changed in cases where the governing parties did not change (or there was no significant ideological distance from the past) (Denmark, the Netherlands). So, the second proposition of H4/1 (if there is no change in governing parties, then there is no policy change) is not validated because changes of policy instruments occurred even if the governing parties did not change (Denmark) or even if there was no significant ideological distance between the new and old governing parties (the Netherlands). This demonstrates that the change of governing parties is not a necessary condition for changes of policy instrument, and that other factors can lead to a change of instrument in the absence of change in the governing parties or their ideologies. Therefore, we suggest to amend H4/1 by amending the second proposition about the link between no change in governing parties and no policy change, because in case of no change in governing parties, we assume that there will not be any change of policy objective, but we cannot predict whether the policy instrument is going to change or not (see figure 4).

Finally, *H4/2*, which assumes that the presence of many ideologically distant and internally non-coherent veto players is expected to prevent the policy changes, is fully validated in this research for both the change of policy objective and instrument (see table 4).

Figure 4: Amended hypothesis H4/1
(political and institutional context)

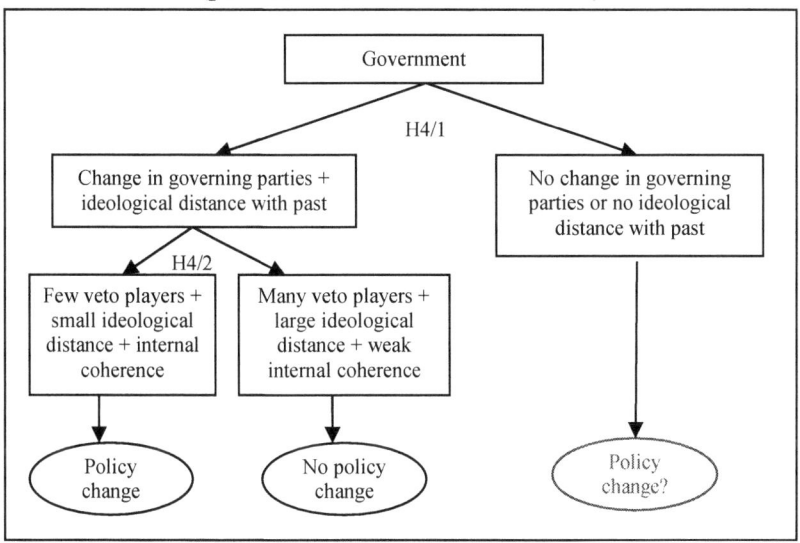

We observe that in all five countries, there are only few institutional veto players: the parliament (one Chamber (Denmark) or two Chambers (Belgium, Germany, the Netherlands and the UK)) and, in the constitutional monarchies, the King or Queen (Belgium, Denmark, the Netherlands and the UK) (see table 12). In the bicameral systems, either the two Chambers share the same majority (Belgium), or one does not have significant veto power (the UK, the Netherlands), or their ideological preference regarding the RES-E policy does not differ significantly, even if they are composed of different majorities (Germany and the Netherlands). In the constitutional monarchies, royal assent is necessary for legislation to enter into force, but this is usually a formality as the King or Queen does not exert any political power anymore. In addition, except for Germany regarding the RES-E issue, those countries are characterised by strong party discipline, which increases the internal coherence of the veto players (see table 12). In Germany, the members of parliament do not always follow the directions of their party leaders, especially regarding the RES-E policy, which explains why some members of parliament of the opposition parties supported the majority in favour of the RES-E policy change in 2000 and 2004. So, the lack of internal coherence in this case was beneficial for the RES-E policy change. What emerges from the comparison is that the most relevant factor that explains the influence of the partisan veto players on the policy change is not the number of veto players but the ideological distance between them and their internal coherence (see table 12). Indeed, in Belgium, Denmark and the Netherlands, many partisan veto players are able to oppose policy changes because of multi-party coalitions with a large number of political parties (Belgium, the Netherlands) or minority government coalitions (Denmark). But, thanks to the coalition agreements in Belgium and the Netherlands and *ad hoc* majorities in Denmark (e.g. the green majority), consensus between the political parties is made easier and reduces the risk of veto. In addition, in these countries, the ideological distance between the political parties regarding the RES-E policy was not significant at the time of the policy change (large consensus in favour of RES-E policy change). So, the assumptions of the veto players theory about the relation between on the one hand the number of veto players, the ideological distance between them and internal coherence, and on the other hand the likeliness of policy change, these assumptions are validated in this research and H4/2 does not need to be amended (see figure 4).

Table 12: Details about the test of H4/2 (political and institutional context)

	Institutional veto players	Partisan veto players	Policy change		H4/2
Belgium	Few and internally coherent with small ideological distance	Many, but small ideological distance over RES-E and strong internal coherence	Policy objective	Yes	V
			Policy instrument	Yes	V
Denmark	Few, and small ideological distance over RES-E and strong internal coherence	Many, but small ideological distance over RES-E and strong internal coherence	Policy objective	No	n.a.
			Policy instrument	Yes	V
Germany	Few and small ideological distance over RES-E, but weak internal coherence	Few and small ideological distance over RES-E and strong internal coherence	Policy objective	Yes	V
			Policy instrument	No	n.a.
Netherlands	Few and internal coherence with small ideological distance over RES-E	Many, but small ideological distance over RES-E and strong internal coherence	Policy objective	No	n.a.
			Policy instrument	Yes	V
UK	Few and internally coherent with small ideological distance	No partisan veto players	Policy objective	Yes	V
			Policy instrument	Yes	V

III. Explanations of Policy Changes

Now that we have tested the validity of the hypotheses and proposed amendments to the original hypotheses when necessary, we still need to evaluate the relevance of the hypotheses to explaining the RES-E policy changes (objective and/or instrument). The analysis of the relevance of the hypotheses has already been done at the end of each of the five case-study chapters, so we will not enter into the details of the case studies anymore, but we will compare the results of the individual analysis in order to identify which hypotheses are the most significant to explain the policy changes (see table 13).

Table 13: Relevance of the hypotheses to explain the RES-E policy changes in the five countries[8]

	Policy change	H1		H2	H3	H4	
		H1/1	H1/2			H4/1	H4/2
Belgium	Objective	++	0	+	++	++	+
	Instrument	0	+	++	++	++	+
Denmark	*Objective*	*+*	*0*	*+*	*+*	*0*	*0*
	Instrument	+	+	++	+	0	+
Germany	Objective	++	++	0	+	++	+
	Instrument	*+*	*+*	*++*	*0*	*0*	*0*
Netherlands	*Objective*	*+*	*0*	*+*	*+*	*0*	*0*
	Instrument	++	0	++	0	0	+
UK	Objective	++	0	++	0	++	+
	Instrument	++	++	++	0	+	+
Comparison	**Objective**	**++**	**0**	**+**	**+**	**++**	**+**
	Instrument	**+**	**+**	**++**	**0**	**0**	**+**

Legend: ++ (major explanation of policy change/no change), + (minor explanation), 0 (does not explain policy change/no change).

Table 13 summarises the results of the five case studies as to the relevance and significance of the hypotheses to explaining the RES-E policy changes. We observe that the patterns of explanation differ not only between the countries, but also between the change of policy objective and instrument. Indeed, the most significant hypotheses to explaining the change of policy objective are not the same as the most significant hypotheses to explaining the change of policy instrument (see table 13), which confirms the relevance of making the distinction between those two types of policy change.

---

[8] The lines in italics correspond to cases of no policy change in which the evaluation of the relevance of the hypothesis could only be done by default.

As far as the change of policy objective is concerned, two explanatory factors emerge from the comparison as the most significant ones: the policy actors (especially the dominant public policy actors, H1/1) and the political context (especially the change in governing parties, H4/1). Those factors were major explanations of the RES-E policy objective change in the three cases of objective change (Belgium, Germany and the UK), and they also appeared significant to explaining the absence of change of policy objective in the other countries (Denmark and the Netherlands). In reality, these factors are closely related, as what explains the change of policy objective in those three cases is the arrival of new political parties in government (H4/1), linked with the fact that the actors who dominate the policy change are the new governing parties (political actors) and not civil servants (H1/1). Table 13 also shows that the other hypotheses of our framework play a minor role (H2, H3, H4/2), or no role at all (H1/2) in the explanation of the RES-E policy objective change. Only in the German case does the influence of the actors of the sector (H1/2) prove to be significant to explaining the change of policy objective, while it is not relevant at all in the other cases. Therefore, it may be deduced that it is not a relevant factor to explain the change of policy objective. The influence of the past policy success/failure on the change of policy objective (H2) turns out to be very different from one country to the other: major explanation in the UK, minor explanation in Belgium (and by default in Denmark and the Netherlands), and not relevant in Germany (see table 13). This means that the past policy is a relevant but not a very significant explanation of the policy objective change. The Europeanisation factor (H3) turns out to be of relevance when explaining the change of policy objective, but not highly significant. The fit/misfit hypothesis is a major explanation of the change in Belgium, but only a minor explanation in Germany (and by default in Denmark and the Netherlands), and it is not relevant at all in the UK case; therefore it only has a minor explanatory value in the light of the comparison. Finally, the absence of veto players (H4/2) opposing the government proved to be significant to explaining the change of the RES-E policy objective in the three countries where it actually changed (and not in the countries where the objective did not change), but to a lesser extent compared to other hypotheses. Therefore, it represents a relevant but only a minor explanatory factor in the explanation of the changes of policy objective (see table 13).

Regarding the change of policy instrument, the most significant explanatory factor is the past policy (H2) (see table 13). In all four cases of change of policy instrument, the failure of the past policy is the main reason why the instrument changed (Belgium, Denmark, the Netherlands and the UK). Moreover, in the German case, the success of the past instrument is the main reason why it did not change. Other factors also contribute to explaining the change of policy instrument, but to a lesser extent (H1/1, H1/2, H4/2); and two hypotheses prove to be irrelevant to explaining the change of policy instrument (H3 and H4/1) (see table 13). The two hypotheses about the policy actors (H1) seem to be relevant to explaining the change of RES-E policy instrument in our cases, but they are only minor explanatory factors. As far as H1/1 is concerned, in the UK and the Netherlands, the fact that the dominant policy actors were civil servants of the administration for energy (and not politicians) is a major explanation of the change of RES-E policy instrument (H1/1). In Denmark (and by default in Germany), this factor is relevant to explaining the change of policy instrument, but only to a smaller extent compared to other explanatory factors. In Belgium, the actors who dominated the change of policy instrument were not civil servants, as would have been expected under H1/1, but political actors (the ministry of Energy), which means that H1/1 is not relevant in this case. As far as H1/2 is concerned, we observe that the influence of the actors of the sector on the change of policy instrument (H1/2) is a significant explanatory factor in all countries, except for the Netherlands, where the policy change process was limited to policy experts' decisions. Therefore it is a relevant explanatory factor to the changes of policy instrument, but not a highly significant one (see table 13). Regarding the Europeanisation factor, we observe that the misfit hypothesis (H3) does not help make sense of the change of policy instrument in the majority of the countries. The Europeanisation hypothesis is a significant explanatory factor of the change of policy instrument in the Belgian and Danish cases, but not in the other countries. So, it may be deduced that the influence of the European policy is not a relevant factor to explaining the change of the policy instrument (see table 13). Similarly, we observe that the change of the governing coalition (H4/1) is a relevant explanatory factor of the RES-E policy instrument change only in a minority of cases (Belgium and the UK), so in most of the cases this hypothesis is not a relevant explanatory factor. Therefore, we may also deduce that this hypothesis is not relevant to explaining the change of policy instrument (see table 13). Finally, the absence of veto players to oppose the change of policy instrument (H4/2) contributes to explaining the policy change, but it is only a minor explanatory factor compared to other factors (see table 13).

In conclusion, all the explanatory factors proved to be relevant to explaining the policy changes in the five countries, but some more than others. In addition, we observe that the factors that are the most significant to explaining the change of policy objective or the change of policy instrument differ; which once again confirms that it is interesting to make the distinction between those two types of policy change.

Our research shows that the main explanatory factors of the change of policy objective are the dominant policy actors (political actors, H1/1) and the governing political parties (new parties, H4/1) (see table 13). This confirms Peter Hall's assumption about the central position of the political actors in case of third order change. On the other hand, we observe that the main explanatory factor for the change of policy instrument is the failure of the past policy (H2) (see table 13). This confirms another assumption of Peter Hall about the past policy failure to be a necessary condition in case of second order change.

IV. Conclusions of the Comparison

As far as the policy changes are concerned, the comparison demonstrates that there is no hierarchical link between the two types of policy change (policy objective and policy instrument) and that the objective of a policy can change while the instrument does not, which questions Peter Hall's assumption about such a hierarchical link. The comparison also shows that the RES-E policies remain very different in Europe, but that they tend to converge on some aspects: convergence of the policy objectives towards the European indicative targets, but not in absolute terms; convergence of the policy instruments regarding the target group (demand side of the market) and the resources (private), but not regarding the incentive (quantity and/or price). Finally, the comparison demonstrates that the indicators we selected to identify the changes in the policy objectives and instruments prove to be relevant and consistent to account for the diversity of policy changes in the five countries. At the end of the day, we recognise that we have not captured all aspects of the complex phenomenon of policy change with our theoretical framework; the cognitive dimension of the policy change and the implementation issues, for instance, have not been addressed; but we believe that this research contributes to understanding and explaining the policy change process from a pragmatic perspective.

As far as the explanation of the policy changes is concerned, the comparison has showed that our theoretical framework is largely validated and thus requires only minor amendments. We have also observed that our hypotheses proved to be relevant, though some more than others, to explaining the RES-E policy changes in the five countries that we analysed. However, it has turned out that distinct patterns of explanation have emerged for the changes of policy objective or the change of policy instrument (see table 14).

Table 14: Conclusions about the validity and the relevance of the hypotheses

	Policy change	H1		H2	H3	H4	
		H1/1	H1/2			H4/1	H4/2
Validity of H	Objective	V	n.a.	V	V	V	V
	Instrument	V	"V"	V	I	I	V
Relevance of H	Objective	++	0	+	+	++	+
	Instrument	+	+	++	0	0	+

As far as the change of policy objective is concerned, two hypotheses stand out as necessary variables (see grey cells in table 14): the dominant policy actors (H1/1) and the change of governing political parties (H4/1). These hypotheses have been validated and appear to be major explanatory factors in all cases. This means that changes of policy objective appear to be very political issues, dependent primarily upon the political actors who dominate the policy-making process and very sensitive to government change.

As far as the change of policy instrument is concerned, one hypothesis stands out as necessary variable (see the grey cells in table 14): the success or failure of the past policy (H2). Therefore, changes of policy instrument appear to be more path-dependent than changes of policy objective, as they heavily depend on the success or failure of the past instruments, and, depending on the politico-administrative context, they can reveal themselves to be either administrative or political issues.

Finally, one of the questions raised in this research is the influence of the European integration on the domestic policies, namely the Europeanisation question, and the result of the comparison is twofold in this case. On the one hand, the influence of the European policy is a relevant factor to explaining the policy changes in almost all countries (except for the UK), which demonstrates the pertinence of the question. However, except for the Belgian case, it is not a significant factor to explaining the policy changes. Thus policy changes prove to be driven mainly by domestic factors and not European integration impulses. This is probably due to the characteristics of the European policy at stake in this research: uncertain and non-constraining. So, we observe that in case of uncertain and non-constraining European policy on issues dominated by domestic interests, the Europeanisation idea is relevant but not significant to explaining domestic policy changes. On the other hand, the comparison also showed that the misfit hypothesis is validated and relevant in case of changes of policy objective, but not in case of change of policy instrument (see table 14). This tends to demonstrate that the European policy more significantly influences the formulation of the policy objectives than the policy instruments in the member states. Indeed, the formulation and implementation of the policy instruments appear to be very sensitive and protectionist issues in most member states regarding the RES-E policy. On the contrary, the adoption of integrated (but still differentiated) policy objectives at the European level seems to be becoming more common, especially in issues with international consequences (e.g. RES-E targets, GHG emissions reduction targets).

Conclusion

The research question addressed in this research is "How and why do public policies change?". In order to answer this question, we have presented a definition of policy changes based on Peter Hall' distinction between three orders of change (Hall, 1993), but we have only considered two dimensions of policy changes: the change of policy objectives and the change of policy instruments. We have developed a theoretical framework to explain the policy change and formulated four sets of hypotheses drawing from different approaches of the policy literature. Then, we confronted the research question and the theoretical framework to an original empirical background: the RES-E policy changes in Belgium, Denmark, Germany, the Netherlands and the UK in the period of 1999 to 2004. The empirical research involved in-depth case studies of each country's RES-E policy changes, followed by a systematic comparison of the individual case studies' findings. The comparison allowed us to draw interesting conclusions about the validity and relevance of our theoretical framework and, more generally, about the contribution of the results of our research to the policy literature.

Now, in this conclusion, we will summarise the main findings of our research from a theoretical point of view and emphasise the contributions of this research to the policy literature. In addition, we will present the main empirical contributions of our research to the analysis of RES-E policies in Europe. And, in the meantime, we will also discuss some normative issues raised by our research for the policy analysts or the policymakers.

Policy changes have been analysed from a very pragmatic perspective in this research, with a focus on empirically observable policy objectives (policy targets) and instruments (target groups, incentives and resources of the policy). This definition of policy change proved to be very useful and consistent, especially in a comparative analysis. The distinction between the policy objectives and the policy instruments, following Peter Hall, also proved to be very relevant from a theoretical point of view, as we observed different patterns for the explanation of those two types of policy change, hence the hypotheses explaining the change of policy objective are not the same as the ones explaining the change of policy instrument. Therefore, our comparative analysis shows that it does make sense (and not only empirically) to theoretically distinguish the changes of policy objective and instrument while trying to explain policy changes.

In order to get a more comprehensive picture of the policy changes, we should also consider the cognitive aspects of the policy changes, namely the motivations of the policy actors that legitimise the policy change. This question has been addressed in the literature by such approaches as the policy paradigm (Hall, 1993), the "referential" (Muller, 2000), or the belief systems (Sabatier, 1993), and it would be interesting to use these approaches to complete this research. However, these approaches usually involve single case studies with more in-depth analysis of the actor's beliefs and preferences, which differs from what we decided to do in this research. Besides, our research focuses on the formulation of the policies and does not address the question of the implementation of the policies, which often turns out to be as (or even more) crucial in the apprehension of policy change phenomena. The analysis of the implementation issues would address the question of the policy change from another angle, which would be complementary to this research, but it was not our initial purpose. Finally, we have looked at policy changes from a dichotomous perspective (change/no change), but it would also have been interesting to consider more gradual modalities of policy change (e.g. no change/minor change/major change), both regarding the policy objectives and the policy instruments. However, we believe that defining clear and unequivocal variables is a key condition for the success of research, especially in comparative analysis. Moreover, it is usually a lot easier to formulate hypotheses to explain variables with two modalities than variables with more than two modalities.

The results of this research regarding the definition of policy change confirm the relevance of distinguishing policy objectives from policy instruments, but invalidate the assumption of Peter Hall about the hierarchical link between these two dimensions of policy change. Indeed, we have observed that changes of policy objective are not necessarily followed by changes of policy instrument, which means that the policy objectives and the policy instruments are more independent than expected. On the contrary, the assumption of Peter Hall about the hierarchical link between the policy instruments and the settings of the instruments is more likely to be observed, though it was not systematically tested in this research. This is due to the fact that the settings of the instruments are not independent from the instruments, as they involve the operational details of the instruments.

Finally, a more normative question emerges from the results of this research: "Which is the most difficult to change, the policy objective or the policy instrument?". According to Peter Hall (1993), one would assume that the policy objectives are the most difficult to change. However, in the light of this research it appears that the policy instruments are more difficult to change than the policy objectives. On the one hand, we observe that policy objectives are political decisions that set the medium and long-term goals of a policy. A change of policy objective thus requires a political consensus among the governing parties, which appeared to be rather easy regarding the RES-E policy because of the absence of strong opposition on this issue. Moreover, the formulation of policy objectives is usually a mere declaration of intent that will need to be formalised through appropriate policy instruments in order to be effective. On the other hand, the policy instruments involve more political, administrative and institutional challenges than policy objectives, as they require the policy actors to make sensitive choices about how to reach the policy objectives: Who are the policy targets? Which incentive is to be mobilised? How is the cost of the policy to be financed? Which institutions are in charge of implementing and controlling the policy? These embarrassing questions can be avoided at the stage of the policy objective, but no more when adopting the policy instruments. Besides, the participation of the actors of the sector in the formulation of the policy instruments also tends to make it more difficult to achieve, especially when a strong opposition rallies against the change of the instrument. In conclusion, unlike Peter Hall, we believe that, for the policymakers, changing the policy objectives is easier than changing the policy instruments. However, we acknowledge that Peter Hall developed its theory primarily to explain policy changes of a larger scope in more conflicting sectors (e.g. macro-economic policies), while we chose to analyse a rather specific and consensual policy issue (RES-E policy).

As far as the explanation of the policy changes is concerned, the confrontation of our theoretical framework with the RES-E policy changes in the five countries allows us to draw some interesting conclusions. They do not bring any theoretical breakthrough in the political science literature, but they merely contribute to the consolidation and extension of the policy science.

Firstly, policy changes cannot be explained without looking at the policy actors. Our approach is deeply actor-based. We observe that, either directly or indirectly, the hypotheses that explain the policy changes involve the policy actors. This legitimises the central position of the policy actors in our theoretical framework. For instance, the influence of the past policy or the European policy on the policy changes can only be explained by looking at the perceptions of the policy actors.

Moreover, we demonstrated that the dominant actors of the policy change process must be political actors (versus civil servants) in case of change of the policy objective (a necessary condition). This confirms

Peter Hall's hypothesis about the dominant position of the political actors in the change of the policy objective. Regarding the change of policy instruments, we observed, like Peter Hall, that the dominant policy change actors are usually administrative actors, though this is not necessary always the case.

Finally, the assumption of Peter Hall that, when the actors who dominate the policy change process are political actors, the policy change involves a large debate with the actors of the sector, while when the policy change process is dominated by civil servants, it remains closed within a small group of policy experts, this assumption is not fully validated by our research because our results are mixed on this question.

Secondly, this research validates Peter Hall's assumption and the path-dependency idea that policy changes are dependent upon the outcomes of the past policies. This is especially true for the changes of policy instrument, in which case the failure of the past policy is a necessary condition for policy changes. Besides, as we already mentioned, we observe that what matters in the evaluation of the outcomes of the past policies, is not the objective information, but the perceptions of the policy actors.

Thirdly, we have demonstrated in this research that the main factors of policy change are domestic factors, and that the influence of the European policies is only marginally significant to explaining the policy changes. This finding will question the relevance of the Europeanisation idea as a driving force in the European Union (and beyond). Nevertheless, this conclusion should be qualified. We observe that the European policy proves to be more relevant to explaining the change of policy objectives (more European integration) than the change of policy instruments (less European integration). In addition, our research confirms the assumption in the Europeanisation literature that the Europeanisation of the domestic policies requires not only that the European and domestic policies misfit, but also that the domestic policy actors perceive a pressure to adapt. As we said above, to be effective, this pressure to adapt does not need to be an objective constraint, but merely a perception of the policy actors.

Fourthly, the change of governing political parties prior to policy change appears to be a necessary condition for the change of policy objective in this research. This confirms the idea of the partisan theory that political parties matter to public policies, but only regarding the formulation of policy objectives and not the formulation of policy instruments. This is not surprising, given the fact that we have already observed that political actors are determining, primarily in cases of change of policy objective and less in cases of change of policy instrument. Besides, our research also confirms what the partisan theory assumes and other empirical research have already observed, viz. that the most relevant factor to explaining policy change is not the change of the governing parties, but the ideological distance between the new and the previous governing parties. In addition, our research shows that the hypothesis about the influence of the governing parties on the policy changes is not relevant in non-majoritarian democracies, which confirms what the partisan theory had already observed.

Finally, our analysis of the influence of the veto players on the policy changes confirms the veto players theory, especially the idea that what matters to policy change is not so much the number of veto players, but the ideological distance between them and their internal coherence.

Apart from the theoretical contributions of the research, we also want to discuss some empirical contributions to the analysis of the RES-E policies in Europe. The in-depth historical case studies presented in this book are an undeniable value of this research. They provide a rigorous description of the RES-E policies from the 1970s that was necessary to understand the recent policy changes and explain how and why they happened. Moreover, the systematic comparison of the case studies based on a common analytical framework demonstrates the value of going beyond individual case studies to draw more general conclusions.

The comparison of the RES-E policy changes in the Belgium, Denmark, Germany, the Netherlands and the UK shows that these countries experienced different types of RES-E policy changes (policy objective and/or policy instrument change), but at the same period of time (1999-2004). In most countries, this period coincided with the liberalisation of the electricity market process as well as the formulation of the first climate change policies, following the adoption of the Kyoto Protocol. These two common factors probably explain to a large extent why policy changes driven primarily by domestic factors occurred at the same period of time. These factors have not been addressed directly in this research because we believe that their influence on the RES-E policy changes is indirect, and we described these indirect influences in the case studies. For instance, the liberalisation of the electricity market contributes to explain why some RES-E policy instruments failed or succeeded (actually or potentially) in a competitive market, or why the actors of the RES-E sector faced new challenges or opportunities that modified their influence on the policy-making process. The specific question of the relation between the liberalisation and the RES-E policy could have been the central question of the research, but we decided to look at the larger picture in which the liberalisation is an important but indirect element. Indeed, at first sight, the liberalisation of the electricity market may appear to be a threat for the future development of the RES-

E because of its commercial, financial and physical disadvantages compared to the traditional electricity technologies. However, we have observed in this research that the liberalisation of the electricity market has turned out to be the opportunity for the emergence of new RES-E policies or the improvement of the existing RES-E policies in most countries. By making the competitive disadvantage of the RES-E on the electricity market more visible and more serious, the liberalisation process has increased the attention of policymakers towards this policy problem and led them to increase their overall support to RES-E. Eventually, the threat has become an opportunity for the RES-E sector and the market has not been the last to realise it, as the recent boom in new RES-E business demonstrates.

Our research did not specifically address the traditional question of the convergence/divergence between the RES-E policies in Europe. However, we can draw some interesting comments on this issue from the findings of this research. Firstly, we did not observe a significant convergence of the RES-E policies between the five countries that we analysed, neither regarding the policy objective, nor the policy instrument. However, we still notice that one dimension of the policy instruments tends to converge: the resources used to finance the instrument (the private funding through the electricity bills). Moreover, the most common target group of the policy instruments seems to be the electricity suppliers, but the convergence on this dimension is not generalised. Secondly, so far the European RES-E policy has refrained from harmonising the RES-E policy instruments, which legitimises and reinforces the diverging patterns between the member states' policies. Regarding the RES-E policy objectives, the European directive of 2001 formulated indicative targets for each member state in accordance with the European target. But the redistribution of the European target was not based on an equality criterion (the same target for all member states), but was based on the current RES-E market of the countries and the evaluation of their potential in the future, which reinforces the divergence between the policy objectives of the member states. Thirdly, besides these diverging trends, we observe some attempts to encourage the transfer of policy instruments between the European countries, leading to increasing policy convergence. On the one hand, we noticed in our research that some countries carried on a benchmarking of the best existing policy instruments in the other member states during the process of policy change (Belgium, the Netherlands). On the other hand, a private initiative in favour of the spread of the TGC system started in 1999 with the ambition to bring the RES-E policy instruments of the European countries to converge towards the TGC model (RECs). Thus, both at the public and private levels, signs of policy transfer appear, but they are still marginal.

During the last decade, there has been a lively controversy within the RES-E sector between the proponents of the so-called feed-in tariff (FIT) system and the proponents of the so-called tradable green certificate (TGC) system, to demonstrate which the "best" instrument is. This research does not take any interest in this discussion, but we made two observations in relation with this controversy. Firstly, we observed in our case studies that the performance of a policy instrument depends not so much on the characteristics of the instrument itself, but on the settings of the instrument (e.g. level of the quota, tariff), its implementation arrangements (whether the instrument is properly implemented or not) and the policy mix that surrounds it (e.g. RD&D subsidies, industrial incentives, planning guidelines). Therefore, the discussion about the performances of the ideal-type policy instruments seems rather vain, as it does not address the right questions. Secondly, these two types of policy instrument (FIT and TGC) are not mutually exclusive, as the Belgian RES-E policy, which involves elements of both types of instrument, demonstrates. Two conclusions for the policymakers can be drawn from these observations: (1) there is no such thing as a "best" policy instrument. T he performance of an instrument depends on its settings, the way it is implemented and the policy mix that surrounds it; and (2) the more interesting RES-E policy instrument might be a mix of FIT and TGC characteristics, which would combine the advantages of both types of instrument (e.g. quota with tradable certificates combined with minimum prices for the TGC differentiated by RES-E technologies).

Today, the energy challenges are global, but the energy policies remain local (national or even sub-national), as this research illustrates perfectly. The European Union is working on the definition of a European energy policy that ought to be sustainable, competitive and secure. But the energy issues are still very sensitive from a nationalist perspective and the achievement of an integrated European energy policy is still out of reach. Even if the European leaders agreed on general European policy objectives (for instance regarding the RES), their compliance by the member states are questioned. In terms of policy instruments, the Commission has been working for a long time on an integrated European energy tax system. But the design of the instruments used to attain the policy objectives is even more sensitive than the policy objectives for the member states. Eventually, even if we observe an increased effort to integrate the national energy policy objectives in Europe, there is still a long way to go for the convergence of the energy policy instruments.

References

Aendekerk, J-M., C. Bourtembourg, J.-J. Grodent, and B. Haubert (1982). In IIAS (ed.), *The adjustment of administration to the energy crisis*. National reports of the international colloquium, Brussels, 9-11 June 1982.

Agerholm, B. and et al. (2002). "Promoting renewables through market mechanisms", *Energy & Environment*, 13, 4&5, 763-778.

Agterbosch, S., W. Vermeulen, and P. Glasbergen (2004). "Implementation of wind energy in the Netherlands: the importance of the social-institutional setting", *Energy Policy*, 32, 2049-2066.

Andersen, M. S. (1997). "Denmark: The shadow of the green majority". In M. S. Andersen and D. Liefferink (eds.), *European Environmental Policy: The Pioneers*. Manchester: Manchester University Press.

Andeweg, R. B. (2004). "Parliamentary democracy in the Netherlands", *Parliamentary Affairs*, 57, 3, 568-580.

Andrews, G. (2001).*Market based instruments: Australia's experience with trading renewable energy certificates*. Workshop on Good Practices in Policies and Measures, Copenhagen, 8-10 October 2001.

Araujo, L. and G. Easton (1996). "Networks in Socioeconomic Systems. A Critical Review". In D. Iacobucci (ed.), *Networks in Marketing*. Thousand Oaks: Sage.

Arentsen, M. J., R. W. Kunneke, and H. C. Moll (1998). "The Dutch electricity reform: reorganisation by negotiation". In A. Midttun (ed.), *European electricity systems in transition*. Elsevier.

Arkesteijn, K. and L. Oerlemans (2005). "The early adoption of green power by Dutch households. An empirical exploration of factors influencing the early adoption of green electricity for domestic purposes", *Energy Policy*, 33, 2, 183-196.

Asano, K. (2004). "Denmark: early promoter of renewable energy". In I. de Lovinfosse and F. Varone (eds.), *Renewable Electricity Policies in Europe. Tradable Green Certificates in Competitive Markets*. Louvain-la-Neuve: Presses universitaires de Louvain.

Bardach, E. (1976). "Policy termination as a political process", *Policy Sciences*, 7, 123-131.

Bartle, I. (2002). "When institutions no longer matter: reform of telecommunications and electricity in Germany, France and Britain", *Journal of Public Policy*, 22, I, 1-27.

Baumgartner, F. R. and B. D. Jones (1993). *Agendas and Instability in American Politics*. Chicago: University of Chicago Press.

Bechberger, M. and D. Reiche (2004). "Renewable energy policy in Germany: pioneering and exemplary regulations", *Energy for Sustainable Development*, VIII, 1, 47-57.

Becker, U. (2005). "An example of competitive corporatism? The Dutch political economy 1983-2004 in critical examination", *Journal of European Public Policy*, 12, 6, 1078-1102.

Bennett, C. J. (1988). "Different processes, one result: the convergence of data protection policy in Europe and the United States", *Governance*, 1, 4, 415-441.

Bennett, C. J. (1988). "Regulating the computer: comparing policy instruments in Europe and the United States", *European Journal of Political Research*, 16, 437-466.

Bennett, C. J. (1997). "Understanding ripple effects: the cross-national adoption of policy instruments for bureaucratic accountability", *Governance*, 10, 3, 213-233.

Bennett, C. J. (1991). "What is policy convergence and what causes it?", *British Journal of Political Science*, 21, 215-233.

Bennett, C. J. and M. Howlett (1992). "The lessons of learning: reconciling theories of policy learning and policy change", *Policy Sciences*, 25, 3, 275-294.

Berry, D. (2002). "The market for tradable renewable energy credits", *Ecological Economics*, 42, 369-379.

Berry F.S. and Berry W.D. (1999). "Innovation and diffusion models in policy research". In P. A. Sabatier (ed.), *Theories of the policy process*. Colorado: Westview Press.

Berry, T. and M. Jaccard (2001). "The renewable portfolio standard: design considerations and an implementation survey", *Energy Policy*, 29, 4, 263-277.

Birkland, T. A. (1997). *After disaster: agenda setting, public policy, and focusing events*. Washington: Georgetown University Press.

Birkland, T. A. (2004). "'The world changed today': agenda-setting and policy change in the wake of the September 11 terrorist attacks", *Review of Policy Research*, 21, 2, 179-200.

Bobrow D. B. and J. S. Dryzek (1987). *Policy Analysis by Design*. Pittsburgh: University of Pittsburgh Press.

Boots, M. G. and et al. (2001). *The interaction of tradable instruments in renewable energy and climate change markets*. Amsterdam: ECN-C-01-048.

Boussaguet, L., S. Jacquot, and P. Ravinet (2004). *Dictionnaire des politiques publiques*. Paris: Presses de Sciences Po.

Brendstrup, S. (2003). *Sectoral autonomy as a barrier to Europeanisation – Misfit revisited in light of utility liberalisations*. ECPR Conference, Marburg, 18-21 September 2003.

Brennand, T. P. (2004). "Renewable energy in the United Kingdom: policies and prospects", *Energy for Sustainable Development*, VIII, 1, 82-92.

Brewer G.D. and P. de Leon (1983). *Foundations of Policy Analysis*. Homewood: The Dorsey Press.

Budge, I. and H. Keman (1990). *Parties and democracies. Coalition formation and government functionning in 20 States*. Oxford: Oxford University Press.

Buelens, J. and K. Deschouwer (2002). Belgium. In F. Müller-Rommel and T. Poguntke (eds.), *Green Parties in National Governments*. London: Frank Cass.

Burgers, J., V. Ducos, H. C. Schneider, and R. M. Breugem (2004). *SETREC Digest Report of RES-E in the Netherlands*. SETREC/GO project.

Börzel, T. A. (1999). "Towards convergence in Europe? Institutional adaptation to Europeanisation in Germany and Spain", *Journal of Common Market Studies*, 37, 4, 573-596.

Börzel, T. A. and T. Risse (2003). "Conceptualizing the Domestic Impact of Europe". In K. Featherstone and C. M. Radaelli (eds.), *The Politics of Europeanisation*. Oxford: Oxford University Press.

Castles, F. (1982). *The impact of parties. Politics and policies in democratic capitalist states*. London: Sage.

Castles, F. G. (1998). *Comparative public policy: patterns of post-war transformation*. Cheltenham: Edward Elgar.

Castles, F. G. (1998). *Comparative Public Policy. Patterns of Post-war Transformation*. Cheltenham: Edward Elgar.

Chabal, P. M. (2003). "Les ministres 'font'-ils une 'différence'? Le style individuel des ministres dans le changement programmé de politiques publiques", *Revue Internationale des Sciences Administratives*, 69, 1, 31-53.

Cobb, R. W. and C. D. Elder (1993). *Participation in American politics: the dynamics of agenda-building*. Baltimore: John Hopkins University Press.

Cobb, R. W. and C. D. Elder (1983). *The Political Uses of Symbols*. New York: Longman.

Coen, D. H. A. and D. Bollhoff (2002). *Regulating the Utilities: UK and Germany Compared*. Anglo-German Foundation for the Study of Industrial Society.

Cohen, M. D., J. G. March, and J. P. Olsen (1991). "Le modèle du 'garbage can' dans les anarchies organisées". In J. G. March (ed.), *Décision et organisations*. Paris: Editions d'Organisation.

Connor, P. (2004). "Renewable electricity in the United Kingdom: developing policy in an evolving electricity market". In I. de Lovinfosse and F. Varone (eds.), *Renewable Electricity Policies in Europe. Tradable Green Certificates in Competitive Markets*. Louvain-la-Neuve: Presses universitaires de Louvain.

Corbin, J. and A. Strauss (1990). "Grounded Theory Research: Procedures, Canons, and Evaluative Criteria", *Qualitative Sociology*, 13, 1, 3-21.

Cowles, M. G., J. Caporaso, and T. Risse (2001). *Transforming Europe. Europeanisation and Domestic Change*. Ithaca: Cornell University Press.

Cox, H. (1999). "Regulations versus public property: a comparative analysis", *Annals of Public and Cooperative Economics*, 70, 2.

Dahl, R. (1961). *Who Governs?: Democracy and Power in an American City.* New Haven: Yale University Press.

Dahl, R. and C. E. Lindblom (1976). *Politics, Economics and Welfare.* New York: Harpers.

De Busquey, L. (1976). "Les centrales électriques en Belgique et les projets de développement", *La Revue Nouvelle*, 9, No. spécial "Énergie nucléaire. Un choix sage?", 119-123.

de Lovinfosse, I. and V. Dinica (2004). "Between success and failure: renewable electricity policies in the Netherlands". In I. de Lovinfosse and F. Varone (eds.), *Renewable Electricity Policies in Europe. Tradable Green Certificates in Competitive Markets.* Louvain-la-Neuve: Presses universitaires de Louvain.

de Lovinfosse, I. and F. Varone (2002). "Belgium". In D. Reiche (ed.), *Handbook of Renewable Energies in the European Union.* Berlin: Peter Lang.

de Lovinfosse, I. and F. Varone (2005). "Belgium". In D. Reiche (ed.), *Handbook of Renewable Energies in the European Union.* Berlin: Peter Lang.

de Lovinfosse, I. and F. Varone (2004). "EU policies: market liberalisation, renewable electricity and TGC". In I. de Lovinfosse and F. Varone (eds.), *Renewable Electricity Policies in Europe. Tradable Green Certificates in Competitive Markets.* Louvain-la-Neuve: Presses universitaires de Louvain.

de Lovinfosse, I. and F. Varone (2004). "From private self-regulation to public renewable electricity policy: a paradigmatic change in Belgium". In I. de Lovinfosse and F. Varone (eds.), *Renewable Electricity Policies in Europe. Tradable Green Certificates in Competitive Markets.* Louvain-la-Neuve: Presses universitaires de Louvain.

de Lovinfosse, I. and F. Varone (2002). *Renewable electricity policies in Europe: patterns of change in the liberalized market.* Université catholique de Louvain, Prix Tractebel 2001, Working Paper No.1.

de Lovinfosse, I. and F. Varone (2004). *Renewable Electricity Policies in Europe. Tradable Green Certificates in Competitive Markets.* Louvain-la-Neuve: Presses universitaires de Louvain.

de Radigues, B. and M. Huart (2000). *Mise en place d'un marché des certificats verts en Belgique.* Brussels: Report realised for the Cabinet of the State Secretary for Energy and Sustainable Development.

De Visscher C., le Bussy G. & Eymeri J.M. (2004). *La relation entre l'autorité politique et la haute administration. Mise en perspective de la situation au niveau fédéral en Belgique.* Gent: Academia Press, Série Modernisation de l'Administration.De Winter, L. (1996). "Party Encroachment on the Executive and Legislative Branch in the Belgian Polity", *Res Publica*, 38, 2, 325-352.

De Winter, L. and P. Dumont (2003). "Belgium: Delegation and Accountability under Partitocratic Rule". In K. Strom and *et al.* (eds.), *Delegation and Accountability in Parliamentary Democracies.* Oxford: Oxford University Press.

Declercq, C. (2000). *L'ouverture du marché de l'électricité. La nouvelle organisation du secteur.* Brussels: CRISP Courrier Hebdomadaire No.1689-1690.

References

Declercq, C. and A. Vincent (2000). *L'ouverture du marché de l'électricité. Le cadre institutionnel*. Brussels: CRISP Courrier Hebdomadaire No.1684.

Dehousse, F. and T. Iotsova (2000). *L'Europe de l'énergie: un projet toujours reporté*, Brussels: CRISP Courrier Hebdomadaire No.1698-1699.

DeLeon, P. (1999). "The stages approach to the policy process". In P. A. Sabatier (ed.), *Theories of the policy process*. Colorado: Westview Press.

DeLeon, P. (1978). "A theory of policy termination". In J. V. May and A. B. Wildavsky (eds.), *The policy cycle*. Beverly Hills: Sage.

Delwit, P. (1999). "Ecolo: les défis du 'plus grand' des partis verts en Europe". In P. Delwit and J.-M. De Waele (eds.), *Les partis verts en Europe*. Brussels: Editions Complexe.

Delwit, P. and J.-M. De Waele (1999). *Les partis verts en Europe*. Brussels: Editions Complexe.

Dente, B., P. Fareri, and J. Ligteringen (1998). "A theoretical framework for case study analysis". In B. Dente, P. Fareri, and J. Ligteringen (eds.), *The waste and the backyard*. Dordrecht: Kluwer.

Dinica, V. (2003). *Sustained diffusion of renewable energy*. Enschede: Twente University Press.

Dinica, V. (2005). "United Kingdom". In D. Reiche (ed.), *Handbook of Renewable Energies in the European Union*. Berlin: Peter Lang.

Dinica, V. and M. Arentsen (2003). "Green certificate trading in the Netherlands in the prospect of the European electricity market", *Energy Policy*, 31, 609-620.

Dinica, V. and M. Arentsen (2001). *Green electricity in the Netherlands*. Report No.2/2001: Center for Energy and Environment, Norwegian School of Management, Sandvika.

Dolowitz, D. P. (2000). "Introduction of the special issue on policy transfer", *Governance*, 13, 1, 1-4.

Dolowitz, D. P. and D. Marsh (2000). "Learning from Abroad: The Role of Policy Transfer in Contemporary Policy-Making", *Governance*, 13, 1, 5-24.

Dolowitz, D. P. and D. Marsh (1996). "Who learns what from whom: a review of the policy transfer literature", *Political Studies*, 44, 2, 343-357.

Dow, S. R. (2001). "Energy law in the United Kingdom". In M. M. Roggenkamp and *et al.* (eds.), *Energy Law in Europe. National, EU and International Law and Institutions*. Oxford: Oxford University Press.

Dowding, K. (1995). "Model or metaphor? A critical review of the policy network approach", *Political Studies*, 43, 1, 136-158.

Dudley, G. (2003). "Ideas, bargaining and flexible policy communities: policy change and the case of the Oxford transport strategy", *Public Administration*, 81, 3, 433-458.

Dunn, W. N. (1994). *Public Policy Analysis. An Introduction*. Upper Saddle River: Prentice Hall.

Duverger, M. (1964). *Méthodes des sciences sociales*. Paris: PUF.

Dye, T. R. (1998). *Understanding Public Policy*. Upper Saddle River: Prentice Hall.

Eising, R. (2002). "Policy Learning in Embedded Negotiations: Explaining EU Electricity Liberalisation", *International Organization*, 56, 1, 85-120.
Eliadis, P., M. M. Hill, and M. Howlett (2005). *Designing Government. From Instruments to Governance*. Montreal & Kingston: McGill-Queen's University Press.
Elliott, D. A. (1994). "UK renewable energy strategy. The need for longer-term support", *Energy Policy*, 22, 12, 1067-1074.
Espey, S. (2001). "Renewables portfolio standards: a mean for trade with electricity from renewable energy sources?", *Energy Policy*, 29, 557-566.
Evans, M. and J. Davies (1999). "Understanding policy transfer: a multi-level, multi-disciplinary perspective", *Public Administration*, 77, 2, 361-385.
Evans, P. B., D. Rueschemeyer, and T. Skocpol (1985). *Bringing the State Back in*. Cambridge: Cambridge University Press.
Falkner, G. (2003). "Comparing Europeanisation Effects: From Methaphor to Operationalisation", *European Integration online Papers*, 7, 13.
Featherstone, K. and C. M. Radaelli (2003). "A Conversant Research Agenda". In K. Featherstone and C. M. Radaelli (eds.), *The Politics of Europeanisation*. Oxford: Oxford University Press.
Featherstone, K. and C. M. Radaelli (2003). *The Politics of Europeanisation*. Oxford: Oxford University Press.
Finnemore, M. and K. Sikkink (1998). "International norm dynamics and political change", *International Organization*, 52, 4, 887-917.
Frankland, E. G. (1995). "Germany. The rise, fall and recovery of Die Grünen". In D. Richardson and C. Rootes (eds.), *The Green Challenge. The development of green parties in Europe*. London: Routledge.
Fristrup, P. (2003). "Some challenges related to introducing tradable green certificates", *Energy Policy*, 31, 15-19.
Fuchs, D. and M. J. Arentsen (2001). *Green electricity in the market place: the policy challenge*. ECPR 2001 Conference, Canterburry, September 2001.
Genoud, C. (2004). "Libéralisation et régulation des industries de réseau: diversité dans la convergence?", *Revue Internationale de Politique Comparée*, 11, 2, 187-204.
Genoud, C. and M. Finger (2004). Electricity Regulation in Europe. In A. Midttun (ed.), *Reshaping the Energy Industry: Regulation, Markets, Business Strategies*. London: Elsevier.
Genoud, C. and M. Finger (2003). "Marchés de l'électricité et stratégie des opérateurs: vers une régulation technocratique multi-niveaux", *Politiques et Management Public*, 21, 3.
Genoud, C. and F. Varone (2002). "Does privatisation matter? Liberalisation and regulation: the case of European electricity", *Public Management Review*, 4, 2, 231-256.
Gerring, J. (2004). "What is a case study and what is it good for?", *American Political Science Review*, 98, 2, 341-354.

Giuliani, M. (2003). "Europeanisation in Comparative Perspective: Institutional Fit and National Adaptation". In K. Featherstone and C. M. Radaelli (eds.), *The Politics of Europeanisation*. Oxford: Oxford University Press.

Glachant, J-M. (2003). "The making of competitive electricity markets in Europe: no single way and no 'single market'". In J.-M. Glachant and D. Finon (eds.), *Competition in European electricity markets*. Cheltenham: Edward Elgar.

Glachant, J.-M. and D. Finon (2003). *Competition in European electricity markets*. Cheltenham: Edward Elgar.

Glaser, B. G. and A. L. Strauss (1967). *The discovery of grounded theory: strategies for qualitative research*. Chicago: Aldine Publishers.

Grohnheit, P. E. (2002). "Denmark: long-term planning with different objectives". In C. Vrolijk (ed.), *Climate change and power – Economic instruments for European electricity*. London: Earthscan.

Grohnheit, P. E. and O. J. Olsen (2001). "Organisation and regulation of the electricity supply industry in Denmark". In L. De Paoli (ed.), *The electricity industry in transition: Organization, Regulation and Ownership in EU Member states*. Milan: FrancoAngeli.

Grossman, E. (2004). "Acteur". In L. Boussaguet, S. Jacquot, and P. Ravinet (eds.), *Dictionnaire des politiques publiques*. Paris: Presses de Sciences Po.

Grotz, C. (2005). "Germany". In D. Reiche (ed.), *Handbook of Renewable Energies in the European Union*. Berlin: Peter Lang.

Haas, E. B. (1990). *When Knowledge is Power: Three Models of Change in International Organisations*. Berkeley: University of California Press.

Haas, R. *et al.* (2004). "How to promote renewable energy systems successfully and effectively", *Energy Policy*, 32, 833-839.

Hadjilambrinos, C. (2000). "Understanding technology choice in electricity industries: a comparative study of France and Denmark", *Energy Policy*, 28, 1111-1126.

Hall, P. A. (1990). *Governing the economy. The politics of state intervention in Britain and France*. Oxford: Oxford University Press.

Hall, Peter A. (1990). "Policy paradigms, experts, and the State: the case of macroeconomic policy-making in Britain". In S. Brooks and A.-G. Gagnon (eds.), *Social scientists, policy, and the State*. New York: Praeger.

Hall, Peter A. (1993). "Policy paradigms, social learning, and the State. The case of economic policymaking in Britain", *Comparative Politics*, 25, 3, 275-296.

Hall, Peter A. (1997). "The role of interests, institutions, and ideas in the comparative political economy of the industrialized nations". In M. I. Lichbach and A. S. Zuckerman (eds.), *Comparative politics: rationality, culture, and structure*. Cambridge-New York: Cambridge University Press.

Hall, Peter A. and R. C. Taylor (1997). "La science politique et les trois néo-institutionnalismes", *Revue française de science politique*, 47, 3-4, 469-496.

Hamel, J. (1997). *Études de cas et sciences sociales*. Paris: L'Harmattan.

Hassenteufel, P. (2000). "Deux ou trois choses que je sais d'elle. Remarques à propos d'expériences de comparaisons européennes". In CURAPP (ed.), *Les méthodes au concret.* Paris: PUF.

Haverland, M. (2003). "The Impact of the European Union on Environmental Policies". In K. Featherstone and C. M. Radaelli (eds.), *The Politics of Europeanisation.* Oxford: Oxford University Press.

Heclo, H. (1974). *Modern Social Politics in Britain and Sweden.* New Haven: Yale University Press.

Heclo, H. (1972). "Review article: Policy analysis", *British Journal of Political Science,* 2, 83-108.

Heidenheimer, A. J., H. Heclo, and C. T. Adams (1990). *Comparative Public Policy. The Politics of Social Choice in America, Europe and Japan.* New York: St Martin's Press.

Helm, D. (2003). *Energy, the State, and the Market. British Energy Policy since 1979.* Oxford: Oxford University Press.

Helm, D., J. Kay, and D. Thompson (1989). *The Market for Energy.* Oxford: Clarendon Press.

Hennicke, P. (2004). "Scenarios for a robust policy mix: the final report of the German study commission on sustainable energy supply", *Energy Policy,* 32, 1673-1678.

Herbiet, M. and I. Gabriel (2002). "Présentation des enjeux non juridiques des secteurs de l'eau et de l'énergie et tour d'horizon de la législation", *Droit Communal,* 1, 2-44.

Héritier, A. (2001). "Differential Europe: The European Union Impact on National Policymaking". In A. Héritier *et al.* (eds.), *Differential Europe. The European Union Impact on National Policymaking.* Lanham: Rowman & Littlefield.

Héritier, A. (2002). "Public-interest services revisited", *Journal of European Public Policy,* 9, 6, 995-1019.

Héritier, A. and C. Knill (2000). *Differential responses to European policies: a comparison.* Bonn: Max-Planck-Projecktgruppe Recht der Gemeinschaftsgüter.

Héritier, Adrienne and C. Knill (2001). "Differential Responses to European Policies: A Comparison". In A. Héritier *et al.* (eds.), *Differential Europe. The European Union Impact on National Policymaking.* Lanham: Rowman & Littlefield.

Herzog, T. (1996). *Research methods in the social sciences.* New York: HarperCollins College Publishers.

Hibbs, D. A. (1992). "Partisan theory after fifteen years", *European Journal of Political Economy,* 8, 361-373.

Hix, S. and K. Goetz (2000). "Introduction: European Integration and National Political Systems", *West European Politics,* 23, 4, 1-26.

Hober, George Jr. (1986). "Technology, political structure, and social regulation", *Comparative Politics,* 18, 3, 357-376.

Hoberg, G. (1991). "Sleeping with an elephant: the American influence on Canadian environmental regulation", *Journal of Public Policy*, 2, 1, 107-132.

Hogwood, B. and B. G. Peters (1983). *Policy Dynamics*. Brighton: Wheatsheaf Books.

Hogwood, B. W. and L. A. Gunn (1984). *Policy Analysis for the Real World*. London: Oxford University Press.

Hood, C. C. (2000). *The tools of government*. London: MacMillan.

Howlett, M. (2005). "What is a Policy Instrument? Tools, Mixes and Implementation Styles". In P. Eliadis, M. M. Hill, and M. Howlett (eds.), *Designing Government. From Instruments to Governance*. Montreal & Kingston: McGill and Queen's University Press.

Howlett, M. and M. Ramesh (1993). "Patterns of Policy Instrument Choice. Policy Style, Policy Learning and the Privatisation Experience", *Policy Studies Review*, 12, 1/2, 3-24.

Howlett, M. C. and M. Ramesh (1995). *Studying Public Policy: Policy Cycle and Policy subsystems*. Oxford: Oxford University Press.

Howlett, Michael (1994). "The judicialization of Canadian environmental policy, 1980-1990: a test of Canada-United States convergence thesis", *Canadian Journal of Political Science*, 27, 1, 99-127.

Huber, E., C. Ragin, and J. D. Stephens (1993). "Social democracy, christian democracy, constitutional structure, and the welfare state", *American Journal of Sociology*, 99, 711-749.

Hvelplund, F. (2005). Denmark. In D. Reiche (ed.), *Handbook of Renewable Energies in the European Union*. Berlin: Peter Lang.

Hvelplund, F. (2001). *Renewable Energy Governance Systems*. Aalborg: Aalborg University, Institute for Development and Planning.

Ingram, H. and A. Schneider (1991). "The Choice of Target Populations", *Administration and Society*, 23, 333-356.

Irondell, B. (2003). "Europeanisation without the European Union? French military reforms 1991-96", *Journal of European Public Policy*, 10, 2, 208-226.

Jacobsson, S. and V. Lauber (2005). "Germany: From a Modest Feed-in Law to a Framework for Transition". In V. Lauber (ed.), *Switching to Renewable Power*. London: Earthscan.

Jacquot, S. and C. Woll (2003). "Usage of European Integration – Europeanisation from a Sociological Perspective", *European Integration online Papers*, 7, 12.

Jahn, D. (1992). "Nuclear power, energy policy and new politics in Sweden and Germany", *Environmental Politics*, 1, 3, 383-417.

James, O. and M. Lodge (2003). "The Limitations of 'Policy Transfer' and 'Lesson Drawing' for Public Policy Research", *Political Studies Review*, 1, 179-193.

Jansen, J. C. (2003). *Policy support for renewable energy in the European Union. A review of the regulatory framework and suggestions for adjustment*. Amsterdam: ECN-C--03-113.

Jensen, S. G. and K. Skytte (2002). "Interaction between the power and green certificate markets", *Energy Policy*, 30, 425-435.
Jensen, S. G. and K. Skytte (2003). "Simultaneous attainment of energy goals by means of green certificates and emission permits", *Energy Policy*, 31, 63-71.
Jobert, B. (1994). *Le tournant néo-libéral en Europe. Idées et recettes dans les pratiques gouvernementales*. Paris: L'Harmattan.
Jobert, B. and P. Muller (1987). *L'État en action: politiques publiques et corporatismes*. Paris: PUF.
Jordan, A. (2005). *Environmental Policy in the European Union*. London: Earthscan.
Jordan, A. (2001). *'New' environmental policy instruments in the UK: policy innovation or 'muddling through'?* ECPR Joint Sessions of Workshops, Grenoble, April 2001.
Jordan, A. and D. Liefferink (2004). *Environment Policy in Europe. The Europeanisation of National Environmental Policy*. Routledge.
Jordan, A., R. Wurzel, A. R. Zito, and L. Brückner (2003). "European governance and the transfer of 'new' environmental policy instruments (NEPIs) in the European Union", *Public Administration*, 18, 3, 555-574.
Junginger, M., S. Agterbosch, A. Faaij, and W. Turkenburg (2004). "Renewable electricity in the Netherlands", *Energy Policy*, 32, 1053-1073.
Kamp, L. M. (2002). *Learning in wind turbine development. A comparison between the Netherlands and Denmark*. Utrecht: Proefschrift Universiteit Utrecht, Faculteit Ruimtelijke Wetenschappen.
Kamp, L. M., R. E. H. M. Smits, and C. D. Andriesse (2004). "Notions on learning applied to wind turbine development in the Netherlands and Denmark", *Energy Policy*, 32, 1625-1637.
Karnoe, P. (1990). "Technological innovation and industrial organisation in the Danish wind industry", *Entrepreneurship & Regional Development*, 2, 105-123.
Kassim, H. (1994). "Policy networks, networks and EU policy making. A sceptical view", *West European Politics*, 17, 4, 17-27.
Keman, H. (2002). *Comparative Democratic Politics*. London: Sage.
Kern, K., H. Jörgens, and M. Jänicke (2001). *The diffusion of environmental policy innovations: a contribution to the globalisation of environmental policy*. Discussion Paper FS II 01-302, Wissenschaftszentrum Berlin für Sozialforschung.
Kingdon, J. W. (1994). *Agendas, Alternatives and Public Policy*. New York: Harper Collins.
Klingemann, H.-D., R. I. Hofferbert, and I. Budge (1994). *Parties, Policies and Democracy*. Boulder: Westview Press.
Knill, C. (1998). "European policies: the impact of national administrative traditions", *Journal of Public Policy*, 18, 1, 1-28.

References

Knill, C. and D. Lehmkuhl (2002). "The national impact of European Union regulatory policy: three Europeanisation mechanisms", *European Journal of Political Research*, 41, 255-280.

Knoepfel, P., C. Larrue, and F. Varone (2001). *Analyse et pilotage des politiques publiques*. Genève: Helbing & Lichtenhahn.

Koster, J. M. (1998). "Organizing for competition: an economic analysis of electricity policy in the Netherlands", *Energy Policy*, 26, 661-668.

Lagasse, C.-E. (1999). *Les nouvelles institutions politiques de la Belgique et de l'Europe*. Namur: Artel.

Landry, R. and F. Varone (2005). "The Choice of Policy Instruments: Confronting the Deductive and the Interactive Approaches". In P. Eliadis, M. M. Hill, and M. Howlett (eds.), *Designing Government. From Instruments to Governance*. Montreal & Kingston: McGill and Queen's University Press.

Lascoumes, P. and P. Le Galès (2004). *Gouverner par les instruments*. Paris: Presses de Sciences Po.

Lasswell, H. D. (1971). *A Pre-view of Policy Sciences*. New York: American Elsevier Publishing.

Lauber, V. (2002). *The Different Concepts of Promoting Res-Electricity and their Political Careers*. Proceedings of the 2001 Berlin Conference on the Human Dimensions of Global Environmental Change: Postdam Institute for Climate Impact Research.

Lauber, V. (2004). "REFIT and RPS: options for a harmonised Community framework", *Energy Policy*, 32, 1405-1414.

Lauber, V. (2002). "REFIT v. RPS: Regulatory competition between support schemes in the EU", *Paper delivered at the Global Windpower Conference in Paris*.

Lauber, V. (2002). "Renewable energy at the EU level". In D. Reiche (ed.), *Handbook of Renewable Energies in the European Union*. Berlin: Peter Lang.

Lauber, V. and L. Mez (2006). "Renewable Electricity Policy in Germany, 1975-2005", *Bulletin of Science and Technology and Society*, 26, 2, 105-120.

Lauber, V. and L. Mez (2004). "Three decades of renewable electricity policies in Germany", *Energy & Environment*, 15, 4, 599-623.

Lauber, V. and D. Pesendorfer (2004). "Success through continuity: renewable electricity policies in Germany". In I. de Lovinfosse and F. Varone (eds.), *Renewable Electricity Policies in Europe. Tradable Green Certificates in Competitive Markets*. Louvain-la-Neuve: Presses universitaires de Louvain.

Le Gales, P. and M. Thatcher (1995). *Les réseaux de politique publique: débat autour des policy networks*. Paris: L'Harmattan.

Lees, C. (2000). *The Red-Green coalition in Germany. Politics, personalities and power*. Manchester: Manchester University Press.

Lehmbruch, G. and P. C. Schmitter (1982). *Patters of Corporatist Policymaking*. London: Sage.

Lemming, J. (2003). "Financial risks for green electricity investors and producers in a tradable green certificate market", *Energy Policy*, 31, 21-32.

Lerner, D. and H. D. Lasswell (1951). *The Policy Sciences*. Stanford: Stanford University Press.

Lieberman, R. C. (2002). "Ideas, institutions, and political order: explaining political change", *American Political Science Review*, 96, 4, 697-712.

Lijphart, A. (1984). *Democracies: Patterns of Majoritarian and Consensual Government in Twenty-One Countries*. New Haven: Yale University Press.

Lijphart, A. (1999). *Patterns of Democracy: Government Form and Performance in Thirty-Six Countries*. New Haven: Yale University Press.

Lijphart, A. (1968). *The politics of accommodation: pluralism and democracy in the Netherlands*. Berkeley: University of California Press.

Lindblom, C. (1959). "The science of muddling through", *Public Administration Review*, 19, 78-88.

Linder, S. and B. G. Peter (1989). "Instruments of Government: Perceptions and Contexts", *Journal of Public Policy*, 9, 1, 35-58.

Louis, F. and A. Vallery (2004). "Ferring Revisited: the Altmark Case and State Financing of Public Service Obligations", *World Competition*, 27, 1, 53-74.

Lowi, T. J. (1964). "American Business, Public Policy, Case Studies and Political Theory", *World Politics*, 16, 677-693.

Lowi, T.J. (1972). "Four systems of policy? Politics and choice", *Public Administration Review*, 32, 298-310.

Lucardie, P. and G. Voerman (2004). "The Netherlands", *European Journal of Political Research*, 43, 1084-1092.

Lucas, N. (1985). *Western European Energy Policies. A Comparative Study*. Oxford: Oxford University Press.

Macatangay, R. F. A. (2001). "Market definition and dominant position abuse under the new electricity trading arrangement in England and Wales", *Energy Policy*, 29, 337-340.

Madlener, R. and S. Stagl (2000). "Promoting Renewable Electricity Generation through Guaranteed Feed-in Tariffs vs Tradable Green Certificates: a co-evolutionary perspective", *Paper presented at the European Society of Ecological Economics in Vienna*.

March, J. G. and J. P. Olsen (1984). "The New Institutionalism: Organizational Factors in Political Life", *The American Political Science Review*, 78, 3, 734-749.

Marsh, D. (1998). *Comparing policy networks*. Buckingham: Open University Press.

Marsh, D. and R. A. W. Rhodes (1992). *Policy networks in British government*. Oxford: Clarendon.

Marsh, D. and M. Smith (2000). "Understanding Policy Networks: towards a Dialectical Approach", *Political Studies*, 48, 4-21.

Marsh, J. G. and J. P. Olsen (1989). *Rediscovering institutions: the organisational basis of politics*. Free Press.

Matlary, J. H. (1997). *Energy Policy in the European Union*. Basingstoke: Macmillan.

Matthew, B. M. and A. M. Huberman (1994). *Qualitative Data Analysis*. Thousand Oaks: Sage.

May, J. V. and A. Wildavsky (1978). *The Policy Cycle*. Beverly Hills: Sage.

May, P. J. (1992). "Policy learning and failure", *Journal of Public Policy*, 12, 4, 331-354.

McCormick, J. (2001). *Environmental policy in the European Union*. Hampshire/New York: Palgrave.

McGowan, F. (1996). "Ideology and Expediency in British Energy Policy". In F. McGowan (ed.), *European Energy Policies in a Changing Environment*. Heidelberg: Physica-Verlag.

Menanteau, P., D. Finon, and M.-L. Lamy (2003). "Prices versus quantities: choosing policies for promoting the development of renewable energy", *Energy Policy*, 31, 799-812.

Menges, R. (2003). "Supporting renewable energy on liberalised markets: green electricity between additionality and consumer sovereignty", *Energy Policy*, 31, 583-596.

Meny, Y. and J.-C. Thoenig (1989). *Politiques publiques*. Paris: PUF.

Meyer, N. I. (2003). "European schemes for promoting renewables in liberalised markets", *Energy Policy*, 31, 665-676.

Meyer, N. I. (2004). "Renewable energy policy in Denmark", *Energy for Sustainable Development*, VIII, 1, 25-35.

Meyer, N. I. and A. L. Koefoed (2003). "Danish energy reform: policy implementations for renewables", *Energy Policy*, 31, 597-607.

Mez, L. (2003). "New corporate strategies in the German electricity supply industry". In J.-M. Glachant and D. Finon (eds.), *Competition in European Electricity Markets*. Cheltenham: Edward Elgar.

Midttun, A. (1998). *European electricity systems in transition*. Elsevier.

Midttun, A. (2004). *Reshaping the Energy Industry: Regulation, Markets, Business Strategies*. London: Elsevier.

Midttun, A. and A. L. Koefoed (2003). "Greening of electricity in Europe: challenges and developments", *Energy Policy*, 31, 677-687.

Mitchell, C. (2000). "The England and Wales Non-Fossil Fuel Obligation: History and Lessons", *Annual Review of Energy and the Environment*, 25, 285-312.

Mitchell, C. (1995). "The renewable NFFO". A review, *Energy Policy*, 23, 12, 1077-1091.

Mitchell, C. and P. Connor (2004). "Renewable energy policy in the UK 1990-2003", *Energy Policy*, 32, 1935-1947.

Morthorst, P. E. (2000). "The development of a green certificate market", *Energy Policy*, 28, 1085-1094.

Morthorst, P. E. (2003). "A green certificate market combined wit a liberalised power market", *Energy Policy*, 31, 1393-1402.

Morthorst, P. E. (2003). "National environmental targets and international emission reduction instruments", *Energy Policy*, 31, 73-83.

Mossberger, K. and H. Wolman (2003). "Policy transfer as a form of prospective policy evaluation: challenges and recommendations", *Public Administration Review*, 63, 4, 428-440.

Muller, P. (2000). "L'analyse cognitive des politiques publiques: vers une sociologie politique de l'action publique", *Revue française de science politique*, 50, 2, 189-207.

Muller, P. (2000). *Les politiques publiques*. Paris: PUF.

Muller, P. and Y. Surel (1998). *L'analyse des politiques publiques*. Paris: Montchrestien.

Mörth, U. (2003). "Europeanisation as Interpretation, Translation, and Editing of Public Policies". In K. Featherstone and C. M. Radaelli (eds.), *The Politics of Europeanisation*. Oxford: Oxford University Press.

Müller-Rommel, F. and T. Poguntke (2002). *Green Parties in National Governments*. London: Frank Cass.

Nielsen, K. H. (2002). *Translating Wind Power Policies. The construction of frames of meaning for wind technologies in Denmark, 1976-2002*. Salzburg: Presentation in the International Summer School on the Politics and Economics of Renewable Energy, Salzburg.

Nielsen, K. H. and L. Backer (2003). *Framing, overflowing, failure. Framing forums and princing devices for the green certificate market in Denmark*. 19th EGOS Colloquium, Copenhagen, 3-5 July 2003.

Nielsen, L. and T. Jeppesen (2003). "Tradable green certificates in selected European countries-overview and assessment", *Energy Policy*, 31, 3-14.

Olsen, O. J. and K. Skyte (2001). "Consumer ownership in liberalized electricity markets. The case of Denmark", *Annals of Public and Cooperative Economics*, 73, 1, 1-18.

Padioleau, J. (1982). *L'État au concret*. Paris: PUF.

Page, E. C. (2000). *Future Governance and the Literature on Policy Transfer and Lesson Drawing*. Paper prepared for the ESCR Future Governance Programme Workshop on Policy Transfer, London.

Palier, B. (2002). *Gouverner la sécurité sociale. Les réformes du système français de protection sociale depuis 1945*. Paris: PUF.

Palier, B. (2004). Path dependence. In L. Boussaguet, S. Jacquot, and P. Ravinet (eds.), *Dictionnaire des politiques publiques*. Paris: Presses de Sciences Po.

Parsons, W. (1995). *Public Policy. An Introduction to the Theory and Practice of Policy Analysis*. Cheltenham: Edward Elgar.

Patterson, M. J. and I. H. Rowlands (2002). "Beauty in the eye of the beholder: a comparison of 'green power' certification programs in Australia, Canada, the United Kingdom and the United States", *Energy & Environment*, 13, 1, 1-25.

Pehle, H. (1997). "Germany: domestic obstacles to an international forerunner". In M. S. Andersen and D. Liefferink (eds.), *European environmental policy. The pioneers*. Manchester: Manchester University Press.

Pennings, P. (2005). "Parties, voters and policy priorities in the Netherlands", 1971-2002, *Party Politics*, 11, 1, 29-45.

Peters, B. G. (2002). "The Politics of Tool Choice". In L. M. Salamon (ed.), *The Tools of Government. A guide to the new governance*. Oxford: Oxford University Press.

Peters, B. G. (1997). "Shouldn't row, can't steer: What's a government to do?", *Public Policy and Administration*, 12, 2, 51-61.

Pielow, J-C. and H.-M. Koopmann (2001). "Energy Law in Germany". In M. Roggenkamp, A. Ronne, C. Redgwell, and I. Del Guayo (eds.), *Energy Law in Europe. National, EU and International Law and Institutions*. Oxford: Oxford University Press.

Pierson, P. (2000). "Increasing returns, path dependence and the study of politics", *American Political Science Review*, 94, 251-267.

Pierson, P. (1993). "When Effects become Cause. Policy Feedback and Political Change", *World Politics*, 45, 4, 595-628.

Poli, S. (2002). "National Schemes Supporting the Use of Electricity Produced from Renewable Energy Sources and the Community Legal Framework", *Journal of Environmental Law*, 14, 2, 209-232.

Radaelli, C. M. (2004). "Europeanisation: Solution or problem?", *European Integration online Papers*, 8, 16.

Radaelli, C. M. (2003). "The Europeanisation of Public Policy". In K. Featherstone and C. M. Radaelli (eds.), *The Politics of Europeanisation*. Oxford: Oxford University Press.

Radaelli, C. M. (2000). "Policy transfer in the European Union: institutional isomorphism as a source of legitimacy", *Governance*, 13, 1, 25-43.

Radaelli, C. M. (2000). "Whither Europeanisation? Concept stretching and substantive change", *European Integration online Papers*, 4, 8.

Rader, N. (2000). "The hazards of implementing renewable portfolio standards", *Energy & Environment*, 11, 4, 391-405.

Reiche, D. (2002). *Handbook of Renewable Energies in the European Union*. Berlin: Peter Lang.

Reiche, D. (2002). *The Meaning of Vertical and Horizontal Policies for Renewable Energies*. Proceedings of the 2001 Berlin Conference on the Human Dimensions of Global Environmental Change: Postdam Institute for Climate Impact Research.

Reiche, D. (2005). "The Netherlands". In D. Reiche (ed.), *Handbook of Renewable Energies in the European Union*. Berlin: Peter Lang.

Reiche, D. and M. Bechberger (2004). "Policy differences in the promotion of renewable energies in the EU member states", *Energy Policy*, 32, 843-849.

Reijnders, L. (2002). "Imports as a major complication: liberalisation of the green electricity market in the Netherlands", *Energy Policy*, 30, 723-726.

Rhodes, R. A. W. and D. Marsh (1995). "Les réseaux d'action publique en Grande-Bretagne". In P. Le Gales and M. Thatcher (eds.), *Les réseaux de politique publique: débat autour des policy networks*. Paris: L'Harmattan.

Richardson, D. and C. Rootes (1995). *The green challenge. The development of green parties in Europe*. London: Routledge.

Richardson, J. (2000). "Government, Interest Groups and Policy Change", *Political Studies*, 48, 1006-1025.

Rihoux, B. (1995). "Belgium. Greens in a divided society". In D. Richardson and C. Rootes (eds.), *The green challenge. The development of green parties in Europe*. London: Routledge.

Ringeling, A. B. (2002). "European Experience with Tools of Government". In L. M. Salamon (ed.), *The Tools of Government. A guide to the new governance*. Oxford: Oxford University Press.

Risse, T., M. G. Cowles, and J. Caporaso (2001). "Europeanisation and Domestic Change: Introduction". In M. G. Cowles, J. Caporaso, and T. Risse (eds.), *Transforming Europe. Europeanisation and Domestic Change*. Ithaca: Cornell University Press.

Roberts, G. K. (1999). "Les verts allemands: changements et développements". In P. Delwit and J.-M. De Waele (eds.), *Les partis verts en Europe*. Brussels: Complexe.

Rochefort, D. A. and R. W. Cobb (1994). *The politics of problem definition. Shaping the policy agenda*. Lawrence: University Press of Kansas.

Roggenkamp, M. M. (2001). "Energy Law in the Netherlands". In M. Roggenkamp, A. Ronne, C. Redgwell, and I. Del Guayo (eds.), *Energy Law in Europe. National, EU and International Law and Institutions*. Oxford: Oxford University Press.

Ronne, A. (2001). "Energy Law in Denmark". In M. Roggenkamp, A. Ronne, C. Redgwell, and I. Del Guayo (eds.), *Energy Law in Europe. National, EU and International Law and Institutions*. Oxford: Oxford University Press.

Rose, R. (1990). Inheritance before choice in public policy, *Journal of Theoretical Politics*, 2, 263-291.

Rose, R. (1993). *Lesson drawing in public policy. A guide to learning across time and space*. Chatham: Chathma House Publishers.

Rose, R. and P. L. Davies (1994). *Inheritance in Public Policy. Change without Choice in Britain*. Chatham: Chatham House Publishers.

Ross, C. (2000). "The Promotion of Renewable Energy in England and Wales: the Use of the Non-Fossil Fuel Obligation". In P. Jasinski and W. Pfaffenberger (eds.), *Energy and Environment: Multiregulation in Europe*. Aldershot: Ashgate.

Rüdig, W. (2002). "Germany". In F. Müller-Rommel and T. Poguntke (eds.), *Green parties in national governments*. London: Frank Cass.

Sabatier, P. A. (1988). "An advocacy coalition framework of policy change and the role of policy-oriented learning therein", *Policy Sciences*, 21, 2-3, 129-168.

Sabatier, P. A. and H. C. Jenkins-Smith (1999). "The Advocacy Coalition Framework: An Assessment". In P. A. Sabatier (ed.), *Theories of the policy process*. Colorado: Westview Press.

Sabatier, P. A. and H. C. Jenkins-Smith (1993). *Policy change and learning: an advocacy coalition approach*. Boulder: Westview Press.

Sabatier, P. A. and E. Schlager (2000). "Les approches cognitives des politiques publiques: perspectives américaines", *Revue française de science politique*, 50, 2, 209-234.

Salamon, L. M. (2002). *The Tools of Government. A guide to the new governance.* Oxford: Oxford University Press.

Sapir, M. and A. Carton (1976). "L'énergie, le nucléaire et l'opinion dans différents pays", *La Revue Nouvelle*, 9, No. spécial "Énergie nucléaire. Un choix sage?", 225-239.

Schaeffer, G. J. *et al.* (2000). *Options for design of tradable green certificate systems.* Amsterdam: ECN-C--00-032.

Scharpf, F. W. (1997). *Games Real Actors Play. Actor-Centered Insitutionalism in Policy Research.* Boulder: Westview Press.

Schmidt, M. G. (2002). "The impact of political parties, constitutional structures and veto players on public policy". In H. Keman (ed.), *Comparative Democratic Politics*. London: Sage.

Schmidt, M. G. (1996). "When parties matter: a review of the possibilities and limits of partisan influence on public policy", *European Journal of Political Research*, 30, 155-183.

Schmidt, V. A. (2002). "Europeanisation and the mechanics of economic policy adjustment", *Journal of European Public Policy*, 9, 6, 894-912.

Schmidt, V. A. and C. M. Radaelli (2004). "Policy Change and Discourse in Europe: Conceptual and Methodological Issues", *West European Politics*, 27, 2, 183-210.

Schmithüsen, F. (2003). *Understanding Cross-Sectoral Policy Impacts: Policy and Legal Aspects.* FAO, Forestry Paper 142, Cross-Sectoral Policy Impacts Between Forestry and Other Sectors.

Schmitt, D. (1983). "West German Energy Policy". In W. L. Kohl (ed.), *After the second oil crisis.* Lexington: Lexington Books.

Schmitter, P. C. (1977). "Modes of Interest Intermediation and Models of Societal Change in Western Europe", *Comparative Political Studies*, 10, 1, 7-38.

Schneider, A. L. and H. Ingram (1990). "Behavioral Assumptions of Policy Tools", *Journal of Politics*, 52, 2, 510-529.

Schneider, A. L. and H. Ingram (1997). *Policy Design for Democracy.* Lawrence KA: University Press of Kansas.

Schneider, A. L. and H. Ingram (1988). "Systematic Pinching Ideas: A Comparative Approach to Policy Design", *Journal of Public Policy*, 8, 1, 61-80.

Seeliger, R. (1996). "Conceptualizing and researching policy convergence", *Policy Studies Journal*, 24, 2, 287-306.

Simon, H. A. (1957). *Models of Man: Social and Rational.* New York: John Wiley.

Sinnaeve, A. (2003). "State Financing of Public Services: The Court's Dilemna in the Altmark Case", *European State Aid Law Quarterly*, 3, 3, 351-363.

Sioshansi, F. P. (2001). "Opportunities and perils of the newly liberalized European electricity markets", *Energy Policy*, 29, 419-427.

Slingerland, S. (1997). "Energy conservation and organisation of electricity supply in the Netherlands", *Energy Policy*, 25, 193-203.

Smith, A. (2000). "L'analyse comparée des politiques publiques: une démarche pour dépasser le tourisme intelligent?", *Revue Internationale de Politique Comparée*, 7, 1.

Smith, A. and J. Watson (2002). *The Challenge for Tradable Green Certificates in the UK*. ENER Forum 3. Successfully Promoting Renewable Energy Sources in Europe, Budapest, June 2002.

Sorrell, S. (2003). "Who owns the carbon? Interaction between the EU emissions trading scheme and the UK renewable obligation and energy efficiency commitment", *Energy & Environment*, 14, 5, 677-703.

Staiss, F. (2000). *Jahrbuch Erneuerbare Energien 2000*. Radebeul: Bieberstein.

Staiss, F. (2003). *Jahrbuch Erneuerbare Energien 2002/2003*. Radebeul: Bieberstein.

Steinmo, S. (2004). "Néo-institutionnalismes". In L. Boussaguet, S. Jacquot, and P. Ravinet (eds.), *Dictionnaire des politiques publiques*. Paris: Presses de Sciences Po.

Steinmo, S., K. A. Thelen, and F. Longstreth (1992). *Structuring Politics: Historical Institutionalism in Comparative Analysis*. Cambridge: Cambridge University Press.

Stone, D. (1999). "Learning lessons and transferring policy across time, space and disciplines", *Politics*, 19, 1, 51-59.

Stone, D. (2000). "Non-governmental policy transfer: the strategies of independent policy institutes", *Governance*, 13, 1, 45-62.

Strauss, A. and J. Corbin (1990). *Basics of qualitative research: grounded theory procedures and techniques*. Newbury Park: Sage Publications.

Surel, Y. (1995). "Les politiques publiques comme paradigmes". In A. Faure, G. Pollet, and P. Warin (eds.), *La construction du sens dans les politiques publiques. Débats autour de la notion de référentiel*. Paris: L'Harmattan.

Surel, Yves (1997). "Quand la politique change les politiques. La loi Lang du 10 août 1981 et les politiques du livre", *Revue française de science politique*, 47, 2, 147-172.

Tchouate Heteu, P. M. (2004). *Les certificats verts et l'électricité renouvelable*. Louvain-la-Neuve: Presses universitaires de Louvain.

Tews, K., P.-O. Busch, and H. Jörgens (2003). "The diffusion of new environmental policy instruments", *European Journal of Political Research*, 42, 569-600.

Thatcher, M. (2002). "Analysing regulatory reform in Europe", *Journal of European Public Policy*, 9, 6, 859-872.

Thatcher, M. (2002). "Regulation after delegation: independent regulatory agencies in Europe", *Journal of European Public Policy*, 9, 6, 954-972.

Thatcher, M. (2004). "Winners and Losers in Europeanisation: Reforming the National Regulation of Telecommunications", *West European Politics*, 27, 2, 284-309.

References

Thelen, K. (2003). "Comment les institutions évoluent: perspectives de l'analyse comparative historique". In Association recherche et régulation (ed.), *L'année de la régulation. Economie, Institutions, Pouvoirs*. Paris: Presses de Science Po.

Thill, G. (1976). "Le travail de la commission des sages", *La Revue Nouvelle*, 9, No. spécial "Énergie nucléaire. Un choix sage?", 118.

Thiriot, C., M. Marty, and E. Nada (2004). *Penser la politique comparée. Un état des savoirs théoriques et méthodologiques*. Paris: Editions Karthala.

Thoenig, J. C. (2004). "Politique publique". In L. Boussaguet, S. Jacquot, and P. Ravinet (eds.), *Dictionnaire des politiques publiques*. Paris: Presses de Sciences Po.

Timmermans, A. I. (2003). *High Politics in the Low Countries. An Empirical Study of Coalition Agreements in Belgium and the Netherlands*. Ashgate: Aldershot and Burlington.

Trebilcock, M. J. and et al. (1982). *The Choice of Governing Instruments*. Ottawa: Canadian Government Publishing Center.

True, J. L., B. D. Jones, and F. R. Baumgartner (1999). "Punctuated-Equilibrium Theory: Explaining stability and change in American policy-making". In P. A. Sabatier (ed.), *Theories of the policy process*. Colorado: Westview Press.

Tsebelis, G. (1995). "Decision making in political systems: veto players in presidentialism, parlementarism, multicameralism and multipartism", *British Journal of Political Science*, 25, 289-325.

Tsebelis, G. (1990). *Nested Games: Rational Choice in Comparative Politics*. Berkeley: University of California Press.

Tsebelis, G. (1999). "Veto players and law production in parliamentary democracies: the empirical analysis", *American Political Science Review*, 93, 591-608.

Tsebelis, G. (2002). *Veto players. How political institutions work*. Princeton: Princeton University Press.

Töller, A. E. (2004). "The Europeanisation of Public Policies – Understanding Idiosyncratic Mechanisms and Contingent Results", *European Integration online Papers*, 8, 9.

van Rooijen, S. N. M. and M. T. van Wees (2006). "Green electricity policies in the Netherlands: an analysis of policy decisions", *Energy Policy*, 34, 60-71.

Van Sambeek, A. J. W. (2002). *The European dimension of national renewable electricity policy making*. ECN-RX-02-060.

Van Sambeek, A. J. W. and E. Van Thuijl (2003). *The Dutch renewable electricity market in 2003. An overview and evaluation of current changes in renewable electricity policy in the Netherlands*. ECN-C03-037.

Varone, F. (2000). "Le choix des instruments de l'action publique: analyse comparée des politiques énergétiques en Europe et en Amérique du Nord", *Revue internationale de politique comparée*, 7, 1.

Varone, F. (1998). *Le choix des instruments des politiques publiques: une analyse comparée des politiques d'efficience énergétique du Canada, du Danemark, des États-Unis, de la Suède et de la Suisse.* Bern: Haupt.

Varone, F. and B. Aebischer (2001). "Energy Efficiency: The Challenges of Policy Design", *Energy Policy*, 29, 8, 615-629.

Varone, F. and C. Genoud (2001). "Libéralisation des services de réseau et redistribution des responsabilités politique et managériale: le cas de l'électricité", *Politiques et Management Public*, 19, 3, 191-212.

Verbong, G. et al. (2001). *Een kwestie van lange adem.* Boxtel: AEneas.

Verbong, G., A. Van Selm, R. Knoppers, and R. Raven (2001). *Een kwestie van lange adem. De geschiedenis van duurzame energie in Nederland.* Boxtel: Aeneas.

Verbruggen, A. (2004). "Tardable green certificates in Flanders (Belgium)", *Energy Policy*, 32, 165-176.

Verbruggen, A. and E. Vanderstappen (2003). "Electricity sector restructuring in Belgium". In J.-M. Glachant and D. Finon (eds.), *Competition in European Electricity Markets.* Cheltenham: Edward Elgar.

Verbruggen, A. and E. Vanderstappen (1999). "Régulation et marché de l'électricité dans le Benelux", *Reflets et Perspectives de la vie économique*, 38, 2.

Vincent, A. and C. Declercq (2000). *L'ouverture du marché de l'électricité. Organisation et stratégie des acteurs.* Brussels: CRISP Courrier Hebdomadaire No.1695.

Vink, M. (2003). "What is europeanisation? And other questions on a new research agenda", *European Political Science*, 3, 1, 63-74.

Vlassopoulou, C. A. (2000). "'Ideas matter too': éléments d'une analyse postpositiviste de la lutte contre la pollution de l'air en France et en Grèce", *Revue Internationale de Politique Comparée*, 7, 1, 113-133.

Vlassopoulou, C. A. (2000). "Politiques publiques comparées. Pour une approche définitionnelle et diachronique". In CURAPP (ed.), *Les méthodes au concret.* Paris: PUF.

Voogt, M., M. G. Boots, G. J. Schaeffer, and J. W. Martens (2000). "Renewable electricity in a liberalised market – the concept of green certificates", *Energy & Environment*, 11, 1, 65-79.

Walker, J. L. (1969). "The diffusion of innovations among the American States", *American Political Science Review*, 63, 3, 881-889.

Weerts, S. (2001). "L'évolution des acteurs du service public dans les secteurs de l'eau et de l'électricité", *Droit Communal*, 1, 45-79.

Weimer, D. L. and A. R. Vining (1999). *Policy Analysis. Concepts and Practice.* Upper Saddle River: Prentice-Hall.

Wildavsky, A. (1979). *Speaking Truth to Power: The Art and Craft of Policy Analysis.* Boston: Little Brown.

Wilson, C. A. (2000). "Policy regimes and policy change", *Journal of Public Policy*, 20, 3, 247-274.

Wolman, H. (1992). "Understanding cross national policy transfers: the case of Britain and the US", *Governance*, 5, 1, 27-45.

Wood, B. D. and A. Doan (2003). "The politics of problem definition: applying and testing threshold models", *American Journal of Political Science*, 47, 4, 640-653.

Yesilkagit, K. (2004). "Bureaucratic autonomy, organizational culture, and habituation. Politicians and independent administrative bodies in the Netherlands", *Administration & Society*, 36, 5, 528-552.

Yin, R. K. (2003). *Case study research: design and methods*. Thousands Oaks: Sage Publications.

Zahariadis, N. (1999). Ambiguity, Time and Multiple Streams. In P. A. Sabatier (ed.), *Theories of the policy process*. Colorado: Westview Press.

"Public Action"

The series "Public Action" studies the State and how it works from a variety of perspectives. It focuses on public policy analysis (environment, health, employment culture, etc.) as well as the institutional and organisational study of public administrations. The mutations that characterise nowadays public action (deregulation, externalisation, contractualisation and networking) are at the heart of these preoccupations.

Consequently, the series "Public Action" also places further emphasis on the relationship the State maintains with its environment, – not only the political apparatus and, of course, civil society, but also with broader social changes such as individualism and globalisation.

Multidisciplinary work is an essential characteristic of contemporary research in the field. Given the essential need for such approaches to the topic, this series encourages economic, historical, judicial, politological, sociological and philosophical perspectives.

Series Editors :
Jean-Louis GENARD, Professor at the Université Libre de Bruxelles
Steve JACOB, Professor at the Université Laval

Series titles

No.3 – Isabelle DE LOVINFOSSE, *How and Why Do Policies Change? A Comparison of Renewable Electricity Policies in Belgium, Denmark, Germany, the Netherlands and the UK*, 2008, 317 p.

N° 2 – Fabrizio CANTELLI, *L'État à tâtons. Pragmatique de l'action publique face au sida*, 2007, 268 p.

N° 1 – Steve JACOB, Frédéric VARONE et Jean-Louis GENARD (dir.), *L'évaluation des politiques au niveau régional*, 2007, 218 p.

Peter Lang – The website

Discover the general website of the Peter Lang publishing group:

www.peterlang.com